Graph Algorithms in the Language of Linear Algebra

SOFTWARE • ENVIRONMENTS • TOOLS

The SIAM series on Software, Environments, and Tools focuses on the practical implementation of computational methods and the high performance aspects of scientific computation by emphasizing in-demand software, computing environments, and tools for computing. Software technology development issues such as current status, applications and algorithms, mathematical software, software tools, languages and compilers, computing environments, and visualization are presented.

Software, Environments, and Tools

Daniel J. Bates, Jonathan D. Hauenstein, Andrew J. Sommese, Charles W. Wampler, *Numerically Solving Polynomial Systems with Bertini*

Uwe Naumann, *The Art of Differentiating Computer Programs: An Introduction to Algorithmic Differentiation*

C. T. Kelley, *Implicit Filtering*

Jeremy Kepner and John Gilbert, editors, *Graph Algorithms in the Language of Linear Algebra*

Jeremy Kepner, *Parallel MATLAB for Multicore and Multinode Computers*

Michael A. Heroux, Padma Raghavan, and Horst D. Simon, editors, *Parallel Processing for Scientific Computing*

Gérard Meurant, *The Lanczos and Conjugate Gradient Algorithms: From Theory to Finite Precision Computations*

Bo Einarsson, editor, *Accuracy and Reliability in Scientific Computing*

Michael W. Berry and Murray Browne, *Understanding Search Engines: Mathematical Modeling and Text Retrieval, Second Edition*

Craig C. Douglas, Gundolf Haase, and Ulrich Langer, *A Tutorial on Elliptic PDE Solvers and Their Parallelization*

Louis Komzsik, *The Lanczos Method: Evolution and Application*

Bard Ermentrout, *Simulating, Analyzing, and Animating Dynamical Systems: A Guide to XPPAUT for Researchers and Students*

V. A. Barker, L. S. Blackford, J. Dongarra, J. Du Croz, S. Hammarling, M. Marinova, J. Wasniewski, and P. Yalamov, *LAPACK95 Users' Guide*

Stefan Goedecker and Adolfy Hoisie, *Performance Optimization of Numerically Intensive Codes*

Zhaojun Bai, James Demmel, Jack Dongarra, Axel Ruhe, and Henk van der Vorst, *Templates for the Solution of Algebraic Eigenvalue Problems: A Practical Guide*

Lloyd N. Trefethen, *Spectral Methods in MATLAB*

E. Anderson, Z. Bai, C. Bischof, S. Blackford, J. Demmel, J. Dongarra, J. Du Croz, A. Greenbaum, S. Hammarling, A. McKenney, and D. Sorensen, *LAPACK Users' Guide, Third Edition*

Michael W. Berry and Murray Browne, *Understanding Search Engines: Mathematical Modeling and Text Retrieval*

Jack J. Dongarra, Iain S. Duff, Danny C. Sorensen, and Henk A. van der Vorst, *Numerical Linear Algebra for High-Performance Computers*

R. B. Lehoucq, D. C. Sorensen, and C. Yang, *ARPACK Users' Guide: Solution of Large-Scale Eigenvalue Problems with Implicitly Restarted Arnoldi Methods*

Randolph E. Bank, *PLTMG: A Software Package for Solving Elliptic Partial Differential Equations, Users' Guide 8.0*

L. S. Blackford, J. Choi, A. Cleary, E. D'Azevedo, J. Demmel, I. Dhillon, J. Dongarra, S. Hammarling, G. Henry, A. Petitet, K. Stanley, D. Walker, and R. C. Whaley, *ScaLAPACK Users' Guide*

Greg Astfalk, editor, *Applications on Advanced Architecture Computers*

Roger W. Hockney, *The Science of Computer Benchmarking*

Françoise Chaitin-Chatelin and Valérie Frayssé, *Lectures on Finite Precision Computations*

Graph Algorithms in the Language of Linear Algebra

Edited by

Jeremy Kepner
MIT Lincoln Laboratories
Lexington, Massachusetts

John Gilbert
University of California at Santa Barbara
Santa Barbara, California

siam.
Society for Industrial and Applied Mathematics
Philadelphia

10 9 8 7 6 5 4 3

This work is sponsored by the Department of the Air Force under Air Force Contract FA8721-05-C-0002. Opinions, interpretations, conclusions, and recommendations are those of the authors and are not necessarily endorsed by the United States Government.

Library of Congress Cataloging-in-Publication Data

Kepner, Jeremy V., 1969-
Graph algorithms in the language of linear algebra / Jeremy Kepner, John Gilbert.
p. cm. – (Software, environments, and tools)
Includes bibliographical references and index.
ISBN 978-0-898719-90-1
1. Graph algorithms. 2. Algebras, Linear. I. Gilbert, J. R. (John R.), 1953- II. Title.
QA166.245.K47 2011
511'.6–dc22 2011003774

for

Dennis Healy

whose vision

allowed us all

to see further

List of Contributors

David A. Bader
College of Computing
Georgia Institute of Technology
Atlanta, GA 30332
bader@cc.gatech.edu

Nadya Bliss
MIT Lincoln Laboratory
244 Wood Street
Lexington, MA 02420
nt@ll.mit.edu

Robert Bond
MIT Lincoln Laboratory
244 Wood Street
Lexington, MA 02420
rbond@ll.mit.edu

Aydın Buluç
High Performance Computing Research
Lawrence Berkeley National Laboratory
1 Cyclotron Road
Berkeley, CA 94720
abuluc@lbl.gov

Daniel M. Dunlavy
Computer Science and Informatics Department
Sandia National Laboratories
Albuquerque, NM 87185
dmdunla@sandia.gov

Alan Edelman
Mathematics Department
MIT
77 Massachusetts Avenue
Cambridge, MA 02139
edelman@mit.edu

Christos Faloutsos
School of Computer Science
Carnegie Mellon University
Pittsburgh, PA 15213
christos@cs.cmu.edu

Jeremy T. Fineman
School of Computer Science
Carnegie Mellon University
Pittsburgh, PA 15213
jfineman@cs.cmu.edu

John Gilbert
Computer Science Department
University of California at Santa Barbara
Santa Barbara, CA 93106
gilbert@cs.ucsb.edu

Christine E. Heitsch
School of Mathematics
Georgia Institute of Technology
Atlanta, GA 30332
heitsch@math.gatech.edu

Bruce Hendrickson
Discrete Algorithms and Mathematics Department
Sandia National Laboratories
Albuquerque, NM 87185
bahendr@sandia.gov

W. Philip Kegelmeyer
Informatics and Decision Sciences Department
Sandia National Laboratories
Livermore, CA 94551
wpk@sandia.gov

Jeremy Kepner
MIT Lincoln Laboratory
244 Wood Street
Lexington, MA 02420
kepner@ll.mit.edu

Tamara G. Kolda
Informatics and Decision Sciences Department
Sandia National Laboratories
Livermore, CA 94551
tgkolda@sandia.gov

Jure Leskovec
Computer Science Department
Stanford University
Stanford, CA 94305
jure@cs.stanford.edu

Kamesh Madduri
Computational Research Division
Lawrence Berkeley National Laboratory
Berkeley, CA 94720
kmadduri@lbl.gov

Sanjeev Mohindra
MIT Lincoln Laboratory
244 Wood Street
Lexington, MA 02420
smohindra@ll.mit.edu

Huy Nguyen
MIT Computer Science and Artificial Intelligence Laboratory (CSAIL)
32 Vassar Street
Cambridge, MA 02139
huy2n@mit.edu

Charles M. Rader
MIT Lincoln Laboratory
244 Wood Street
Lexington, MA 02420
charlesmrader@verizon.net

Steve Reinhardt
Microsoft Corporation
716 Bridle Ridge Road
Eagan, MN 55123
steve.reinhardt@microsoft.com

Eric Robinson
MIT Lincoln Laboratory
244 Wood Street
Lexington, MA 02420
erobinson@ll.mit.edu

Viral B. Shah
82 E Marine Drive
Badrikeshwar, Flat No. 25
Mumbai 400 002
India
viral@mayin.org

Contents

List of Figures

List of Tables

List of Algorithms

Preface

Graphs are among the most important abstract data structures in computer science, and the algorithms that operate on them are critical to modern life. Graphs have been shown to be powerful tools for modeling complex problems because of their simplicity and generality. For this reason, the field of graph algorithms has become one of the pillars of theoretical computer science, informing research in such diverse areas as combinatorial optimization, complexity theory, and topology. Graph algorithms have been adapted and implemented by the military and commercial industry, as well as by researchers in academia, and have become essential in controlling the power grid, telephone systems, and, of course, computer networks.

The increasing preponderance of computer and other networks in the past decades has been accompanied by an increase in the complexity of these networks and the demand for efficient and robust graph algorithms to govern them. To improve the computational performance of graph algorithms, researchers have proposed a shift to a parallel computing paradigm. Indeed, the use of parallel graph algorithms to analyze and facilitate the operations of computer and other networks is emerging as a new subdiscipline within the applied mathematics community.

The combination of these two relatively mature disciplines—graph algorithms and parallel computing—has been fruitful, but significant challenges still remain. In particular, the tasks of implementing parallel graph algorithms and achieving good parallel performance have proven especially difficult.

In this monograph, we address these challenges by exploiting the well-known duality between the canonical representation of graphs as abstract collections of vertices with edges and a sparse adjacency matrix representation. In so doing, we show how to leverage existing parallel matrix computation techniques as well as the large amount of software infrastructure that exists for these computations to implement efficient and scalable parallel graph algorithms. In addition, and perhaps more importantly, a linear algebraic approach allows the large pool of researchers trained in fields other than computer science, but who have a strong linear algebra background, to quickly understand and apply graph algorithms.

Our treatment of this subject is intended formally to complement the large body of literature that has already been written on graph algorithms. Nevertheless, the reader will find several benefits to the approaches described in this book.

(1) *Syntactic complexity.* Many graph algorithms are more compact and are easier to understand when presented in a sparse matrix linear algebraic format. An algorithmic description that assumes a sparse matrix representation of the graph, and operates on that matrix with linear algebraic operations, can be readily

understood without the use of additional data structures and can be translated into a program directly using any of a number of array-based programming environments (e.g., MATLAB®).

(2) *Ease of implementation.* Parallel graph algorithms are notoriously difficult to implement. By describing graph algorithms as procedures of linear algebraic operations on sparse (adjacency) matrices, all the existing software infrastructure for parallel computations on sparse matrices can be used to produce parallel and scalable programs for graph problems. Moreover, much of the emerging Partitioned Global Address Space (PGAS) libraries and languages can also be brought to bear on the parallel computation of graph algorithms.

(3) *Performance.* Graph algorithms expressed by a series of sparse matrix operations have clear data-access patterns and can be optimized more easily. Not only can the memory access patterns be optimized for a procedure written as a series of matrix operations, but a PGAS library could exploit this transparency by ordering global communication patterns to hide data-access latencies.

This work represents the first of its kind on this interesting topic of linear algebraic graph algorithms, and represents a collection of original work on the topic that has historically been scattered across the literature. This is an edited volume and each chapter is self-contained and can be read independently. However, the authors and editors have taken great care to unify their notation and terminology to present a coherent work on this topic.

The book is divided into three parts: (I) Algorithms, (II) Data, and (III) Computation. Part I presents the basic mathematical framework for expressing common graph algorithms using linear algebra. Part II provides a number of examples where a linear algebraic approach is used to develop new algorithms for modeling and analyzing graphs. Part III focuses on the sparse matrix computations that underlie a linear algebraic approach to graph algorithms. The book concludes with a discussion of some outstanding questions in the area of large graphs.

While most algorithms are presented in the form of pseudocode, when working code examples are required, these are expressed in MATLAB, and so a familiarity with MATLAB is helpful, but not required.

This book is suitable as the primary book for a class on linear algebraic graph algorithms. This book is also suitable as either the primary or supplemental book for a class on graph algorithms for engineers and scientists outside of the field of computer science. Wherever possible, the examples are drawn from widely known and well-documented algorithms that have already been identified as representing many applications (although the connection to any particular application may require examining the references).

Finally, in recognition of the severe time constraints of professional users, each chapter is mostly self-contained and key terms are redefined as needed. Each chapter has a short summary and references within that chapter are listed at the end of the chapter. This arrangement allows the professional user to pick up and use any particular chapter as needed.

Jeremy Kepner
Cambridge, MA

John Gilbert
Santa Barbara, CA

Acknowledgments

There are many individuals to whom we are indebted for making this book a reality. It is not possible to mention them all, and we would like to apologize in advance to those we may not have mentioned here due to accidental oversight on our part.

The development of linear algebraic graph algorithms has been a journey that has involved many colleagues who have made important contributions along the way. This book marks an important milestone in that journey: the broad availability and acceptance of linear algebraic graph algorithms.

Our own part in this journey has been aided by numerous individuals who have directly influenced the content of this book. In particular, our collaboration began in 2002 during John's sabbatical at MIT, and we are grateful to our mutual friend Prof. Alan Edelman for facilitating this collaboration. This early work was also aided by the sponsorship of Mr. Zachary Lemnios. Subsequent work was supported by Mr. David Martinez, Dr. Ken Senne, Mr. Robert Graybill, and Dr. Fred Johnson. More recently, the idea of exploiting linear algebra for graph computations found a strong champion in Prof. Dennis Healy who made numerous contributions to this work.

In addition to those folks who have helped with the development of the technical ideas, many additional folks have helped with the development of this book. Among these are the SIAM Software Environments and Tools series editor Prof. Jack Dongarra, our book editor at SIAM, Ms. Elizabeth Greenspan, the copyeditor at Lincoln Laboratory, Ms. Dorothy Ryan, and the students of MIT and UCSB. Finally, we would like to thank several anonymous reviewers whose comments enhanced this book (in particular, the one who gave us the idea for the title of the book).

Part I

Algorithms

Chapter 1

Graphs and Matrices

*Jeremy Kepner**

Abstract

A linear algebraic approach to graph algorithms that exploits the sparse adjacency matrix representation of graphs can provide a variety of benefits. These benefits include syntactic simplicity, easier implementation, and higher performance. Selected examples are presented illustrating these benefits. These examples are drawn from the remainder of the book in the areas of algorithms, data analysis, and computation.

1.1 Motivation

The duality between the canonical representation of graphs as abstract collections of vertices and edges and a sparse adjacency matrix representation has been a part of graph theory since its inception [Konig 1931, Konig 1936]. Matrix algebra has been recognized as a useful tool in graph theory for nearly as long (see [Harary 1969] and the references therein, in particular [Sabadusi 1960, Weischel 1962, McAndrew 1963, Teh & Yap 1964, McAndrew 1965, Harary & Trauth 1964, Brualdi 1967]). However, matrices have not traditionally been used for practical computing with graphs, in part because a dense 2D array is not an efficient representation of a sparse graph. With the growth of efficient data structures and algorithms for *sparse* arrays and

*MIT Lincoln Laboratory, 244 Wood Street, Lexington, MA 02420 (kepner@ll.mit.edu).

This work is sponsored by the Department of the Air Force under Air Force Contract FA8721-05-C-0002. Opinions, interpretations, conclusions, and recommendations are those of the author and are not necessarily endorsed by the United States Government.

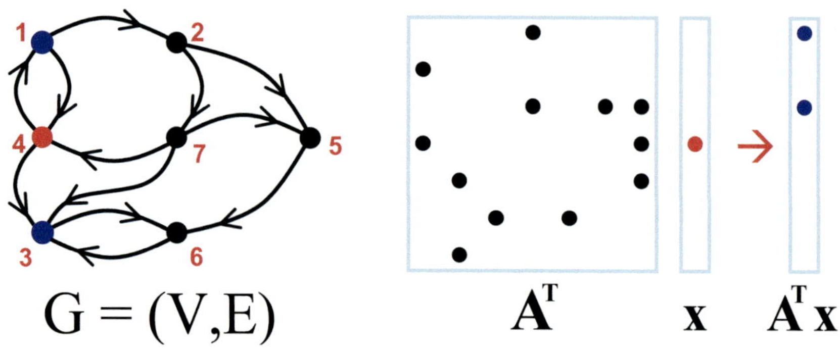

Figure 1.1. Matrix graph duality.
Adjacency matrix **A** is dual with the corresponding graph. In addition, vector matrix multiply is dual with breadth-first search.

matrices, it has become possible to develop a practical array-based approach to computation on large sparse graphs.

There are several benefits to a linear algebraic approach to graph algorithms. These include:

1. *Syntactic complexity.* Many graph algorithms are more compact and are easier to understand in an array-based representation. In addition, these algorithms are accessible to a new community not historically trained in canonical graph algorithms.

2. *Ease of implementation.* Array-based graph algorithms can exploit the existing software infrastructure for parallel computations on sparse matrices.

3. *Performance.* Array-based graph algorithms more clearly highlight the data-access patterns and can be readily optimized.

The rest of this chapter will give a brief survey of some of the more interesting results to be found in the rest of this book, with the hope of motivating the reader to further explore this interesting topic. These results are divided into three parts: (I) Algorithms, (II) Data, and (III) Computation.

1.2 Algorithms

Linear algebraic approaches to fundamental graph algorithms have a variety of interesting properties. These include the basic graph/adjacency matrix duality, correspondence with semiring operations, and extensions to tensors for representing multiple-edge graphs.

1.2.1 Graph adjacency matrix duality

The fundamental concept in an array-based graph algorithm is the duality between a graph and its adjacency representation (see Figure 1.1). To review, for a graph

$G = (V, E)$ with N vertices and M edges, the $N \times N$ adjacency matrix $\mathbf{A}$ has the property $\mathbf{A}(i, j) = 1$ if there is an edge e_{ij} from vertex v_i to vertex v_j and is zero otherwise.

Perhaps even more important is the duality that exists with the fundamental operation of linear algebra (vector matrix multiply) and a breadth-first search (BFS) step performed on G from a starting vertex s

$$BFS(G, s) \quad \Leftrightarrow \quad \mathbf{A}^T\mathbf{v}, \;\; \mathbf{v}(s) = 1$$

This duality allows graph algorithms to be simply recast as a sequence of linear algebraic operations. Many additional relations exist between fundamental linear algebraic operations and fundamental graph operations (see chapters in Part I).

1.2.2 Graph algorithms as semirings

One way to employ linear algebra techniques for graph algorithms is to use a broader definition of matrix and vector multiplication. One such broader definition is that of a semiring (see Chapter 2). In this context, the basic multiply operation becomes (in MATLAB notation)

$$\mathbf{A} \; op_1.op_2 \; \mathbf{v}$$

where for a traditional matrix multiply $op_1 = +$ and $op_2 = *$ (i.e., $\mathbf{Av} = \mathbf{A} \; +.* \; \mathbf{v}$). Using such notation, canonical graph algorithms such as the Bellman–Ford shortest path algorithm can be rewritten using the following semiring vector matrix product (see Chapters 3 and 5)

$$\mathbf{d} = \mathbf{d} \;\; +.\min \;\; \mathbf{A}$$

where the $N \times 1$ vector $\mathbf{d}$ holds the length of the shortest path from a given starting vertex s to all the other vertices.

More complex algorithms, such as betweenness centrality (see Chapter 6), can also be effectively represented using this notation. In short, betweenness centrality tries to measure the "importance" of a vertex in a graph by determining how many shortest paths the vertex is on and normalizing by the number of paths through the vertex. In this instance, we see that the algorithm effectively reduces to a variety of matrix matrix and matrix vector multiplies.

Another example is subgraph detection (see Chapter 8), which reduces to a series of "selection" operations

Row selection:	$\mathbf{A} \; \mathrm{diag}(\mathbf{v})$
Col selection:	$\mathrm{diag}(\mathbf{u}) \; \mathbf{A}$
Row/Col selection:	$\mathrm{diag}(\mathbf{u}) \; \mathbf{A} \; \mathrm{diag}(\mathbf{v})$

where $\mathrm{diag}(\mathbf{v})$ is a diagonal matrix with the values of the vector $\mathbf{v}$ along the diagonal.

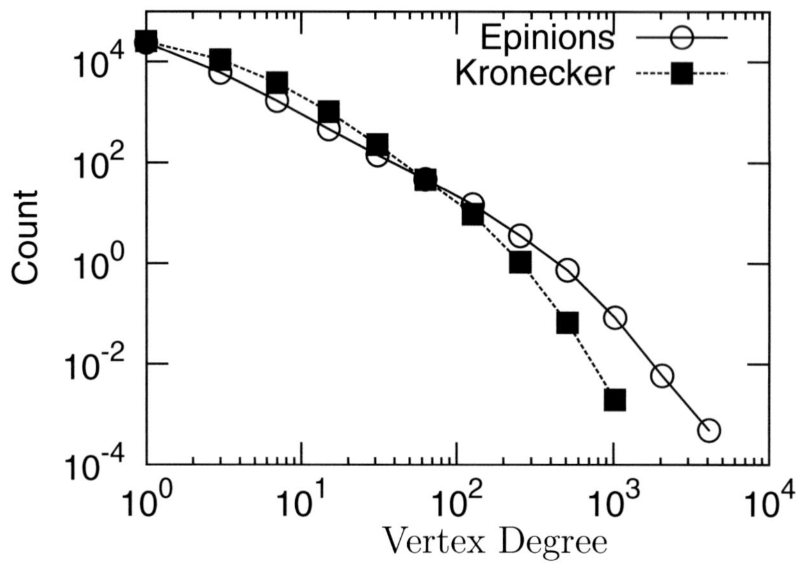

Figure 1.2. Power law graph.
Real and simulated in-degree distribution for the Epinions data set.

1.2.3 Tensors

In many domains (e.g., network traffic analysis), it is common to have multiple edges between vertices. Matrix notation can be extended to these graphs using tensors (see Chapter 7). For example, consider a graph with at most N_k edges between any two vertices. This graph can be represented using the $N \times N \times N_k$ tensor $\mathcal{X}$ where $\mathcal{X}(i,j,k)$ is the kth edge going from vertex i to vertex j.

1.3 Data

A matrix-based approach to the analysis of real-world graphs is useful for the simulation and theoretical analysis of these data sets.

1.3.1 Simulating power law graphs

Power law graphs are ubiquitous and arise in the Internet, the web, citation graphs, and online social networks. Power law graphs have the general property that the histograms of their degree distribution $Deg()$ fall off with a power law and are approximately linear in a log-log plot (see Figure 1.2). Mathematically, this observation can be stated as

$$Slope[\log(Count[Deg(g)])] \approx -constant$$

Efficiently generating simulated data sets that satisfy this property is difficult. Interestingly, an array-based approach using Kronecker products naturally produces graphs of this type (see Chapters 9 and 10). The Kronecker product graph generation algorithm can be described as follows. First, let $\mathbf{A} : \mathbb{R}^{M_B M_C \times N_B N_C}$, $\mathbf{B} : \mathbb{R}^{M_B \times N_B}$, and $\mathbf{C} : \mathbb{R}^{M_C \times N_C}$. Then the Kronecker product is defined as follows:

$$\mathbf{A} = \mathbf{B} \otimes \mathbf{C} = \begin{pmatrix} b_{1,1}\mathbf{C} & b_{1,2}\mathbf{C} & \cdots & b_{1,M_B}\mathbf{C} \\ b_{2,1}\mathbf{C} & b_{2,2}\mathbf{C} & \cdots & b_{2,M_B}\mathbf{C} \\ \vdots & \vdots & & \vdots \\ b_{N_B,1}\mathbf{C} & b_{N_B,2}\mathbf{C} & \cdots & b_{N_B,M_B}\mathbf{C} \end{pmatrix}$$

Now let $\mathbf{G} : \mathbb{R}^{N \times N}$ be an adjacency matrix. The Kronecker exponent to the power k is as follows

$$\mathbf{G}^{\otimes k} = \mathbf{G}^{\otimes k-1} \otimes \mathbf{G}$$

which generates an $N^k \times N^k$ adjacency matrix.

1.3.2 Kronecker theory

It would be very useful if it were possible to analytically compute various centrality metrics for power law graphs. This is possible (see Chapter 10), for example, for Kronecker graphs of the form

$$(\mathbf{B}(n, m) + \mathbf{I})^{\otimes k}$$

where $\mathbf{I}$ is the identity matrix and $\mathbf{B}(n, m)$ is the adjacency matrix of a complete bipartite graph with sets of n and m vertices. For example, the degree distribution (i.e., the histogram of the degree centrality) of the above Kronecker graph is

$$Count[Deg = (n+1)^r (m+1)^{k-r}] = \binom{k}{r} n^{k-r} m^r$$

for $r = 0, \ldots, k$.

1.4 Computation

The previous sections have given some interesting examples of the uses of array-based graph algorithms. In many cases, these algorithms reduce to various sparse matrix multiply operations. Thus, the effectiveness of these algorithms depends upon the ability to efficiently run such operations on parallel computers.

1.4.1 Graph analysis metrics

Centrality analysis is an important tool for understanding real-world graphs. Centrality analysis deals with the identification of critical vertices and edges (see Chapter 12). Example centrality metrics include

Degree centrality is the in-degree or out-degree of the vertex. In an array formulation, this is simply the sum of a row or a column of the adjacency matrix.

Closeness centrality measures how close a vertex is to all the vertices. For example, one commonly used measure is the reciprocal of the sum of all the shortest path lengths.

Stress centrality computes how many shortest paths the vertex is on.

Betweenness centrality computes how many shortest paths the vertex is on and normalizes this value by the number of shortest paths to a given vertex.

Many of these metrics are computationally intensive and require parallel implementations to compute them on even modest-sized graphs (see Chapter 12).

1.4.2 Sparse matrix storage

An array-based approach to graph algorithms depends upon efficient handling of sparse adjacency matrices (see Chapter 13). The primary goal of a sparse matrix is efficient storage that is a small multiple of the number of nonzero elements in the matrix M. A standard storage format used in many sparse matrix software packages is the Compressed Storage by Columns (CSC) format (see Figure 1.3). The CSC format is essentially a dense collection of sparse column vectors. Likewise, the Compressed Storage by Rows (CSR) format is essentially a dense collection of sparse row vectors. Finally, a less commonly used format is the "tuples" format, which is simply a collection of row, column, and value 3-tuples of the nonzero

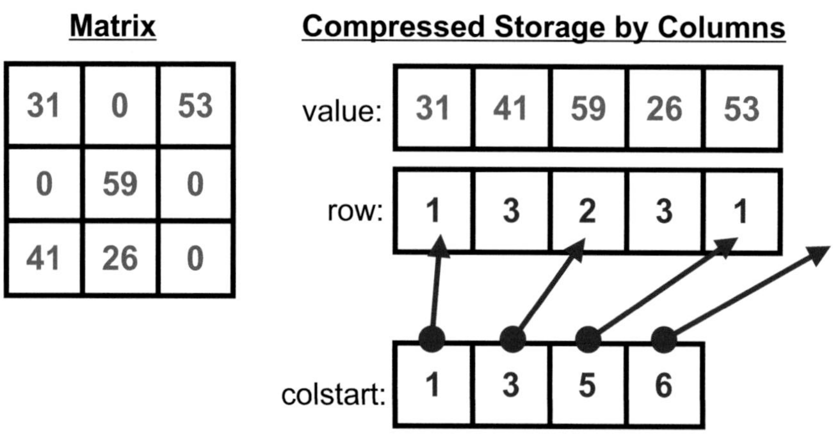

Figure 1.3. Sparse matrix storage.
The CSC format consists of three arrays: `colstart`, `row`, and `value`. `colstart` is an N-element vector that holds a pointer into `row` which holds the row index of each nonzero value in the columns.

elements. Mathematically, the following notation can be used to differentiate these different formats

$$\mathbf{A}: \mathbb{R}^{S(N)\times N} \quad \text{sparse rows (CSR)}$$

$$\mathbf{A}: \mathbb{R}^{N\times S(N)} \quad \text{sparse columns (CSC)}$$

$$\mathbf{A}: \mathbb{R}^{S(N\times N)} \quad \text{sparse rows and columns (tuples)}$$

1.4.3 Sparse matrix multiply

In addition to efficient sparse matrix storage, array-based algorithms depend upon an efficient sparse matrix multiply operation (see Chapter 14). Independent of the underlying storage representation, the amount of useful computation done when two random $N \times N$ matrices with M nonzeros are multiplied together is approximately $2M^2/N$. By using this model, it is possible to quickly estimate the computational complexity of many linear algebraic graph algorithms. A more detailed model of the useful work in multiplying two specific sparse matrices $\mathbf{A}$ and $\mathbf{B}$ is

$$flops(\mathbf{A}\cdot\mathbf{B}) = 2\sum_{k=1}^{N_v} nnz(\mathbf{A}(:,k)) \cdot nnz(\mathbf{B}(k,:))$$

where $M = nnz()$ is the number of nonzero elements in the matrix. Sparse matrix matrix multiply is a natural primitive operation for graph algorithms but has not been widely studied by the numerical sparse matrix community.

1.4.4 Parallel programming

Partitioned Global Address Space (PGAS) languages and libraries are the natural environment for implementing array-based algorithms. PGAS approaches have been implemented in C, Fortran, C++, and MATLAB (see Chapter 4 and [Kepner 2009]). The essence of PGAS is the ability to specify how an array is decomposed on a parallel processor. This decomposition is usually specified in a structure called a "map" (or layout, distributor, distribution, etc.). Some typical maps are shown in Figure 1.4.

The usefulness of PGAS can be illustrated in the following MATLAB example, which creates two distributed arrays `A` and `B` and then performs a data redistribution via the assignment "=" operation

```
Amap = map([Np 1],{},0:Np-1);     % Row map.
Bmap = map([1 Np],{},0:Np-1);     % Column map.
A = rand(N,N,Amap);               % Distributed array.
B = zeros(N,N,Bmap);              % Distributed array.
B(:,:) = A;                       % Redistribute A to B.
```

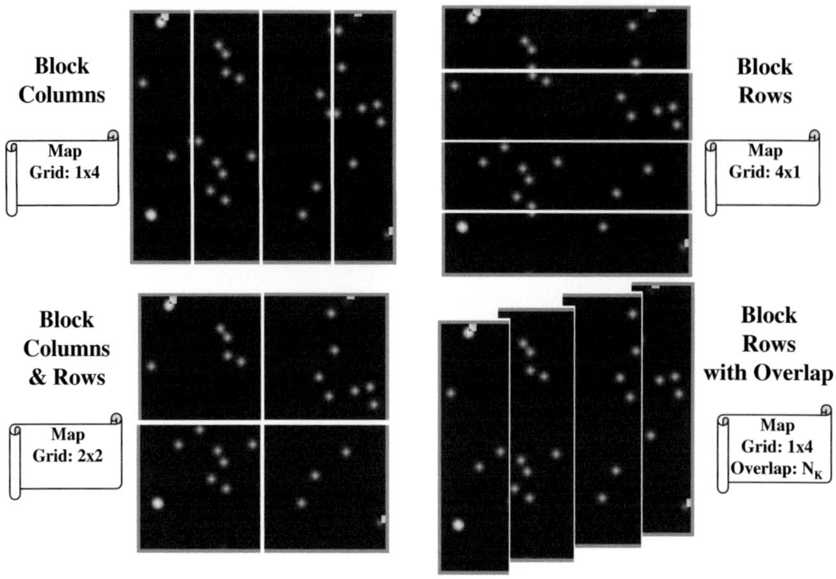

Figure 1.4. Parallel maps.
A selection of maps that are typically supported in PGAS programming environments.

Mathematically, we can write the same algorithm as follows

$$\mathbf{A} : \mathbb{R}^{P(N) \times N}$$

$$\mathbf{B} : \mathbb{R}^{N \times P(N)}$$

$$\mathbf{B} = \mathbf{A}$$

where $P()$ is used to denote the dimension of the array that is being distributed across multiple processors.

1.4.5 Parallel matrix multiply performance

The PGAS notation allows array algorithms to be quickly transformed into graph algorithms. The performance of such algorithms can then be derived from the performance of parallel sparse matrix multiply (see Chapter 14), which can be written as

$$\mathbf{A}, \mathbf{B}, \mathbf{C} : \mathbb{R}^{P(N \times N)}$$

$$\mathbf{A} \leftarrow \mathbf{BC}$$

The computation and communication times of such an algorithm for *random* sparse matrices are

$$T_{comp}(N_P) \propto (M/N)M/N_P$$

$$T_{comm}(N_P) \propto MN_P^{1/2}$$

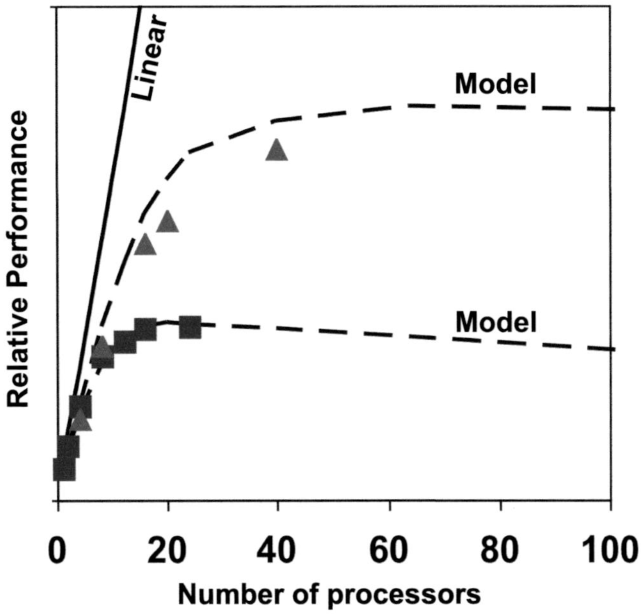

Figure 1.5. Sparse parallel performance.
Triangles and squares show the measured performance of a parallel betweenness centrality code on two different computers. Dashed lines show the performance predicted from the parallel sparse matrix multiply model showing the implementation achieved near the theoretical maximum performance the computer hardware can deliver.

The resulting performance speedup (see Figure 1.5) on a typical parallel computing architecture then shows the characteristic scaling behavior empirically observed (see Chapter 11).

Finally, it is worth mentioning that the above performance is for *random* sparse matrices. However, the adjacency matrices of power law graphs are far from random, and the parallel performance is dominated by the large load imbalance that occurs because certain processors hold many more nonzero values than others. This has been a historically difficult problem to address in parallel graph algorithms. Fortunately, array-based algorithms combined with PGAS provide a mechanism to address this issue by remapping the matrix. One such remapping is the two-dimensional cyclic distribution that is commonly used to address load balancing in parallel linear algebra. Using $P_c()$ to denote this distribution, we have the following algorithm

$$\mathbf{A}, \mathbf{B}, \mathbf{C} : \mathbb{R}^{P_c(N \times N)}$$

$$\mathbf{A} = \mathbf{BC}$$

Thus, with a very minor algorithmic change: $P() \rightarrow P_c()$, the distribution of nonzero values can be made more uniform across processors.

More optimal distributions for sparse matrices can be discovered using automated parallel mapping techniques (see Chapter 15) that exploit the specific distribution of non-zeros in a sparse matrix.

1.5 Summary

This chapter has given a brief survey of some of the more interesting results to be found in the rest of this book, with the hope of motivating the reader to further explore this fertile area of graph algorithms. The book concludes with a final chapter discussing some of the outstanding issues in this field as it relates to the analysis of large graph problems.

References

[Brualdi 1967] R.A. Brualdi. Kronecker products of fully indecomposable matrices and of ultrastrong digraphs. *Journal of Combinatorial Theory,* 2:135–139, 1967.

[Harary & Trauth 1964] F. Harary and C.A. Trauth Jr. Connectedness of products of two directed graphs. *SIAM Journal on Applied Mathematics,* 14:250–254, 1966.

[Harary 1969] F. Harary. *Graph Theory.* Reading: Addison–Wesley. 1969.

[Kepner 2009] J. Kepner. *Parallel MATLAB for Multicore and Multinode Computers.* Philadelphia: SIAM. 2009.

[Konig 1931] D. Konig. Graphen und Matrizen (Graphs and matrices). *Matematikai Lapok,* 38:116–119, 1931.

[Konig 1936] D. Konig. *Theorie der endlichen und unendlichen graphen* (Theory of Finite and Infinite Graphs). Leipzig: Akademie Verlag M.B.H. 1936. See Richard McCourt (Birkhauser 1990) for an English translation of this classic work.

[McAndrew 1963] M.H. McAndrew. On the product of directed graphs. *Proceedings of the American Mathematical Society,* 14:600–606, 1963.

[McAndrew 1965] M.H. McAndrew. On the polynomial of a directed graph. *Proceedings of the American Mathematical Society,* 16:303–309, 1965.

[Sabadusi 1960] G. Sabadusi. Graph multiplication. *Mathematische Zeitschrift,* 72:446–457, 1960.

[Teh & Yap 1964] H.H. Teh and H.D. Yap. Some construction problems of homogeneous graphs. *Bulletin of the Mathematical Society of Nanying University,* 1964:164–196, 1964.

[Weischel 1962] P.M. Weischel. The Kronecker product of graphs. *Proceedings of the American Mathematical Society,* 13:47–52, 1962.

Chapter 2

Linear Algebraic Notation and Definitions

Eric Robinson, Jeremy Kepner*, and John Gilbert*†

Abstract

This chapter presents notation, definitions, and conventions for graphs, matrices, arrays, and operations upon them.

2.1 Graph notation

For the most part, this book will not distinguish between a graph and its adjacency matrix and will move freely between vertex/edge notation and matrix notation. Thus, a graph G can be written either as $G = (V, E)$, where V is a set of N vertices and E is a set of M edges (directed edges unless otherwise stated), or as $G = \mathbf{A}$, where $\mathbf{A}$ is an $N \times N$ matrix with M nonzeros, namely $\mathbf{A}(i, j) = 1$ whenever (i, j) is an edge. This representation will allow many standard graph algorithms to be expressed in a concise linear algebraic form.

*MIT Lincoln Laboratory, 244 Wood Street, Lexington, MA 02420 (erobinson@ll.mit.edu, kepner@ll.mit.edu).

†Computer Science Department, University of California, Santa Barbara, CA 93106-5110 (gilbert@cs.ucsb.edu).

This work is sponsored by the Department of the Air Force under Air Force Contract FA8721-05-C-0002. Opinions, interpretations, conclusions, and recommendations are those of the authors and are not necessarily endorsed by the United States Government.

Usually N will be the number of vertices and M the number of edges in a graph. There are several other equivalent notations

Parameter	Variable	Adjacency Matrix Notation	Vertex/Edge Notation
number of vertices	N	$\|\mathbf{A}\|$	$\|V\|$
number of directed edges	M	$nnz(\mathbf{A})$	$\|E\|$

2.2 Array notation

Most of the arrays (including vectors, matrices, and tensors) in this book have elements that are either boolean (from $\mathbb{B}$), integer (from $\mathbb{Z}$), or real (from $\mathbb{R}$). The notation $A : \mathbb{R}^{5\times 6\times 7}$, for example, indicates that A is a 3D array of 210 real numbers, of size 5 by 6 by 7.

Scalars, vectors, matrices, and tensors are considered arrays; we use the following typographical conventions for them.

Dimensions	Name	Typeface	Example
0	scalar	italic lowercase	s
1	vector	boldface lowercase	$\mathbf{v}$
2	matrix	boldface capital	$\mathbf{M}$
3 or more	tensor	boldface script	$\boldsymbol{\mathfrak{T}}$

The ith entry of a vector $\mathbf{v}$ is denoted by $\mathbf{v}(i)$. An individual entry of a matrix $\mathbf{M}$ or a three-dimensional tensor $\boldsymbol{\mathfrak{T}}$ is denoted by $\mathbf{M}(i,j)$ or $\boldsymbol{\mathfrak{T}}(i,j,k)$. We also allow indexing on expressions; for example, $[(\mathbf{I}-\mathbf{A})^{-1}](i,j)$ is an entry of the inverse of the matrix $\mathbf{I}-\mathbf{A}$.

We will often use the MATLAB notation for subsections and indexes of arrays with any number of dimensions. For example, $\mathbf{A}(1\!:\!5,[3\ 1\ 4\ 1])$ is a 5×4 array containing the elements in the first five rows of columns 3, 1, 4, and 1 (again) in that order. If I is an index or a set of row indices, then $\mathbf{A}(I,:)$ is the submatrix of $\mathbf{A}$ with those rows and all columns.

2.3 Algebraic notation

Here we describe the common algebraic structures and operations on arrays that are used throughout the book. Some individual chapters also define notation specific to their topics; for example, Chapter 7 introduces a number of additional types of matrix and tensor multiplication.

2.3.1 Semirings and related structures

A *semiring* is a set of elements with two binary operations, sometimes called "addition" and "multiplication," such that

- Addition and multiplication have identity elements, sometimes called 0 and 1, respectively.

- Addition and multiplication are associative.
- Addition is commutative.
- Multiplication distributes over addition from both left and right.
- The additive identity is a multiplicative annihilator, $0 * a = a * 0 = 0$.

Both $\mathbb{R}$ and $\mathbb{Z}$ are semirings under their usual addition and multiplication operations. The booleans $\mathbb{B}$ are a semiring under $\wedge$ and $\vee$, as well as under $\vee$ and $\wedge$. If $\mathbb{R}$ and $\mathbb{Z}$ are augmented with $+\infty$, they become semirings with min for "addition" and $+$ for "multiplication." Linear algebra on this $(\min, +)$ semiring is often useful for solving various types of shortest path problems.

We often write semiring addition and multiplication using ordinary notation as $a + b$ and $a * b$ or just ab. When this could be ambiguous or confusing, we sometimes make the semiring operations explicit.

Most of matrix arithmetic and much of linear algebra can be done in the context of a general semiring. Both more and less general settings are sometimes useful. We will see examples that formulate graph algorithms in terms of matrix vector and matrix matrix multiplication over structures that are semiring-like except that addition is not commutative. We will also see algorithms that require a semiring to be *closed*, which means that the equation $x = 1 + ax$ has a solution for every a. Roughly speaking, this corresponds to saying that the sequence $1 + a + a^2 + \cdots$ converges to a limit.

2.3.2 Scalar operations

Scalar operations like $a + b$ and ab have the usual interpretation. An operation between a scalar and an array is applied pointwise; thus $a + \mathbf{M}$ is a matrix the same size as $\mathbf{M}$.

2.3.3 Vector operations

We depart from the convention of numerical linear algebra by making no distinction between row and column vectors. (In the context of multidimensional tensors, we prefer not to deal with notation for a different kind of vector in each dimension.) For vectors $\mathbf{v} : \mathbb{R}^M$ and $\mathbf{w} : \mathbb{R}^N$, the *outer product* of $\mathbf{v}$ and $\mathbf{w}$ is written as $\mathbf{v} \circ \mathbf{w}$, which is the $M \times N$ matrix whose (i, j) element is $\mathbf{v}(i) * \mathbf{w}(j)$. If $M = N$, the *inner product* $\mathbf{v} \cdot \mathbf{w}$ is the scalar $\sum_i \mathbf{v}(i) * \mathbf{w}(i)$.

Given also a matrix $\mathbf{M} : \mathbb{R}^{M \times N}$, the products $\mathbf{vM}$ and $\mathbf{Mw}$ are both vectors, of dimension N and M, respectively.

When we operate over semirings other than the usual $(+, *)$ rings on $\mathbb{R}$ and $\mathbb{Z}$, we will sometimes make the semiring operations explicit in matrix vector (and matrix matrix) multiplication. For example, $\mathbf{M}(\min.+)\mathbf{w}$, or $\mathbf{M} \min.+ \mathbf{w}$, is the M-vector whose ith element is $\min(\mathbf{M}(i, j) + \mathbf{w}(j) : 1 \leq j \leq N)$. The usual matrix vector multiplication $\mathbf{Mw}$ could also be written as $\mathbf{M} +.* \mathbf{w}$.

2.3.4 Matrix operations

Three kinds of matrix "multiplication" arise frequently in graph algorithms. All three are defined over any semiring.

If $\mathbf{A}$ and $\mathbf{B}$ are matrices of the same size, the *pointwise product* (or *Hadamard product*) $\mathbf{A} .* \mathbf{B}$ is the matrix $\mathbf{C}$ with $\mathbf{C}(i,j) = \mathbf{A}(i,j) * \mathbf{B}(i,j)$. Similar notation applies to other pointwise binary operators; for example, $\mathbf{C} = \mathbf{A}./\mathbf{B}$ has $\mathbf{C}(i,j) = \mathbf{A}(i,j)/\mathbf{B}(i,j)$, and $\mathbf{A} .+ \mathbf{B}$ is the same as $\mathbf{A} + \mathbf{B}$.

If $\mathbf{A}$ is $M \times N$ and $\mathbf{B}$ is $N \times P$, then $\mathbf{AB}$ is the conventional $M \times P$ matrix product. We sometimes make the semiring explicit by writing, for example, $\mathbf{A} +.* \mathbf{B}$ or $\mathbf{A} \min.+ \mathbf{B}$.

Finally, if $\mathbf{A}$ is $M \times N$ and $\mathbf{B}$ is $P \times Q$, the *Kronecker product* $\mathbf{A} \otimes \mathbf{B}$ is the $MP \times NQ$ matrix $\mathbf{C}$ with $\mathbf{C}(i,j) = \mathbf{A}(s,t) * \mathbf{B}(u,v)$, where $i = (s-1)P + u$ and $j = (t-1)Q + v$. One can think of $\mathbf{A} \otimes \mathbf{B}$ as being obtained by replacing each element $\mathbf{A}(s,t)$ of $\mathbf{A}$ by its pointwise product with a complete copy of $\mathbf{B}$. The Kronecker power $\mathbf{A}^{\otimes k}$ is defined as the k-fold Kronecker product $\mathbf{A} \otimes \mathbf{A} \otimes \cdots \otimes \mathbf{A}$.

It is useful to extend the "dotted" notation to represent matrix scalings. If $\mathbf{A}$ is an $M \times N$ matrix, $\mathbf{v}$ is an M-vector, and $\mathbf{w}$ is an N-vector, then $\mathbf{v} .* \mathbf{A}$ scales the rows of $\mathbf{A}$; that is, the result is the matrix whose (i,j) entry is $\mathbf{v}(i) * \mathbf{A}(i,j)$. Similarly, $\mathbf{A} .* \mathbf{w}$ scales columns, yielding the matrix whose (i,j) entry is $\mathbf{w}(j) * \mathbf{A}(i,j)$. In MATLAB notation, these could be written $\mathrm{diag}(\mathbf{v}) * \mathbf{A}$ and $\mathbf{A} * \mathrm{diag}(\mathbf{w})$.

2.4 Array storage and decomposition

Section 2.2 defined multidimensional arrays as mathematical objects, without reference to how they are stored in a computer. When presenting algorithms, we sometimes need to talk about the representation used for storage. This section gives our notation for describing sparse and distributed array storage.

2.4.1 Sparse

An array whose elements are mostly zeros can be represented compactly by storing only the nonzero elements and their indices. Many different sparse data structures exist; Chapter 13 surveys several of them.

It is often useful to view sparsity as an attribute attached to one or more dimensions of an array. For example, the notation $A : \mathbb{R}^{500 \times S(600)}$ indicates that A is a 500×600 array of real numbers, which can be thought of as a dense array of 500 rows, each of which is a sparse array of 600 columns. Figure 2.1 shows two possible data structures for an array $A : \mathbb{Z}^{4 \times S(4)}$. A data structure for $A : \mathbb{R}^{S(500) \times 600}$ would interchange the roles of rows and columns. An array $A : \mathbb{R}^{S(500) \times S(600)}$, or equivalently $A : \mathbb{R}^{S(500 \times 600)}$, is sparse in both dimensions; it might be represented simply as an unordered sequence of triples (i, j, a) giving the positions and values of the nonzero elements. A three-dimensional array $A : \mathbb{R}^{500 \times 600 \times S(700)}$ is a dense two-dimensional array of 500×600 sparse 700-vectors.

Sparse representations generally trade off ease of access for memory. Most data structures support constant-time random indexing along dense dimensions, but not

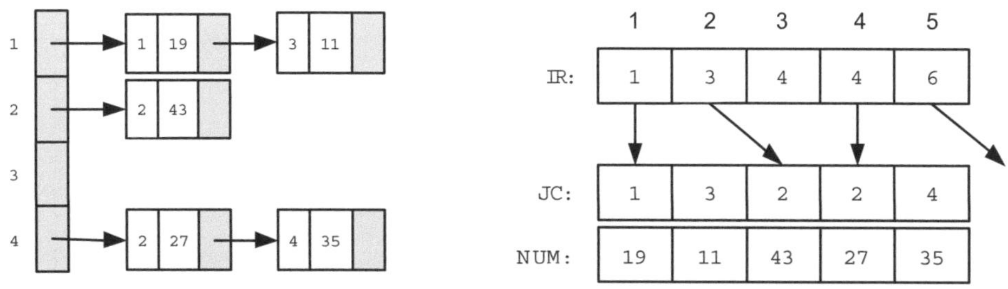

Figure 2.1. Sparse data structures.
Two data structures for a sparse array $A : \mathbb{Z}^{4\times S(4)}$. Left: adjacency lists. Right: compressed sparse rows.

along sparse dimensions. The memory requirement is typically proportional to the number of nonzeros times the number of sparse dimensions, plus the product of the sizes of the dense dimensions.

2.4.2 Parallel

When analyzing the parallel performance of the algorithms described in this book, it is important to consider three things: the number of instances of the program used in the computation (denoted N_P), the unique identifier of each instance of the program (denoted $P_{ID} = 0, \ldots, N_P - 1$), and the distribution of the arrays used in those algorithms over those P_{ID}s. Consider a nondistributed array $\mathbf{A} : \mathbb{R}^{N\times N}$. The corresponding distributed array is given in "P notation" as $\mathbf{A} : \mathbb{R}^{P(N)\times N}$, where the first dimension is distributed among N_P program instances. Figure 2.2 shows $\mathbf{A} : \mathbb{R}^{P(16)\times 16}$ for $N_P = 4$. Likewise, Figure 2.3 shows $\mathbf{A} : \mathbb{R}^{16\times P(16)}$ for $N_P = 4$.

Block distribution

A block distribution is the default distribution. It is used to represent the grouping of adjacent columns/rows, planes, or hyperplanes on the same P_{ID}. A parallel dimension is declared using $P(N)$ or $P_b(N)$. For $\mathbf{A} : \mathbb{R}^{P(N)\times N}$, each row $\mathbf{A}(i,:)$ is assumed to reside on $P_{ID} = \lceil i/\lceil N/N_P \rceil \rceil$. Some examples of block distributions for matrices are provided. Figure 2.2 shows a block distribution over the rows of a matrix. Figure 2.3 shows a block distribution over the columns of a matrix.

Cyclic distribution

A cyclic distribution is used to represent distributing adjacent items in a distributed dimension onto different P_{ID}s. For $\mathbf{A} : \mathbb{R}^{P_c(N)\times N}$, each row $\mathbf{A}(i,:)$ is assumed to reside on $P_{ID} = (i-1) \bmod N_P$.

Some examples of cyclic distributions for matrices are provided. Figure 2.4 shows a cyclic distribution over the rows of a matrix. Figure 2.5 shows a cyclic distribution over the columns of a matrix.

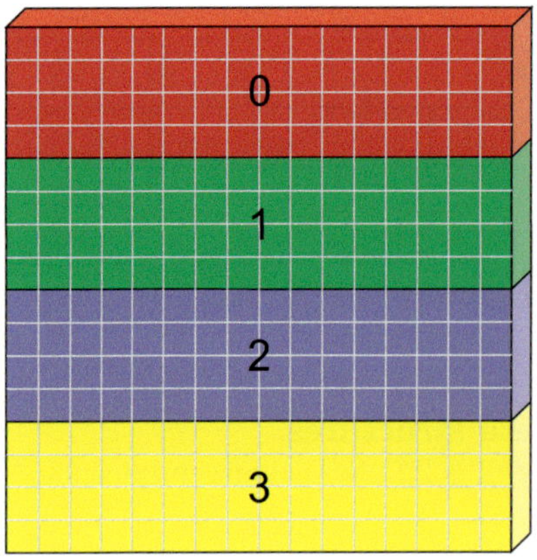

Figure 2.2. A row block matrix.

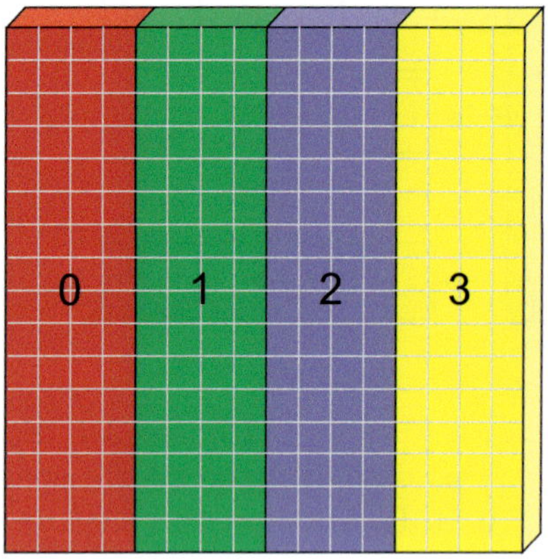

Figure 2.3. A column block matrix.

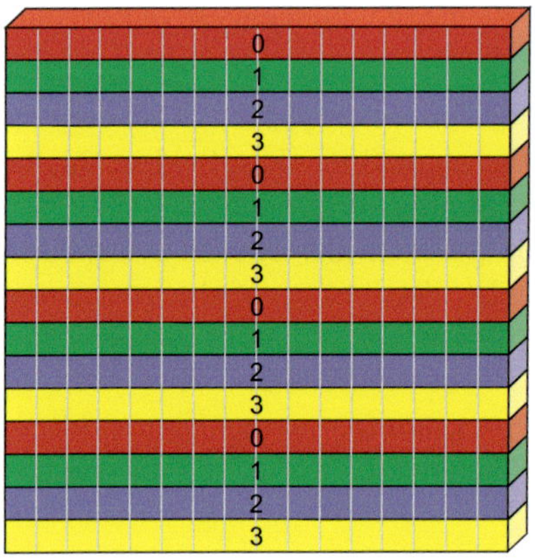

Figure 2.4. A row cyclic matrix.

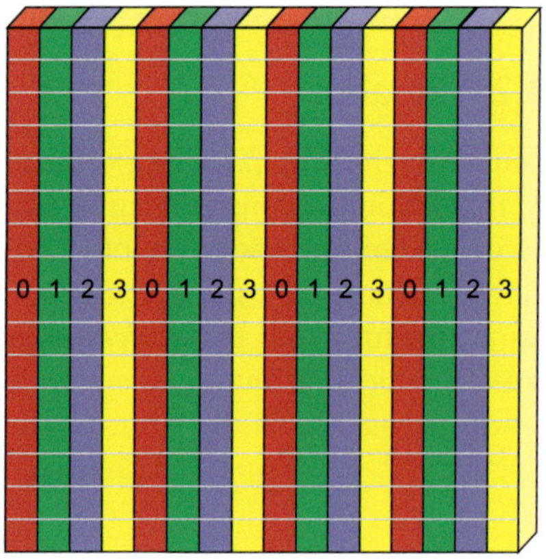

Figure 2.5. A column cyclic matrix.

Chapter 3

Connected Components and Minimum Paths

*Charles M. Rader**

Abstract

A familiarity with matrix algebra is useful in understanding and inventing graph algorithms. In this chapter, two very different examples of graph algorithms based on linear algebra are presented. Strongly connected components are obtained via efficient computation of infinite powers of the adjacency matrix. Shortest paths are computed using a modification of matrix exponentiation.

3.1 Introduction

Any graph can be represented by its adjacency matrix $\mathbf{A}$. Familiarity with matrix operations and properties could therefore be a useful asset for solving some problems in graph theory. In this chapter, we consider two of the classical graph theory problems, finding *strongly connected components*, and finding *minimum path lengths*, from the point of view of linear algebra.

To find strongly connected components, we will rely on a relationship in linear algebra

$$(\mathbf{I} - \mathbf{A})^{-1} = \sum_{n=0}^{\infty} \mathbf{A}^n$$

which is the matrix analogy to the series identity $1/(1-x) = 1 + x + x^2 + \cdots$.

*MIT Lincoln Laboratory, 244 Wood Street, Lexington, MA 02420 (charlesmrader@verizon.net).

This work is sponsored by the Department of the Air Force under Air Force Contract FA8721-05-C-0002. Opinions, interpretations, conclusions, and recommendations are those of the author and are not necessarily endorsed by the United States Government.

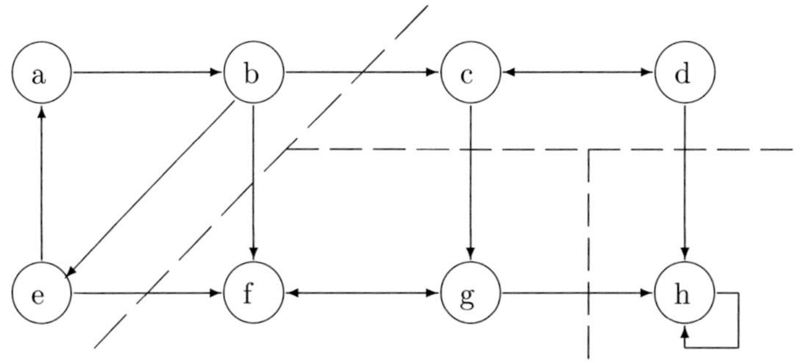

Figure 3.1. Strongly connected components.
An eight-node graph illustrating strongly connected components. In the adjacency matrix of the graph row/col 1 is node a, row/col 2 is node b, etc.

To compute minimum path lengths, we will need an invented operation that looks like matrix exponentiation, except that the two operations + and × are replaced by min and +, respectively. A well-known shortcut algorithm for computing a power of a matrix will still work even after making that change.

3.2 Strongly connected components

For any directed graph, it is possible to group the nodes of the graph into maximal sets such that within each set of nodes there exist paths connecting any node to any other node in the same set. For example, for the eight-node graph in Figure 3.1, those sets are nodes a,b,e, nodes c,d, nodes f,g, and node h. Segregating the nodes of a graph into such sets is called finding *strongly connected components.*

The graph in Figure 3.1 is taken from [Cormen et al. 2001], page 553.* We will use it as an example in this chapter.

For our example graph, the incidence matrix is

$$\mathbf{A} = \begin{bmatrix} 0 & 1 & 0 & 0 & 0 & 0 & 0 & 0 \\ 0 & 0 & 1 & 0 & 1 & 1 & 0 & 0 \\ 0 & 0 & 0 & 1 & 0 & 0 & 1 & 0 \\ 0 & 0 & 1 & 0 & 0 & 0 & 0 & 1 \\ 1 & 0 & 0 & 0 & 0 & 1 & 0 & 0 \\ 0 & 0 & 0 & 0 & 0 & 0 & 1 & 0 \\ 0 & 0 & 0 & 0 & 0 & 1 & 0 & 1 \\ 0 & 0 & 0 & 0 & 0 & 0 & 0 & 1 \end{bmatrix}$$

Identifying strongly connected components can be accomplished using matrix operations.

*T.H. Cormen, C.E. Leiserson, R. Rivest and C. Stein. *Introduction to Algorithms, third edition*, figure, page 553, © 2009 Massachusetts Insitute of Technology, by permission of the MIT Press.

Let's define another matrix $\mathbf{C}$ computed by using the element wise or-function ($\vee$) of $\mathbf{I}, \mathbf{A}, \mathbf{A}^2, \mathbf{A}^3, \ldots,$ where $a \vee b$ is 0 when both a and b are 0, and is 1 if either a or b is nonzero. The matrix $\mathbf{C}$ can be made up of an infinite number of terms (although a large enough finite number of terms would do). For the moment, let's not worry about how many terms we use

$$\mathbf{C} = \mathbf{I} \vee \mathbf{A} \vee \mathbf{A}^2 \vee \mathbf{A}^3 \vee \mathbf{A}^4 \vee \cdots$$

$\mathbf{A}^k$ has the property that $\mathbf{A}^k(i,j) > 0$ if and only if there is a path from node i to node j in exactly k steps. If there are m ways to go from node i to node j in exactly k steps, then $\mathbf{A}^k(i,j) = m$. $\mathbf{A}^k(i,j) = 0$ when there is no such path with exactly k steps.

The resulting matrix $\mathbf{C}$ has a nonzero in position i, j if and only if there are paths from node i to node j in any number of steps—including 0 steps, because we included $\mathbf{I}$ in the series.

Here's what $\mathbf{C}$ looks like in our example

$$\mathbf{C} = \begin{bmatrix} 1 & 1 & 1 & 1 & 1 & 1 & 1 & 1 \\ 1 & 1 & 1 & 1 & 1 & 1 & 1 & 1 \\ 0 & 0 & 1 & 1 & 0 & 1 & 1 & 1 \\ 0 & 0 & 1 & 1 & 0 & 1 & 1 & 1 \\ 1 & 1 & 1 & 1 & 1 & 1 & 1 & 1 \\ 0 & 0 & 0 & 0 & 0 & 1 & 1 & 1 \\ 0 & 0 & 0 & 0 & 0 & 1 & 1 & 1 \\ 0 & 0 & 0 & 0 & 0 & 0 & 0 & 1 \end{bmatrix}$$

Now consider the element-by-element logical and function ($\wedge$) of $\mathbf{C}$ and $\mathbf{C}^T$

$$\mathbf{C} \wedge \mathbf{C}^T = \begin{bmatrix} 1 & 1 & 0 & 0 & 1 & 0 & 0 & 0 \\ 1 & 1 & 0 & 0 & 1 & 0 & 0 & 0 \\ 0 & 0 & 1 & 1 & 0 & 0 & 0 & 0 \\ 0 & 0 & 1 & 1 & 0 & 0 & 0 & 0 \\ 1 & 1 & 0 & 0 & 1 & 0 & 0 & 0 \\ 0 & 0 & 0 & 0 & 0 & 1 & 1 & 0 \\ 0 & 0 & 0 & 0 & 0 & 1 & 1 & 0 \\ 0 & 0 & 0 & 0 & 0 & 0 & 0 & 1 \end{bmatrix}$$

This is essentially the answer we seek. Row i of $\mathbf{C} \wedge \mathbf{C}^T$ has a 1 in column k if and only if node i and node k belong to the same strongly connected set of nodes.

3.2.1 Nondirected links

If the links in the graph are all bidirectional, then $\mathbf{A}$ is a symmetric matrix. Therefore, $\mathbf{C}$ is symmetric. Hence $\mathbf{C} \wedge \mathbf{C}^T$ is the same as $\mathbf{C}$ and that step can be omitted.

3.2.2 Computing C quickly

Computing $\mathbf{C} = \mathbf{I} \vee \mathbf{A} \vee \mathbf{A}^2 \vee \mathbf{A}^3 \vee \mathbf{A}^4 \vee \cdots$ looks complicated. Suppose instead we compute

$$\mathbf{D} = \mathbf{I} + \mathbf{A} + \mathbf{A}^2 + \mathbf{A}^3 + \mathbf{A}^4 + \cdots + \mathbf{A}^K$$

Then $\mathbf{D}$ has 0 where $\mathbf{C}$ has 0, and $\mathbf{D}$ has a nonzero positive entry where $\mathbf{C}$ has 1. So $\mathbf{C}(i,j) = (\mathbf{D}(i,j) > 0)$. We can compute $\mathbf{D}$ and we instantly have $\mathbf{C}$. Maybe $\mathbf{D}$ is easier to compute.

Now let $\mathbf{D}$ be taken for an infinite number of terms. This series almost certainly does not converge, but we will fix that later. For now consider $\mathbf{E} \equiv (\mathbf{I} - \mathbf{A})\mathbf{D}$

$$\begin{aligned}
\mathbf{E} &= \mathbf{D} - \mathbf{A}\mathbf{D} \\
&= \mathbf{I} + \mathbf{A} + \mathbf{A}^2 + \mathbf{A}^3 + \mathbf{A}^4 + \cdots \\
&\qquad - \mathbf{A} - \mathbf{A}^2 - \mathbf{A}^3 - \mathbf{A}^4 - \cdots \\
&= \mathbf{I} \\
&= (\mathbf{I} - \mathbf{A})\mathbf{D}
\end{aligned}$$

Hence $\mathbf{D} = (\mathbf{I} - \mathbf{A})^{-1}$. This is the compact representation of $\mathbf{D}$, which we should be able to compute quickly using sparse matrix algorithms.

Unfortunately, the method will almost always fail. The problem is that the series for $\mathbf{D}$ usually does not converge. But we can fix our convergence problem very easily. Let's introduce α, which is any positive number, replace $\mathbf{A}$ by $\alpha\mathbf{A}$, and redefine $\mathbf{D}$

$$\mathbf{D} = \mathbf{I} + (\alpha\mathbf{A}) + (\alpha\mathbf{A})^2 + (\alpha\mathbf{A})^3 + (\alpha\mathbf{A})^4 + \cdots$$

taken for an infinite number of terms. Once again, we argue that the term $(\alpha\mathbf{A})^k$ has a nonzero positive entry in position i,j if and only if there is a path from node i to node j in k steps. So $\mathbf{D}$ has a nonzero positive entry in position i,j if and only if there is a path from node i to node j.

Now we examine $\mathbf{E} \equiv (\mathbf{I} - \alpha\mathbf{A})\mathbf{D}$

$$\begin{aligned}
\mathbf{E} &= \mathbf{I} + \alpha\mathbf{A} + \alpha^2\mathbf{A}^2 + \alpha^3\mathbf{A}^3 + \alpha^4\mathbf{A}^4 + \cdots \\
&\qquad - \alpha\mathbf{A} - \alpha^2\mathbf{A}^2 - \alpha^3\mathbf{A}^3 - \alpha^4\mathbf{A}^4 - \cdots \\
&= \mathbf{I} \\
&= (\mathbf{I} - \alpha\mathbf{A})\mathbf{D}
\end{aligned}$$

Hence $\mathbf{D} = (\mathbf{I} - \alpha\mathbf{A})^{-1}$.

If we choose α sufficiently small, we are sure that all our infinite series converge, so all our math is valid. In the illustrated case, we used $\alpha = 0.5$ and $\mathbf{D}$ was computed as

$$(\mathbf{I} - \alpha\mathbf{A})^{-1} = \begin{bmatrix} 1.143 & 0.571 & 0.381 & 0.191 & 0.286 & 0.698 & 0.540 & 0.730 \\ 0.286 & 1.143 & 0.762 & 0.381 & 0.571 & 1.397 & 1.079 & 1.460 \\ 0 & 0 & 1.333 & 0.667 & 0 & 0.444 & 0.889 & 1.556 \\ 0 & 0 & 0.667 & 1.333 & 0 & 0.222 & 0.444 & 1.778 \\ 0.571 & 0.286 & 0.191 & 0.095 & 1.143 & 1.016 & 0.603 & 0.698 \\ 0 & 0 & 0 & 0 & 0 & 1.333 & 0.667 & 0.667 \\ 0 & 0 & 0 & 0 & 0 & 0.667 & 1.333 & 1.333 \\ 0 & 0 & 0 & 0 & 0 & 0 & 0 & 2.000 \end{bmatrix}$$

and $\mathbf{D}$ has the same pattern of 0 and nonzero elements as $\mathbf{C}$.

3.3 Dynamic programming, minimum paths, and matrix exponentiation

In this section, we will formulate dynamic programming [Bellman 1952], a method of finding the least cost paths through a graph, by using incidence matrices, and we will define a matrix operation which is isomorphic to matrix multiplication. Then we will extend that to an isomorphism to matrix exponentiation and show how it finds all the shortest paths in a graph.

Suppose we have a directed graph, G, with N nodes and M edges, and an extended incidence matrix $\mathbf{A}$ (whose i, j entry is the cost, nonnegative, of going from node i to node j of the graph G in one step). We are interested in finding another matrix whose i, j entry is the least possible cost of going from node i to node j in whatever number of steps gives the least cost. The entries $\mathbf{A}(i, i)$ on the diagonal of matrix $\mathbf{A}$ will be 0, as we will explain later.

When we multiply a matrix $\mathbf{A}$ by another matrix $\mathbf{B}$, component i, j of the resulting matrix matrix product $\mathbf{C}$ is

$$\mathbf{C}(i,j) = \sum_k \mathbf{A}(i,k) * \mathbf{B}(k,j)$$

Let's define a different way of matrices operating on one another

$$\mathbf{C} = \mathbf{A} \ \min.+ \ \mathbf{B}$$

$$\mathbf{C}(i,j) = \min_k \{ \ \mathbf{A}(i,k) + \mathbf{B}(k,j) \ \} \tag{3.1}$$

The min .+ operation is isomorphic to matrix matrix multiplication. The roles played by scalar multiplication and scalar addition in a conventional matrix matrix multiply are played by addition and minimum in the min .+ operation

$$\sum_k \to \min_k; \qquad * \to +$$

Next we will apply the min .+ operation to finding the costs of shortest paths in G.

When the graph G has no edge from node i to node j, the incidence matrix $\mathbf{A}$ would normally have $\mathbf{A}(i,j) = 0$. But that would mean that there was no cost to go from one node to another node not even connected to it. We would rather have missing edges *left out* of the minimization defined in equation (3.1). Instead of formally changing equation (3.1), it is easiest to use a modified incidence matrix for which a left-out edge has $\mathbf{A}(i,j) = \infty$. Fortunately, it is not necessary to have a computer code representing ∞. Any sufficiently large number, larger than the cost of any $N-1$ step path, would serve as well.

Let's consider $\mathbf{C} = \mathbf{A} \;\min.+\; \mathbf{A}$

$$\mathbf{C}(i,j) = \min_k\{\ \mathbf{A}(i,k) + \mathbf{A}(k,j)\ \}$$

$\mathbf{A}(i,k)$ is the weight along a path from node i to node k and $\mathbf{A}(k,j)$ is the weight along a path from node k to node j. Then $\mathbf{A}(i,k)+\mathbf{A}(k,j)$ is the sum of the weights (hereafter called the *cost*) encountered along the two-step path from node i to node j passing through node k, and $\mathbf{C}(i,j) = \min_k\{\ \mathbf{A}(i,k) + \mathbf{A}(k,j)\ \}$ is the cost along the two-step path from node i to node j for which that cost is minimum.

It follows that $\mathbf{C}$ will have in position i,j the lowest cost of moving from node i to node j in exactly two steps. But recall that earlier we decided that $\mathbf{A}(i,i)$ would be zero. So the minimization in $\mathbf{C}(i,j) = \min_k\{\ \mathbf{A}(i,k) + \mathbf{A}(k,j)\ \}$ includes $k = i$ and $k = j$, and these represent the cost of going from i to j in only one step. Therefore, $\mathbf{C} = \mathbf{A} \;\min.+\; \mathbf{A}$ will have in position i,j the lowest cost of moving from node i to node j in *no more than* two steps. We could consider some path as using two steps even when one of the steps goes from node i to itself. In the remainder of this discussion, when we refer to paths of a given length, that will always include all shorter paths as well.

Now we introduce the notation $\mathbf{A}^{\diamond n}$. We will define

$$\mathbf{A}^{\diamond 1} = \mathbf{A}, \quad \mathbf{A}^{\diamond 2} = \mathbf{A} \;\min.+\; \mathbf{A}$$

and in general

$$\mathbf{A}^{\diamond n} = \mathbf{A} \;\min.+\; (\mathbf{A}^{\diamond(n-1)})$$

$\mathbf{A}(i,j)$ has the minimum cost of all one-step paths from node i to node j because there is only one such path. We have the beginning of a pattern. $\mathbf{A}^{\diamond 1}$ gives the minimum costs for one step, and $\mathbf{A}^{\diamond 2}$ gives the minimum costs for two steps. We will next show that the pattern continues, e.g., $\mathbf{A}^{\diamond 3}$ gives the minimum costs for three steps, etc.

We will later show that $\mathbf{A}^{\diamond n}$ shares an important property with matrix exponentiation; namely, we will prove that for any nonnegative p and q, $\mathbf{A}^{\diamond(p+q)} = \mathbf{A}^{\diamond p} \;\min.+\; \mathbf{A}^{\diamond q}$.

First, we simply explain why the i,j entry of the matrix $\mathbf{A}^{\diamond n}$ is the cost of the least cost path from node i to node j using n steps. We do this by mathematical induction. We assume that the statement is true for n and show that it must then be true for $n+1$. We already know that the statement is true for $n = 1$ since that is how $\mathbf{A}$ was defined.

The smallest one-step path cost from node i to any node k is given in $\mathbf{A}$, and the smallest n step path cost from any node k to node j is given in $\mathbf{B} =$

$\mathbf{A}^{\diamond n}$. But there must be some node k on the smallest cost $n+1$ step path from node i to any node j. Bellman's principle tells us that all subpaths on an optimum path must be optimum subpaths. So if node k is on the optimum path from i to j, the one step subpath from i to k must be (trivially) optimum and its cost is $\mathbf{A}(i,k)$, and the subpath of n steps from k to j must be optimum and its cost is $\mathbf{B}(k,j)$. The computation $\mathbf{A}$ min.+ $\mathbf{B}$ explicitly determines which node k optimizes $\mathbf{A}(i,k)+\mathbf{B}(k,j)$, which is the cost of a path with $n+1$ steps. Q.E.D.

3.3.1 Matrix powers

We could compute $\mathbf{B} = \mathbf{A}^{\diamond K}$ by the following loop

$\mathbf{B} = \mathbf{A}$
for $j = 2 : K$
  do
    $\mathbf{B} = \mathbf{A}\ \text{min}.+\ \mathbf{B}$

As we have seen, $\mathbf{B}(i,j)$ would contain the cost of the least cost K step path from node i to node j. But suppose we are interested in the absolute least cost. It would never require more than $N-1$ steps to get from one node to another at least cost. Therefore, $\mathbf{A}^{\diamond n}$ does not change as n increases beyond $n = N-1$. [Note: it is assumed that it is possible to get from node i to node j. If a graph consists of several disconnected subgraphs, then there are nodes i, j for which the absolute least cost path, with at most $N-1$ steps, would have to include one impossible step, e.g., it would have to include one step with infinite costs.]

$\mathbf{B} = \mathbf{A}^{\diamond K}$ contains the costs of least cost paths, but does not contain the paths themselves. The information about paths was thrown away when, in equation (3.1), we saved only the minimum $\mathbf{A}(i,k)+\mathbf{B}(k,j)$ and did not record the k which achieved it. If we want to identify the optimum paths, whenever we perform a min.+ calculation $\mathbf{B} = \mathbf{A}$ min.+ $\mathbf{B}$, we simply maintain a second matrix $\mathbf{D}$ whose i, j element is the k for which the minimum was achieved. There may sometimes be ties, but if a tie is arbitrarily broken, then *one* of the minimum cost paths will be identified.

Now we return to prove the claim that for any nonnegative p and q, $\mathbf{A}^{\diamond(p+q)} = \mathbf{A}^{\diamond p}$ min.+ $\mathbf{A}^{\diamond q}$. But this is simply another statement of Bellman's principle. For any k, $\mathbf{R} = \mathbf{A}^{\diamond p}$ has the optimum p-step path cost $\mathbf{R}(i,k)$ between node i and node k. $\mathbf{S} = \mathbf{A}^{\diamond q}$ has the optimum q-step path cost $\mathbf{S}(k,j)$ between node k and node j. If an optimum path of $p+q$ steps passes through some node k, the subpaths from node i to node k and from node k to node j must be optimum subpaths. The $\mathbf{R}$ min.+ $\mathbf{S}$ operation simply finds the one node k that optimizes $\mathbf{R}(i,k)+\mathbf{S}(k,j)$. By Bellman's principle, this must be the cost of the least cost path of $p+q$ steps and so it is also $\mathbf{A}^{\diamond(p+q)}$.

It is in that sense that we consider $\mathbf{A}^{\diamond n}$ to be isomorphic to matrix exponentiation.

This gives us other more efficient ways to compute $\mathbf{A}^{\diamond K}$. If we choose K to be the smallest power of two greater than or equal to $N-1$, we can then economically compute the matrix whose i, j element is the absolute optimum path cost from node i to node j

$$
\begin{aligned}
\mathbf{B}^{(0)} &= \mathbf{A} \\
\mathbf{B}^{(1)} &= \mathbf{B}^{(0)} \ \min.+ \ \mathbf{B}^{(0)} = \mathbf{A}^{\diamond 2} \\
\mathbf{B}^{(2)} &= \mathbf{B}^{(1)} \ \min.+ \ \mathbf{B}^{(1)} = \mathbf{A}^{\diamond 4} \\
&\vdots \\
\mathbf{B}^{(r)} &= \mathbf{B}^{(1)} \ \min.+ \ \mathbf{B}^{(1)} = \mathbf{A}^{\diamond 2^r}
\end{aligned}
$$

and stop when we have computed $\mathbf{A}^{\diamond 2^r}$ where $2^r \geq N-1$.

Again, we note that $\mathbf{A}^{\diamond 2^r}$ gives the costs of the optimum paths, not the paths themselves. If we want to recover the paths, we must save, in a set of matrices $\mathbf{D}^{(m)}, m = 1, \ldots, r$, the values of k, which were optimum in each min.+ computation. That is, we redefine the min.+ operation as $[\mathbf{C}, \mathbf{D}] = \mathbf{A} \ \min.+ \ \mathbf{B}$ so that at the same time we compute

$$
\begin{aligned}
\mathbf{C}(i,j) &= \min_k \{ \ \mathbf{A}(i,k) + \mathbf{A}(k,j) \ \} \\
\mathbf{D}(i,j) &= \operatorname{argmin}\{ \ \mathbf{A}(i,k) + \mathbf{A}(k,j) \ \}
\end{aligned}
$$

Now our algorithm is

$$
\begin{aligned}
\mathbf{B}^{(0)} &= \mathbf{A} \\
\left[\mathbf{B}^{(1)}, \mathbf{D}^{(1)}\right] &= \mathbf{B}^{(0)} \ \min.+ \ B^{(0)} \\
\left[\mathbf{B}^{(2)}, \mathbf{D}^{(2)}\right] &= \mathbf{B}^{(1)} \ \min.+ \ B^{(1)} \\
&\vdots
\end{aligned}
$$

If we want to recover the overall optimum path from node i to node j, we look first at $\mathbf{D}^{(r)}(i,j)$ to find which node k is "halfway" from node i to node j. Then in $\mathbf{D}^{(r-1)}$ we look at $\mathbf{D}^{(r-1)}(i,k)$ to find the node halfway between node i and node k, and we also look at $\mathbf{D}^{(r-1)}(k,j)$ to find the node halfway between node k and node j, and so on.

3.4 Summary

If a graph is represented by its incidence matrix, some algorithms that find properties of a graph may be expressible as operations on matrices. So, often a familiarity with matrix algebra will be useful in understanding some graph algorithms, or even in inventing some graph algorithms. In this chapter, we have given two very different examples of graph algorithms based on familiarity with linear algebra.

In the first example, finding strongly connected components of a graph, we reduced the graph problem to computing a sum of powers of the incidence matrix, and then we recognized a matrix identity which helped us compute that sum of powers in a simple closed form. It is worth noting that the Kth power of an incidence matrix gives, in position i, j, the number of different paths that can possibly be taken to traverse from node i to node j in exactly K steps.

In the second example, finding minimum path lengths, the usual definitions of matrix multiplication and addition were not useful, but the operation that was useful was isomorphic to standard matrix multiplication, using the min operation instead of addition, and the $+$ operation in place of multiplication, but the basic steps were ordered just as in matrix multiplication. We then showed that the computation of minimum path lengths was isomorphic to computation of a power of the incidence matrix, and then we were able to see that repeated squaring, the shortcut method of computing a power of a matrix, is isomorphic to a similar shortcut for finding all shortest paths.

References

[Cormen et al. 2001] T.H. Cormen, C.E. Leiserson, R. Rivest, and C. Stein. *Introduction to Algorithms.* Cambridge, Mass.: MIT Press, 2001.

[Bellman 1952] R. Bellman. On the theory of dynamic programming. *Proceedings of the National Academy of Sciences*, 38:716–719, 1952.

Chapter 4

Some Graph Algorithms in an Array-Based Language

Viral B. Shah, John Gilbert†, and Steve Reinhardt‡*

Abstract

This chapter describes some of the foundations of linear algebraic graph algorithms and presents a number of classic graph algorithms using MATLAB style syntax. These algorithms are implicitly parallel, provided the underlying parallel matrix operations are supported in the array-based language.

4.1 Motivation

High performance applications increasingly combine numerical and combinatorial algorithms. Past research on high performance computation has focused mainly on numerical algorithms, and we have a rich variety of tools for high performance numerical computing. On the other hand, few tools exist for large-scale combinatorial computing. Our goal is to allow scientists and engineers to develop applications using both numerical and combinatorial methods with as little effort as possible.

Sparse matrix computations allow structured representation of irregular data structures, decompositions, and irregular access patterns in parallel applications. Sparse matrices are a convenient way to represent graphs. Since sparse matrices are first-class citizens in MATLAB and many of its parallel dialects [Choy & Edelman 2005],

*This author's work was done while at the Computer Science Department, University of California, Santa Barbara, CA 93106-5110 (viral@mayin.org).

†Computer Science Department, University of California, Santa Barbara, CA 93106-5110 (gilbert@cs.ucsb.edu).

‡Microsoft Corporation, 716 Bridle Ridge Road, Eagan, MN 55123 (steve.reinhardt@microsoft.com).

it is natural to use the duality between sparse matrices and graphs to develop a unified infrastructure for numerical and combinatorial computing.

Several researchers are building libraries of parallel graph algorithms: the Parallel Boost Graph Library (PBGL) at Indiana University [Gregor & Lumsdaine 2005], the SNAP library at the Georgia Institute of Technology [Bader 2005], and the Multi-threaded Graph Library (MTGL) at Sandia National Laboratories (see [Fleischer et al. 2000]). PBGL uses MPI for parallelism and builds upon the Boost graph library [Siek et al. 2002]. Both SNAP and MTGL focus on thread-level parallelism.

Our approach relies upon representing graphs with sparse matrices. The efficiency of our graph algorithms thus depends upon the efficiency of the underlying sparse matrix implementation. We use the distributed sparse array infrastructure in STAR-P (see [Shah & Gilbert 2005]).

Parallelism arises from parallel operations on sparse matrices. This yields several advantages. Graph algorithms are written in a high-level language (here, MATLAB), making codes short, simple, and readable. The data-parallel matrix code has a single thread of control, which simplifies writing and debugging programs. The distributed sparse array implementation in STAR-P provides a set of well-tested primitives.

The primitives described in the next section are used to implement several graph algorithms in our "Knowledge Discovery Toolbox" (KDT, formerly GAPDT); see [Gilbert et al. 2007a, Gilbert et al. 2007b]. High performance and interactivity are salient features of this toolbox. KDT was designed from the outset to run interactively with terascale graphs via STAR-P. Much of KDT scales to tens or hundreds of processors.

4.2 Sparse matrices and graphs

Every sparse matrix problem is a graph problem, and every graph problem is a sparse matrix problem.

A graph consists of a set of nodes V, connected by a set E of directed or undirected edges. A graph can be specified by triples: (u, v, w) represents a directed edge of weight w from node u to node v; the edge is a loop if $u = v$. This corresponds to a nonzero w at location (u, v) in a sparse matrix. An undirected graph corresponds to a symmetric matrix. Table 4.1 lists some corresponding matrix and graph operations.

The sparse matrix implementations of both MATLAB and STAR-P (see [Gilbert et al. 1992, Shah & Gilbert 2005]) attempt to follow two rules for complexity.

- Storage for a sparse matrix should be proportional to the number of rows, columns, and nonzero entries.

- A sparse matrix operation should (as nearly as possible) take time proportional to the size of the data accessed and the number of nonzero arithmetic operations.

Table 4.1. Matrix/graph operations.
Many simple sparse matrix operations can be used directly to perform basic operations on graphs.

Matrix operation	Graph operation
`G = sparse(U, V, W)`	Construct a graph from an edge list
`[U, V, W] = find(G)`	Obtain the edge list from a graph
`vtxdeg = sum(spones(G))`	Node degrees for an undirected graph
`indeg = sum(spones(G))`	Indegrees for a directed graph
`outdeg = sum(spones(G), 2)`	Outdegrees for a directed graph
`N = G(i, :)`	Find all neighbors of node i
`Gsub = G(subset, subset)`	Extract a subgraph of G
`G(i, j) = w`	Add or relabel a graph edge
`G(i, j) = 0`	Delete a graph edge
`G(I, I) = []`	Delete graph nodes
`G = G(perm, perm)`	Permute nodes of a graph
`reach = G * start`	Breadth-first search step

Translating to graph language, we see that the storage for a graph is $O(|V| + |E|)$, and elementary operations (ideally) take time linear in the size of the data accessed.

4.2.1 Sparse matrix multiplication

Sparse matrix multiplication can be a basic building block for graph computations. Path problems on graphs, for example, have been studied in terms of matrix operations; see [Aho et al. 1974, Tarjan 1981]. Specifically, sparse matrix multiplication over semirings can be used to implement a wide variety of graph algorithms.

A *semiring* is an algebraic structure $(S, \oplus, \otimes, 0, 1)$, where S is a set of elements with binary operations $\oplus$ ("addition") and $\otimes$ ("multiplication"), and distinguished elements 0 and 1, that satisfies the following properties

1. $(S, \oplus, 0)$ is a commutative monoid with identity 0
   - associative: $a \oplus (b \oplus c) = (a \oplus b) \oplus c$
   - commutative: $a \oplus b = b \oplus a$
   - identity: $a \oplus 0 = 0 \oplus a = a$
2. $(S, \otimes, 1)$ is a monoid with identity 1
   - associative: $a \otimes (b \otimes c) = (a \otimes b) \otimes c$
   - identity: $a \otimes 1 = 1 \otimes a = a$
3. $\otimes$ distributes over $\oplus$
   - $a \otimes (b \oplus c) = (a \otimes b) \oplus (a \otimes c)$
   - $(b \oplus c) \otimes a = (b \otimes a) \oplus (c \otimes a)$
4. 0 is an annihilator under $\otimes$
   - $a \otimes 0 = 0 \otimes a = 0$

Sparse matrix multiplication has the same control flow and data dependencies over any semiring, so it is straightforward to extend any matrix multiplication code to an arbitrary semiring. We have extended STAR-P's sparse matrix arithmetic to operate over semirings.

Here are a few of the semirings useful for graph algorithms

- $(\mathbb{R}, +, \times, 0, 1)$ is the usual real field, which is a semiring.
- $(\{0,1\}, |, \&, 0, 1)$ is the *boolean semiring*. It is useful in graph traversal algorithms such as breadth-first search.
- $(\mathbb{R} \cup \{\infty\}, \min, +, \infty, 0)$, sometimes called the *tropical semiring*, can be used to implement various shortest path algorithms.
- $(\mathbb{R} \cup \{\infty\}, \min, \times, \infty, 1)$ can be used for operations like selecting a subgraph or contracting nodes to form a quotient graph. Typically the graph's adjacency matrix is multiplied by an indexing matrix containing only zeros and ones. The min operator decides the weight of collapsed edges; other operators can be substituted. This semiring can also be used to implement Cohen's algorithm [Cohen 1998] to estimate fill in sparse matrix multiplication.
- Semirings over tuples can be used to compute actual shortest paths (rather than just their lengths); see, for example, Fineman and Robinson's chapter (Chapter 5) in this book.

4.3 Graph algorithms

In this section, we describe algebraic implementations of several computations on graphs that arise in combinatorial scientific computing.

4.3.1 Breadth-first search

A breadth-first search can be performed by multiplying a sparse matrix $\mathbf{G}$ with a sparse vector $\mathbf{x}$. To search from node i, we begin with $\mathbf{x}(i) = 1$ and $\mathbf{x}(j) = 0$ for $j \neq i$. Then $\mathbf{y} = \mathbf{G}^{\mathbf{T}} * \mathbf{x}$ picks out row i of $\mathbf{G}$, which contains the neighbors of node i. Multiplying $\mathbf{y}$ by $\mathbf{G}^{\mathbf{T}}$ gives nodes two steps away, and so on. Figure 4.1 shows an example. (The example uses the matrix $\mathbf{A} = \mathbf{G} + \mathbf{I}$ in place of $\mathbf{G}$, which has the effect of selecting all nodes at distance at most k on the kth step.)

We can perform several independent breadth-first searches simultaneously by using sparse matrix matrix multiplication. Instead of the vector $\mathbf{x}$, we use a matrix $\mathbf{X}$ with a column for each starting node. After $\mathbf{Y} = \mathbf{G}^{\mathbf{T}} * \mathbf{X}$, column j of $\mathbf{Y}$ contains the result of a breadth-first search step from the node (or nodes) specified by column j of $\mathbf{x}$. Using an efficient sparse matrix data structure, the time complexity of the search is the same as it would be with a traditional sparse graph data structure.

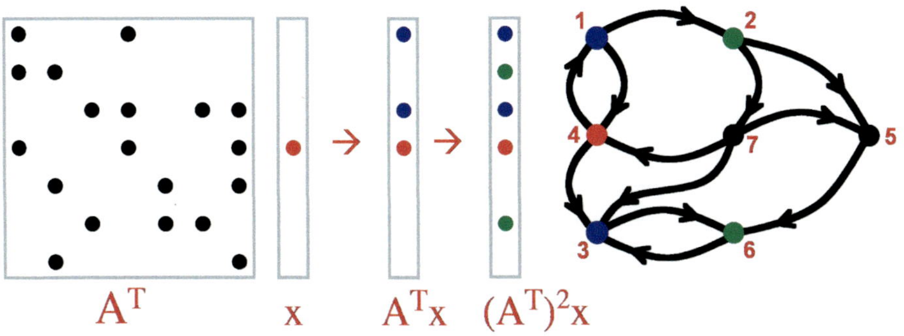

Figure 4.1. Breadth-first search by matrix vector multiplication. A sparse vector is intitialized with a 1 in the position of the start node. Repeated multiplication yields multiple breadth-first steps on the graph. The matrix can be either symmetric or asymmetric.

4.3.2 Strongly connected components

A strongly connected component of a directed graph is a maximal set of nodes all mutually reachable by directed paths. Every node belongs to exactly one such component. Tarjan's seminal paper [Tarjan 1972] uses depth-first search to find strongly connected components in linear time, but the depth-first search does not parallelize well. Instead, we implement a recursive divide-and-conquer algorithm due to Fleischer, Hendrickson, and Pinar [Fleischer et al. 2000] that is efficient in practice on many realistic graphs, although not in the worst case.

Node v is *reachable* from node u if there is a path of directed edges from u to v. The *descendants* of v are those nodes reachable from v. The *predecessors* of v are those nodes from which v is reachable. Descendants and predecessors can be found by breadth-first search, as in Algorithm 4.1.

Algorithm 4.1. Predecessors and descendants.
Predecessors of v are nodes from which v is reachable and are found by breadth-first search in $\mathbf{G}$. Descendants are found by breadth-first search in $\mathbf{G}^\mathbf{T}$.

```
function x = predecessor(G, v)
% Predecessors of a node in a graph

x = sparse(length(G), 1);
xold = x;
x(v) = 1;      % Start BFS from v.

while x ~= xold
   xold = x;
   x = x | G * x;
end
```

The Fleischer–Hendrickson–Pinar algorithm chooses a "pivot" node at random and computes its descendants and predecessors. The set of nodes that are both descendants and predecessors of the pivot (including the pivot itself) forms one strongly connected component. It is easy to show that every remaining component consists either entirely of descendants, entirely of predecessors, or entirely of nodes that are neither descendants nor predecessors. Thus, the algorithm proceeds recursively on these three disjoint subsets of nodes. Algorithm 4.2 gives the code.

Algorithm 4.2. Strongly connected components.

```
function scomponents(G,map)
% Strongly connected components of a graph

global label, count;

if nargin==1; map = 1:length(G); end;     % start recursion
if is empty(G); return; end;              % end recursion

v = 1 + fix(rand * length(G));            % random pivot
pred = predecessor(G,v);
desc = predecessor(G',v);

% Intersection of predecessors+descendants is a component.
scc = pred & desc;
count = count + 1;
label(map(scc)) = count;

% Recurse on subgraph of predecessors
remain = xor(pred,scc);
scomponents(G(remain,remain),map(remain));

% Recurse on subgraph of descendants
remain = xor(desc,scc);
scomponents(G(remain,remain),map(remain));

% Recurse on subgraph of remaining nodes
remain = ~(pred | desc);
scomponents(G(remain,remain),map(remain));
```

4.3.3 Connected components

A *connected component* in an undirected graph is a maximal connected subgraph. Every node belongs to exactly one connected component.

We implement the Awerbuch–Shiloach algorithm [Awerbuch & Shiloach 1987] to find connected components of a graph in parallel. This algorithm builds up rooted trees of nodes, beginning with each node in its own tree and ending with one tree for each component. The final trees are *stars*, which means that every tree node except the root is a child of the root. The set of trees is represented by a vector D that gives the parent of each node. If i is a root, $D(i) = i$. At the end, the root serves as a label for the component; each component consists of nodes with a common $D(i)$.

The method, shown in Algorithm 4.3, iterates a sequence of three steps until the trees stop changing. The first two steps, "conditional hooking" and "unconditional hooking," combine pairs of trees into single trees. The third step, "pointer jumping," collapses each tree into a star. Roughly speaking, the conditional hooking step connects a star with a higher-numbered root to a lower-numbered node of an adjacent tree, and the unconditional hooking step forces every star that is not a complete component to connect to something. Awerbuch and Shiloach show that the algorithm terminates in $O(\log |V|)$ iterations. The total work is therefore $O(|E| \log |V|)$.

4.3.4 Maximal independent set

An *independent set* in an undirected graph is a set of nodes, no two of which are adjacent. An independent set is *maximal* if it is not a subset of any other independent set.

We use Luby's randomized algorithm [Luby 1985] to compute a maximal independent set (MIS), as shown in Algorithm 4.4. We begin by selecting nodes in the graph with probability inversely proportional to their degrees. If we select both endpoints of an edge, we deselect the lower-degree one. We add the remaining selected nodes to the independent set. We then iterate on the subgraph that remains after removing the selected nodes and their neighbors.

Luby shows that, with high probability, each iteration eliminates at least 1/8 of the edges, and therefore the number of iterations is almost certainly $O(\log |V|)$.

4.3.5 Graph contraction

A *contraction* of a graph, also called a *quotient graph*, is obtained by merging subsets of nodes into single nodes. Contraction is a common operation in recursive computations on graphs, appearing, for example, in some graph partitioning algorithms and in numerical multigrid solvers.

As Algorithm 4.5 shows, graph contraction can be implemented in parallel by sparse matrix matrix multiplication. The input is a graph and an integer label for each node; the quotient graph merges nodes with the same label. The key is to form the sparse matrix $\mathbf{S}$, which has a column for every node of the input graph and a row for every node of the quotient graph.

Edges in the input graph merge when their endpoint nodes are contracted together. The code here sums the weights on the merged edges, but by using a different semiring, we could specify other ways to combine weights.

Algorithm 4.3. Connected components.

```
% Label nodes by connected components
function D = components(G)
   D = 1:length(G); [u v] = find (G);
   while true
        D = conditional_hooking (D, u, v);
        star = find_stars(D);
        if nnz(star) == length(G); break; end;
        D = unconditional_hooking (D, star, u, v);
        D = pointer_jumping (D);
   end;
end % components()

% Hook higher roots to lower neighbors
function D = conditional_hooking (D, u, v)
   Du = D(u); Dv = D(v);
   hook = Du == D(Du) & Dv < Du;
   Du = Du(hook); Dv = Dv(hook);
   D(Du) = Dv;
end % conditional_hooking()

% Hook adjacent stars together
function D = unconditional_hooking (D, star, u, v)
   Du = D(u); Dv = D(v);
   hook = star(u) & Dv ~= Du;
   Du = Du(hook); Dv = Dv(hook);
   D(Du) = Dv;
end % unconditional_hooking()

% Determine which nodes are in stars
function star = find_stars (D)
   star = D == D(D);
   star(D) = star(D) & star;
   star(D) = star(D) & star;
end % find_stars()

% Shortcut all paths to point directly to roots
function D = pointer_jumping (D)
   Dold = zeros(1,length(D));
   while any(Dold ~= D)
        Dold = D;
        D = D(D);
   end;
end % pointer_jumping()
```

Algorithm 4.4. Maximal independent set.
Luby's algorithm randomly selects low degree nodes to add to the MIS. Neighbors of selected nodes are then ignored while the computation proceeds on the remaining subgraph. Part of the code is omitted for brevity.

```
function IS = mis (G)
% Maximal independent set of a graph

IS = [];
while length(G) > 0
   % Select vertices with probability 1/(2*degree)
   degree = sum(G,2);
   prob = 1 ./ (2 * degree);
   select = rand(length(G), 1) <= prob;

   % Deselect one of each pair of selected neighbors
   neighbors = select & G * select;
   deselects = ...;          % lower degree neighbors
   if ~isempty(neighbors); select(deselects) = 0; end

   % Add selected nodes to independent set
   IS = [IS find(select)];

   % Exclude neighbors of selected vertices
   remain = not(select | G * select);

   % Iterate on the remaining subgraph
   G = G(remain, remain);
end
```

Algorithm 4.5. Graph contraction.

```
function C = contract(G,labels)
% Contract nodes with the same label

n = length(G);
m = max(labels);
S = sparse(labels, 1:n, 1, m, n);
C = S * G * S';
end
```

4.3.6 Graph partitioning

It is often useful to divide a graph into two (or more) pieces by removing a small number of edges or vertices. Such partitions, usually continued recursively to pieces and subpieces and so on, have been applied to load balancing for parallel computation, VLSI circuit layout, direct solvers for linear systems, and many divide-and-conquer graph algorithms. Finding the best partition for an arbitrary graph is intractable; many heuristic approaches have been developed. Our KDT toolbox includes three partitioning heuristics based on MATLAB codes from the Meshpart toolbox [Gilbert & Teng 2002]. In each case, an efficient parallel implementation follows directly from the parallel array infrastructure of STAR-P.

KDT implements the geometric sphere partitioning algorithm of Miller et al. (see [Miller et al. 1998]) for graphs whose nodes have coordinates in low-dimensional Euclidean space. This algorithm guarantees the quality of partitions for meshes whose elements are "well shaped" in a certain sense that includes most meshes used for numerical finite element methods in 2D and 3D. The algorithm parallelizes naturally; indeed, the sequential MATLAB code from Meshpart [Gilbert et al. 1998] works efficiently in STAR-P with one minor modification to serialize computation on small matrices.

KDT also implements a spectral partitioning algorithm for arbitrary graphs. Spectral partitioning methods have arisen independently in several different fields. The theory is based on ideas of Fiedler [Fiedler 1975]; Pothen, Simon, and Liou suggested applying spectral partitions to parallel computation [Pothen et al. 1990]. Spectral partitioning begins with a graph's Laplacian matrix. The Laplacian $\mathbf{L}$ has off-diagonal elements $\mathbf{L}_{ij} = -1$ if (i, j) is an edge of the graph, and $\mathbf{L}_{ij} = 0$ otherwise. The diagonal element $\mathbf{L}_{ii}$ is the degree of node i; thus the rows of $\mathbf{L}$ sum to zero. The Laplacian is symmetric and positive semidefinite, so it has nonnegative real eigenvalues. The multiplicity of zero as an eigenvalue is equal to the number of connected components of the graph; if the graph is connected, zero is a simple eigenvalue. The smallest nonzero eigenvalue of a connected graph's Laplacian is called its *Fiedler value*, and the corresponding eigenvector is called a *Fiedler vector*.

The simplest spectral partitioning method labels node i of the graph with the ith component of its Fiedler vector, and then partitions the nodes into equal-sized subsets around the median label. This choice can be heuristically justified as representing the continuous relaxation of the NP-complete discrete problem of finding the minimum-sized edge partition of the graph. This method, which is implemented in KDT's `specpart()` routine, tends to find good partitions in practice. A slightly more complicated recursive Fiedler vector method guarantees good partitions for planar graphs and well-shaped finite element meshes [Spielman & Teng 2007]. In STAR-P, KDT uses the built-in eigensolver to obtain the Fiedler vector. This parallelizes well, but it is nonetheless an expensive computation.

KDT also includes the geometric spectral partitioner from the Meshpart toolbox. This algorithm first computes a small number k of eigenvectors of the graph Laplacian and uses them as node coordinates in k-dimensional Euclidian space. It then uses the geometric sphere separator algorithm to find a small cut for the graph as embedded. This method was suggested by Chan, Gilbert, and Teng [Chan et al. 1994], who showed that it worked well on several example problems.

4.4 Graph generators

Our KDT toolbox includes routines to generate three types of graphs that are useful as test cases for algorithms in various domains. The generators are fast and scalable; we use them to generate graphs ranging from a few nodes to billions of nodes.

4.4.1 Uniform random graphs

Our uniform random graph generator produces either directed or undirected graphs in which a specified number or density of edges are chosen uniformly at random, with replacement. This uses the logic of MATLAB `sprand()` routine, which generates a set of edges and builds the sparse adjacency matrix efficiently via `sparse()`; in STAR-P, this in turn uses an efficient parallel sorting routine [Cheng et al. 2006]. For low densities, the resulting graphs have distribution similar to the Erdős–Rényi random graph model [Erdős & Rényi 1959], in which each potential edge in the graph is created independently with fixed probability.

4.4.2 Power law graphs

For graphs with highly variable node degrees, we include a generator due to Kepner [Bader et al. 2004] for Leskovec et al.'s R-MAT graphs [Leskovec et al. 2005]. With appropriate parameters, R-MAT produces graphs with power law degree distributions similar to those occurring in many applications. For an R-MAT graph with 2^k nodes, the MATLAB code uses k iterations of a simple vectorized computation to generate edges, and then calls `sparse()` to create the graph. The code runs efficiently in STAR-P without modification.

Figure 4.2 shows a density spy plot of an R-MAT graph. The recursive structure is visible in the nonrandomized plot. Randomly relabeling the nodes destroys locality and structure. Figure 4.3 shows the degree distribution of an R-MAT graph. The performance plot (Figure 4.4) shows that the parallel R-MAT generator scales well, all the way to a billion nodes. These experiments were performed using 240 processors of a 256-processor shared memory computer.

4.4.3 Regular geometric grids

The KDT toolbox also includes generators for two- and three-dimensional regular grids, which arise in many settings in physical modeling. The routines we provide are the same as those in the Meshpart toolbox [Gilbert & Teng 2002], which generate 5-point, 7-point, and 9-point finite difference meshes on the unit square; equilateral triangular meshes; and cubic and tetrahedral meshes in three dimensions. The mesh generators also return the coordinates of the nodes in Euclidean space.

The original mesh generators use a routine called `blockdiags()` to build sparse matrices with specified diagonal and block diagonal structure. Our STAR-P version of `blockdiags()` uses Kronecker products (via the `kron()` routine) to generate block diagonal matrices. The parallel code for sparse `kron()` is identical to the sequential code.

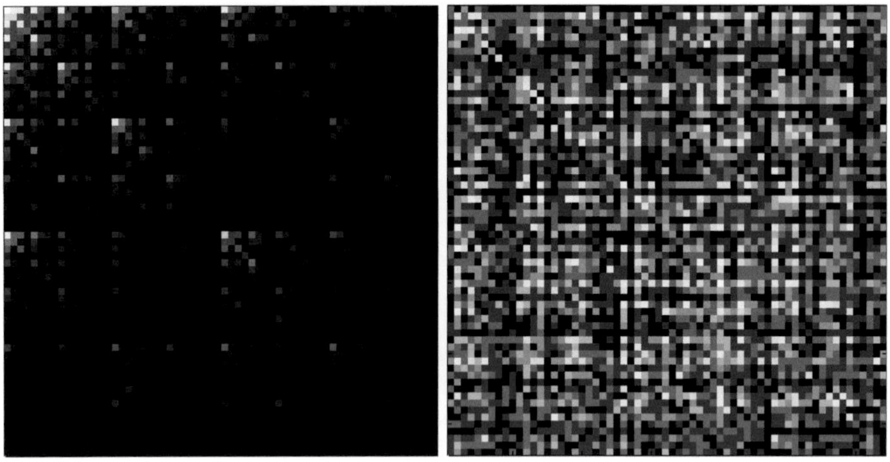

Figure 4.2. Adjacency matrix density of an R-MAT graph. The left image is an R-MAT graph with 1024 nodes and 6671 edges; note the recursive structure. The right image shows the same graph with nodes relabeled randomly.

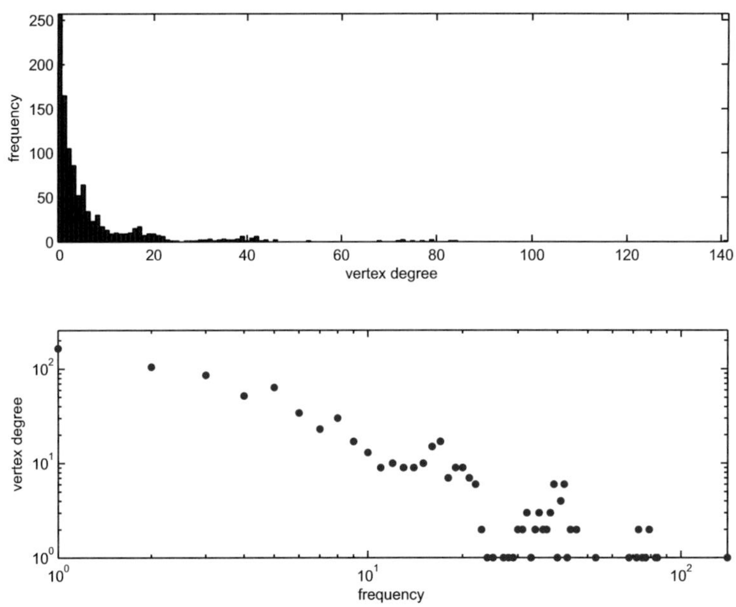

Figure 4.3. Vertex degree distribution in an R-MAT graph.

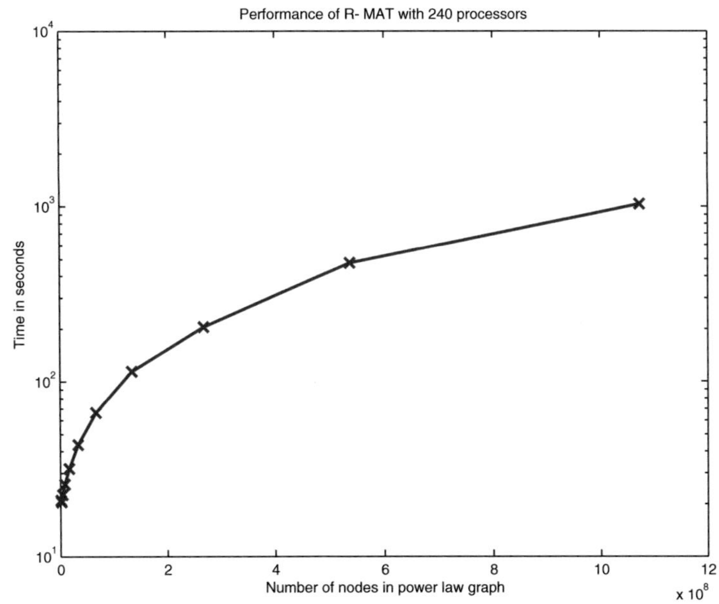

Figure 4.4. Performance of parallel R-MAT generator.

References

[Aho et al. 1974] A.V. Aho, J.E. Hopcroft, and J.D. Ullman. *The Design and Analysis of Computer Algorithms.* Boston: Addison–Wesley, 1974.

[Awerbuch & Shiloach 1987] B. Awerbuch and Y. Shiloach. New connectivity and MSF algorithms for shuffle-exchange network and PRAM. *IEEE Transactions on Computers*, 36:1258–1263, 1987.

[Bader et al. 2004] D.A. Bader, K. Madduri, J.R. Gilbert, V. Shah, J. Kepner, T. Meuse, and A. Krishnamurthy. Designing scalable synthetic compact applications for benchmarking high productivity computing systems. *Cyberinfrastructure Technology Watch*, 2:41–51, 2006.

[Bader 2005] D.A. Bader. High-performance algorithm engineering for large-scale graph problems and computational biology. *Proc. of WEA 2005.* In S. E. Nikoletseas, editor, *Lecture Notes in Computer Science*, 3503:16–21. Heidelberg: Springer-Verlag, 2005.

[Chan et al. 1994] T.F. Chan, J.R. Gilbert, and S.-H. Teng. Geometric spectral partitioning. Technical Report CSL-94-15, Palo Alto Research Center, Xerox Corporation, 1994.

[Cheng et al. 2006] D.R. Cheng, A. Edelman, J.R. Gilbert, and V. Shah. A novel parallel sorting algorithm for contemporary architectures. Technical Report, University of California at Santa Barbara, 2006.

[Choy & Edelman 2005] R. Choy and A. Edelman. Parallel MATLAB: Doing it right. *Proceedings of the IEEE*, 93:331–341, 2005.

[Cohen 1998] E. Cohen. Structure prediction and computation of sparse matrix products. *Journal of Combinatorial Optimization*, 2:307–332, 1998.

[Erdős & Rényi 1959] P. Erdős and A. Rényi. On random graphs. *Publicationes Mathematicae*, 6:290–297, 1959.

[Fiedler 1975] M. Fiedler. A property of eigenvectors of non-negative symmetric matrices and its application to graph theory. *Czechoslovak Mathematical Journal*, 25:619–632, 1975.

[Fleischer et al. 2000] L. Fleischer, B. Hendrickson, and A. Pinar. On identifying strongly connected components in parallel. *Proc. IPDPS Workshops on Parallel and Distributed Processing*, 505–511, Heidelberg: Springer-Verlag, 2000.

[Gilbert et al. 1992] J.R. Gilbert, C. Moler, and R. Schreiber. Sparse matrices in MATLAB: Design and implementation. *SIAM Journal on Matrix Analysis and Applications*, 13:333–356, 1992.

[Gilbert et al. 2007a] J.R. Gilbert, S. Reinhardt, and V. Shah. An interactive environment to manipulate large graphs. *Proc. 2007 IEEE International Conference on Acoustics, Speech, and Signal Processing*, 2007.

[Gilbert & Teng 2002] J.R. Gilbert and S.-H. Teng. MATLAB mesh partitioning and graph separator toolbox, 2002. http://www.cerfacs.fr/algor/Softs/MESHPART/.

[Gilbert et al. 1998] J.R. Gilbert, G.L. Miller, and S.-H. Teng. Geometric mesh partitioning: Implementation and experiments. *SIAM Journal on Scientific Computing*, 19:2091–2110, 1998.

[Gilbert et al. 2007b] J.R. Gilbert, S. Reinhardt, and V. Shah. High performance graph algorithms from parallel sparse matrices. *Proc. PARA06: 8th Workshop on State of the Art in Scientific and Parallel Computing*, 260–269, 2007.

[Gregor & Lumsdaine 2005] D. Gregor and A. Lumsdaine. The parallel BGL: A generic library for distributed graph computations. In *Parallel Object-Oriented Scientific Computing (POOSC)*, July 2005.

[Leskovec et al. 2005] J. Leskovec, D. Chakrabarti, J.M. Kleinberg, and C. Faloutsos. Realistic, mathematically tractable graph generation and evolution, using Kronecker multiplication. *Proc. PKDD 2005: 9th European Conf. on Principles and Practice of Knowledge Discovery in Databases*. In *Lecture Notes in Computer Science*, 3721:133–145. Heidelberg: Springer-Verlag, 2005.

[Luby 1985] M. Luby. A simple parallel algorithm for the maximal independent set problem. *Proc. 17th Annual Symposium on Theory of Computing*, 1–10, New York: ACM, 1985.

[Miller et al. 1998] G.L. Miller, S.-H. Teng, W. Thurston, and S.A. Vavasis. Geometric separators for finite-element meshes. *SIAM Journal on Scientific Computing*, 19:364–386, 1998.

[Pothen et al. 1990] A. Pothen, H.D. Simon, and K.-P. Liou. Partitioning sparse matrices with eigenvectors of graphs. *SIAM Journal on Matrix Analysis and Applications*, 11:430–452, 1990.

[Shah & Gilbert 2005] V. Shah and J.R. Gilbert. Sparse matrices in Matlab*P: Design and implementation. *Proc. HiPC04: 11th International Conference on High Performance Computing.* In *Lecture Notes in Computer Science*, 3296:144–155. Heidelberg: Springer-Verlag, 2005.

[Siek et al. 2002] J.G. Siek, L.-Q. Lee, and A. Lumsdaine. *The Boost Graph Library: User Guide and Reference Manual.* Boston: Addison–Wesley, 2002.

[Spielman & Teng 2007] D.A. Spielman and S.-H.Teng. Spectral partitioning works: Planar graphs and finite element meshes. *Linear Algebra and Its Applications*, 421:284–305, 2007.

[Tarjan 1972] R. Tarjan. Depth-first search and linear graph algorithms. *SIAM Journal on Computing*, 1:146–160, 1972.

[Tarjan 1981] R.E. Tarjan. Fast algorithms for solving path problems. *Journal of the ACM*, 28:594–614, 1981.

Chapter 5

Fundamental Graph Algorithms

Jeremy T. Fineman * ***and Eric Robinson*** †

Abstract

This chapter discusses the representation of several fundamental graph algorithms as algebraic operations. Even though the underlying algorithms already exist, the algebraic representation allows for easily expressible efficient algorithms with appropriate matrix constructs. This chapter gives algorithms for single-source shortest paths, all-pairs shortest paths, and minimum spanning tree.

5.1 Shortest paths

In shortest path problems, we are given weighted, directed graphs $G = (V, E)$ with edge weights $w : E \to \mathbb{R}_\infty$, where $\mathbb{R}_\infty = \mathbb{R} \cup \{\infty\}$, and a weight matrix $\mathbf{W} : \mathbb{R}_\infty^{N \times N}$ where $\mathbf{W}(u, v) = \infty$ if $(u, v) \notin E$. For simplicity, $\mathbf{W}(v, v) = 0$ for all $v \in V$. A path p from v_0 to v_k, denoted $v_0 \overset{p}{\rightsquigarrow} v_k$, is a sequence of vertices $p = \langle v_0, v_1, \ldots, v_k \rangle$ such that $(v_{i-1}, v_i) \in E$. We define the **weight** of the path to be $w(p) = \sum_{i=1}^{k} \mathbf{W}(v_{i-1}, v_i)$. We say that the **size** of the path p is k hops, denoted by $|p| = k$.

The *shortest path distance* (or shortest path weight) from u to v is given by

$$\mathbf{\Delta}(u, v) = \begin{cases} \min\{w(p) : u \overset{p}{\rightsquigarrow} v\} & \text{if a path from } u \text{ to } v \text{ exists} \\ \infty & \text{otherwise} \end{cases}$$

A *shortest path* from u to v is a path $u \overset{p}{\rightsquigarrow} v$ with $w(p) = \mathbf{\Delta}(u, v)$.

---

*School of Computer Science, Carnegie Mellon University, Pittsburgh, PA 15213 (`jfineman@cs.cmu.edu`).

†MIT Lincoln Laboratory, 244 Wood Street, Lexington, MA 02420 (`erobinson@ll.mit.edu`).

This work is sponsored by the Department of the Air Force under Air Force Contract FA8721-05-C-0002. Opinions, interpretations, conclusions, and recommendations are those of the authors and are not necessarily endorsed by the United States Government.

This section considers the Bellman–Ford algorithm for single-source shortest paths and the Floyd–Warshall algorithm for all-pairs shortest paths. The algebraic variants given here of both these algorithms achieve the same running times as the standard graph notations, e.g., as given in [Cormen et al. 2001]. This section does not include Dijkstra's algorithm (which restricts the input to graphs with positive edge weights) since Section 5.2 gives the remarkably similar Prim's algorithm.

5.1.1 Bellman–Ford

The Bellman–Ford algorithm [Bellman 1958, Ford & Fulkerson 1962] solves the single-source shortest paths problem. Given a graph $G = (V, E)$ with edge weights w and a designated source $s \in V$, this algorithm determines whether there is a negative-weight cycle reachable from s. If there is no negative-weight cycle, Bellman–Ford returns the shortest path distances $\mathbf{\Delta}(s, v)$ for all $v \in V$ and the corresponding paths.

One standard presentation of the Bellman–Ford algorithm is given by Algorithm 5.1. Bellman–Ford stores for each vertex v an estimate $\mathbf{d}(v)$ on the shortest path distance, maintaining that $\mathbf{d}(v) \geq \mathbf{\Delta}(s, v)$. The algorithm performs a sequence of "edge relaxations," after which $\mathbf{d}(v) = \mathbf{\Delta}(s, v)$. To *relax* the edge (u, v) simply means that $\mathbf{d}(v) = \min\{\mathbf{d}(v), \mathbf{d}(u) + \mathbf{W}(u, v)\}$. In particular, Bellman–Ford consists of N iterations, relaxing all edges in each iteration (in arbitrary order). lines 8–10 simply give the implementation of an edge relaxation. We use $\pi(v)$ to store the parent of v in the shortest path tree. Upon completion of the algorithm, π can be used to find the shortest paths and not just the distances.

Algorithm 5.1. Bellman–Ford.
A standard implementation of the Bellman–Ford algorithm [Cormen et al. 2001].

```
BELLMAN–FORD(V, E, w, s)
 1  foreach v ∈ V
 2       do d(v) = ∞
 3          π(v) = NIL
 4  d(s) = 0
 5  for k = 1 to N − 1
 6       do foreach edge (u, v) ∈ E
 7              do ▷ Relax (u, v)
 8                 if d(v) > d(u) + W(u, v)
 9                    then d(v) = d(u) + W(u, v)
10                         π(v) = u
11  foreach edge (u, v) ∈ E
12       do if d(v) > d(u) + W(u, v)
13             then return "A negative-weight cycle exists."
```

Our algebraic formulation of Bellman–Ford more closely resembles a dynamic programming interpretation. We define

$$\mathbf{\Delta}_k(u,v) = \begin{cases} \min\{w(p) : u \overset{p}{\leadsto} v, |p| \le k\} & \text{if a } (\le k)\text{-hop path from } u \text{ to } v \text{ exists} \\ \infty & \text{otherwise} \end{cases}$$

to be the shortest path distance from u to v using at most k edges. If $\mathbf{\Delta}_n(s,v) < \mathbf{\Delta}_{N-1}(s,v)$ for any v, then a negative-weight cycle exists. Otherwise, $\mathbf{\Delta}_{N-1}(s,v) = \mathbf{\Delta}(s,v)$.

Computing $\mathbf{\Delta}_k(s,v)$ from $\mathbf{\Delta}_{k-1}(s,v)$ is logically identical to relaxing all edges incident on v. In particular

$$\mathbf{\Delta}_k(s,v) = \min_u\{\mathbf{\Delta}_{k-1}(s,u) + \mathbf{W}(u,v)\} \tag{5.1}$$

Note that $\mathbf{\Delta}_k$ can be computed from only $\mathbf{\Delta}_{k-1}$, so all other $\mathbf{\Delta}_i$ can be discarded.

Algebraic Bellman–Ford

To represent Bellman–Ford as matrix and vector operations, we use the (sparse) $N \times N$ adjacency matrix $\mathbf{A}$ to store the edge weights and a $1 \times N$ vector $\mathbf{d}_k$ to store shortest $(\le k)$-hop path distances $\mathbf{\Delta}_k(s,*)$. Here, the values along the diagonal of the adjacency matrix are zeros.

Translating equation (5.1) to a vector matrix product is relatively straightforward. We have $\mathbf{d}_k(v) = \min_{\forall u \in N}(\mathbf{d}_{k-1}(u) + \mathbf{A}(u,v))$, which is just the product $\mathbf{d}_k(v) = \mathbf{d}_{k-1} \text{ min.+ } \mathbf{A}(:,v)$. Thus, we have $\mathbf{d}_k = \mathbf{d}_{k-1} \text{ min.+ } \mathbf{A}$.

We can represent the shortest path distance as $\mathbf{d} = \mathbf{d}_0\mathbf{A}^{N-1}$, where each multiplication is min.+ , and

$$\mathbf{d}_0(v) = \begin{cases} 0 & \text{if } v = s \\ \infty & \text{otherwise} \end{cases}$$

The expression $\mathbf{d} = \mathbf{d}_0\mathbf{A}^N$ yields two natural algorithms for computing single-source shortest path distances. The first is the faithful algebraic representation of Bellman–Ford given in Algorithm 5.2. This algorithm uses $O(N)$ dense vector, sparse matrix multiplications, so the running time is $\Theta(NM)$ comparisons and additions matching the regular Bellman–Ford.

Algorithm 5.2. Algebraic Bellman–Ford.
An algebraic implementation of the Bellman–Ford algorithm.

```
BELLMAN–FORD(A, s)
d = ∞
d(s) = 0
for k = 1 to N − 1
      do d = d min.+ A
if d ≠ d min.+ A
   then return "A negative-weight cycle exists."
return d
```

The second algorithm is to compute $\mathbf{A}^N$ first by repeated squaring. Note that this approach actually computes all-pairs shortest paths first, and the product d min.+ $(\mathbf{A}^N)$ just selects a single row from the matrix $\mathbf{A}^N$. This algorithm requires $\log N$ dense matrix multiplies that require $\Theta(N^3 \log N)$ work. Note that from the perspective of total work, the algebraic Bellman–Ford given in Algorithm 5.2 asymptotically outperforms the repeated squaring in the worst case, but repeated squaring may allow for more parallelism.

5.1.2 Computing the shortest path tree (for Bellman–Ford)

In addition to computing shortest path distances, Bellman–Ford and other shortest path algorithms find shortest paths. These paths are typically stored using the parent pointers π, comprising a shortest path tree. We now extend our algorithms to compute the shortest path tree. This section describes shortest path trees in the context of Bellman–Ford, but the same techniques can be trivially applied to Floyd–Warshall, for example.

In the absence of 0-weight cycles, computing the shortest path tree is relatively easy. For each $v \neq s$ with $\mathbf{\Delta}(s, v) \neq \infty$, we simply set $\pi(v) = u$ such that $\mathbf{d}(u) + \mathbf{W}(u, v) = \mathbf{d}(v)$ for each $v \neq s$. Since $\mathbf{d}(v) = \mathbf{d}(u) + \mathbf{W}(u, v)$ for some $u \neq v \in V$, and $\mathbf{d}(v) \geq \mathbf{d}(u) + \mathbf{W}(u, v)$ for all $u \in V$, we have $\pi(v) = \text{argmin}_{u \neq v}\{\mathbf{d}(u) + \mathbf{W}(u, v)\}$. We represent this expression algebraically as

$$\pi = \mathbf{d} \text{ argmin.+ } (\mathbf{A} + \text{diag}(\infty)) \tag{5.2}$$

We add $\text{diag}(\infty)$—the matrix having ∞'s along the diagonal—to eliminate self-loops from consideration. Note that $\pi(s)$ has incorrect value, so we must set $\pi(s) = \text{NIL}$. Similarly, for any v with $\mathbf{\Delta}(s, v) = \infty$, the value at $\pi(v)$ is incorrect, which must also be remedied.*

In a graph with 0-weight cycles, using equation (5.2) may not yield a shortest path tree (even after resolving complications induced by unreachable vertices). In particular, the argmin may select vertices forming 0-weight cycles.

We have two approaches for computing the parent pointers on arbitrary graphs (without negative-weight cycles). The first approach entails augmenting our operators to operate on 3-tuples rather than real-valued weights. This 3-tuple approach operationally follows the standard Bellman–Ford algorithm by updating parent pointers as the distances decrease, and it results in only a constant-factor overhead decrease in performance.† The second approach uses ideas from equation (5.2), pushing off the shortest-path-tree computation until the end.

Both approaches have their advantages. The 3-tuple approach uses a single semiring for all operations and needs only to introduce tuples and change the operations used from our algebraic Bellman–Ford. The other approach uses simpler 2-tuples, but we must operate over more than one semiring. Moreover, the second

---

*Assuming every vertex is reachable from s, there is no vertex with $\mathbf{\Delta}(s, v) = \infty$.

†Note that an implementation does not have to explicitly modify the input adjacency matrix. Instead, we can modify the matrix-primitive implementations to treat three same-sized matrices as a single matrix of 3-tuples.

approach involves some additional cleanup to fix some of the incorrect settings to π (as mentioned for equation (5.2)). We give both approaches here because either one may be useful for different problems.

Parent pointers are on smallest size paths

In the original Bellman–Ford algorithm, parent pointers are updated only when edges are relaxed and a strictly shorter (distance) path is discovered. In particular, let Π_k be the possible values assigned to parent pointer π on the kth iteration of Bellman–Ford.* Then

$$\Pi_k(s,v) = \begin{cases} \{\text{NIL}\} & \text{if } k=0,\ v=s \\ \emptyset & \text{if } k=0,\ v\neq s \\ \Pi_{k-1}(s,v) & \text{if } k\geq 1,\ \mathbf{\Delta}_k(s,v) \\ & \quad = \mathbf{\Delta}_{k-1}(s,v) \\ \{u : \mathbf{\Delta}_k(s,v) = \mathbf{\Delta}_{k-1}(s,u) + \mathbf{W}(u,v)\} & \text{otherwise} \end{cases}$$

Consideration of this update rule gives the following lemma, stating that the parent selected is the parent on a shortest ($\leq k$)-hop path having the fewest hops.

Lemma 5.1. *Consider a graph with designated source vertex s and no negative-weight cycles. Then $u \neq$ NIL $\in \Pi_k(s,v)$ if and only if there exists a path $p = \langle s,\ldots,u,v\rangle$ such that $w(p) = \mathbf{\Delta}_k(s,v)$, and there is no path $s \overset{p'}{\leadsto} v$ such that $w(p') = \mathbf{\Delta}_k(s,v)$ and $|p'| < |p|$. Moreover, $\Pi_k(s,s) = \{\text{NIL}\}$.*

Proof. By induction on k. The lemma trivially holds at $k=0$.

If $\mathbf{\Delta}_k(s,v) = \mathbf{\Delta}_{k-1}(s,v)$, then a shortest ($\leq k$)-hop path from s to v has size at most $k-1$, and $\Pi_k(s,v) = \Pi_{k-1}(s,v)$ satisfies the lemma. If $\mathbf{\Delta}_k(s,v) < \mathbf{\Delta}_{k-1}(s,v)$, then all shortest ($\leq k$)-hop paths from s to v have size exactly k, and the lemma holds given the definition of Π_k. Since $\mathbf{\Delta}_k(s,s)=0$, it follows that $\Pi_k(s,s) = \{\text{NIL}\}$. □

This lemma is useful for both of the following shortest path tree approaches.

Computing parent pointers with 3-tuples

Since the original Bellman–Ford algorithm is correct when relaxing edges in an arbitrary order, Lemma 5.1 implies that we can update the parent pointer to be the penultimate vertex along any smallest (size) path, as the final edges on all equally sized paths are relaxed on the same iteration. Thus, as long as we consider path size and weight when performing an update, we do not need the strict inequality used in the parent-pointer update rule.

The goal here is to link the distance and parent-pointer updates by changing our operations to work over 3-tuples. These 3-tuples consist of components corresponding to total path weight, path size, and the penultimate vertex. By cleverly

*An "iteration" here refers to an iteration of the **for** loop in lines 5–10 of Algorithm 5.1.

redesigning operators for these 3-tuples, we can use the same algebra-based algorithms (i.e., $\delta = \mathbf{d}_n = \mathbf{d}_0\mathbf{A}^N$ still applies for a good choice of scalar addition and multiplication operations).

We define our scalars as 3-tuples of the form $\langle w, h, \pi \rangle \in S = (\mathbb{R}_\infty \times \mathbb{N} \times V) \cup \{\langle \infty, \infty, \infty \rangle, \langle 0, 0, \text{NIL} \rangle\}$, where $\mathbb{R}_\infty = \mathbb{R} \cup \{\infty\}$ and $\mathbb{N} = \{1, 2, 3, \ldots\}$. The value $\langle \infty, \infty, \infty \rangle$ corresponds to nonexistence of a path, and the value $\langle 0, 0, \text{NIL} \rangle$ corresponds to a path from a vertex to itself. The first tuple element $w \in \mathbb{R}_\infty$ corresponds to a path weight. The second tuple element $h \in \mathbb{N}$ corresponds to a path size or number of hops. The third element $\pi \in V$ corresponds to the penultimate vertex along a path.

Given this definition of 3-tuples, setting up the entries in the adjacency matrix $\mathbf{A}$ is straightforward. In particular, we have

$$\mathbf{A}(u, v) = \begin{cases} \langle 0, 0, \text{NIL} \rangle & \text{if } u = v \\ \langle \mathbf{W}(u, v), 1, u \rangle & \text{if } u \neq v, (u, v) \in E \\ \langle \infty, \infty, \infty \rangle & \text{if } (u, v) \notin E \end{cases} \tag{5.3}$$

Note that all of the tuple values, except the parent u when the edge exists, are implicit just given an edge weight. Moreover, a natural implementation likely already stores u, either implicitly as a dense matrix index, or explicitly as an (adjacency) list in a sparse matrix. Thus, when implementing this algorithm, it is not strictly necessary to increase the storage used for the adjacency matrix.

Setting up the initial distance vector $\mathbf{d}_0$ is also straightforward. We have

$$\mathbf{d}_0(v) = \begin{cases} \langle 0, 0, \text{NIL} \rangle & \text{if } v = s \\ \langle \infty, \infty, \infty \rangle & \text{otherwise} \end{cases} \tag{5.4}$$

For our addition operation, we use an operator lmin that is defined to be the lexicographic minimum. To perform a lexicographic minimum lmin, compare the first tuple part. If they are equal, compare the second part, and, if they are equal, the third part.

$$\text{lmin}\{\langle w_1, h_1, \pi_1 \rangle, \langle w_2, h_2, \pi_2 \rangle\} = \begin{cases} \langle w_1, h_1, u_1 \rangle & \begin{array}{l} \text{if } w_1 < w_2, \text{ or} \\ \text{if } w_1 = w_2 \text{ and } h_1 < h_2, \text{ or} \\ \text{if } w_1 = w_2 \text{ and } h_1 = h_2 \text{ and} \\ u_1 < u_2 \end{array} \\ \langle w_2, h_2, u_2 \rangle & \text{otherwise} \end{cases}$$

Without loss of generality, the vertices are number $1, 2, \ldots, N$, and $\text{NIL} < v < \infty$ for all $v \in V$.

For our multiplication operation, we define a new binary function called $+_{\text{rhs}}$. This function adds the first two parts of the tuple and retains the third part of the tuple from the right-hand-side argument of the operator. We define this operator as follows

$$\langle w_1, h_1, \pi_1 \rangle +_{\text{rhs}} \langle w_2, h_2, \pi_2 \rangle = \begin{cases} \langle w_1 + w_2, h_1 + h_2, \pi_2 \rangle & \text{if } \pi_1 \neq \infty, \pi_2 \neq \text{NIL} \\ \langle w_1 + w_2, h_1 + h_2, \pi_1 \rangle & \text{otherwise} \end{cases}$$

We introduce the exceptions for $\pi_1 = \infty$ simply to give us a multiplicative identity when this operation is used as multiplication in conjunction with lmin as addition.

The exception when $\pi_2 = \text{NIL}$ is for correctness. Note that unlike regular $+$, this $+_{\text{rhs}}$ operation is not commutative.

Lemma 5.2. *The set* $S = (\mathbb{R}_\infty \times \mathbb{N} \times V) \cup \{\langle\infty,\infty,\infty\rangle, \langle 0,0,\text{NIL}\rangle\}$ *under the operations* $\text{lmin}.+_{\text{rhs}}$ *is a semiring.*

Proof. The lmin function is obviously commutative and associative as it is a natural extension of min. Moreover, S is closed under lmin and $+_{\text{rhs}}$.

The additive identity is $\langle\infty,\infty,\infty\rangle$, and the multiplicative identity is $\langle 0,0,\text{NIL}\rangle$. Note that having the exception for ∞ in $+_{\text{rhs}}$ is necessary to make the additive identity ($\langle\infty,\infty,\infty\rangle$) multiplied by anything be the additive identity. Having the exception for NIL makes $\langle 0,0,\text{NIL}\rangle$ act as a multiplicative identity.

To show associativity of $+_{\text{rhs}}$, consider the multiplication

$$\langle w_1, h_1, \pi_1\rangle +_{\text{rhs}} \langle w_2, h_2, \pi_2\rangle +_{\text{rhs}} \langle w_3, h_3, \pi_3\rangle$$

with any parenthesization. The result is $\langle w_1 + w_2 + w_3, h_1 + h_2 + h_3, \pi'\rangle$. If any $\pi_i = \infty$, then $\pi' = \infty$. Suppose all $\pi_i \neq \infty$. Then the result is $\pi' = \pi_i$, where i is the largest value such that $\pi_i \neq \text{NIL}$.

Finally, we show that $+_{\text{rhs}}$ (implicitly denoted by $ab = a{+_{\text{rhs}}}b$) distributes over lmin. In particular, let $t = \langle w, h, \pi\rangle$, $t_1 = \langle w_1, h_1, \pi_1\rangle$, and $t_2 = \langle w_2, h_2, \pi_2\rangle$. Then we must show left distributivity $t(t_1 \text{lmin} t_2) = tt_1 \text{lmin} tt_2$ and right distributivity $(t_1 \text{lmin} t_2)t$. Without loss of generality (by commutativity of lmin), suppose that $\text{lmin}\{t_1, t_2\} = t_1$. We consider all cases separately.

1. Suppose $t = \langle\infty,\infty,\infty\rangle$. Then $\langle\infty,\infty,\infty\rangle(t_1 \text{lmin} t_2) = \langle\infty,\infty,\infty\rangle t_1 = \langle\infty,\infty,\infty\rangle = \langle\infty,\infty,\infty\rangle t_1 \text{lmin}\langle\infty,\infty,\infty\rangle t_2$. Similarly for right distributivity.
2. Suppose $t = \langle 0,0,\text{NIL}\rangle$. Then $\langle 0,0,\text{NIL}\rangle(t_1 \text{lmin} t_2) = \langle 0,0,\text{NIL}\rangle t_1 = t_1 = t_1 \text{lmin} t_2 = \langle 0,0,\text{NIL}\rangle t_1 \text{lmin}\langle 0,0,\text{NIL}\rangle t_2$. Similarly for right distributivity.
3. Suppose $t_2 = \langle\infty,\infty,\infty\rangle$. Then $t(t_1 \text{lmin}\langle\infty,\infty,\infty\rangle) = tt_1 = tt_1 \text{lmin}\langle\infty,\infty,\infty\rangle = tt_1 \text{lmin} t\langle\infty,\infty,\infty\rangle$. Similarly for right distributivity.
4. Suppose $t_1 = t_2 = \langle 0,0,\text{NIL}\rangle$. Trivial.
5. Suppose $t_1 = \langle 0,0,\text{NIL}\rangle$ and $t, t_2 \in \mathbb{R} \times \mathbb{N} \times V$. Since we suppose $\text{lmin}\{t_1, t_2\} = t_1$, we implicitly have $w_2 \geq 0$. Thus, $\langle w,h,\pi\rangle(\langle 0,0,\text{NIL}\rangle\text{lmin}\langle w_2,h_2,\pi_2\rangle) = \langle w,h,\pi\rangle\langle 0,0,\text{NIL}\rangle = \langle w,h,\pi\rangle\langle 0,0,\text{NIL}\rangle\text{lmin}\langle w,h,\pi\rangle\langle w_2,h_2,\pi_2\rangle$. The last step follows because $w + w_2 \geq w$ and $h + h_2 \geq h + 1$. Similarly for right distributivity.
6. Suppose $t, t_1 \in \mathbb{R} \times \mathbb{N} \times V$ and $t_2 = \langle 0,0,\text{NIL}\rangle$. Again, we implicitly have $w_1 < 0$. Then $\langle w,h,\pi\rangle(\langle w_1,h_1,\pi_1\rangle\text{lmin}\langle 0,0,\text{NIL}\rangle) = \langle w,h,\pi\rangle\langle w_1,h_1,\pi_1\rangle = \langle w,h,\pi\rangle\langle w_1,h_1,\pi_1\rangle\text{lmin}\langle w,h,\pi\rangle\langle 0,0,\text{NIL}\rangle$. The last step follows because $w + w_1 < w$. Similarly for right distributivity.
7. Suppose $t, t_1, t_2 \in \mathbb{R}\times\mathbb{N}\times V$. Then we have $\langle w,h,\pi\rangle(\langle w_1,h_1,\pi_1\rangle\text{lmin}\langle w_2,h_2,\pi_2\rangle) = \langle w,h,\pi\rangle\langle w_1,h_1,\pi_1\rangle = \langle w,h,\pi\rangle\langle w_1,h_1,\pi_1\rangle\text{lmin}\langle w,h,\pi\rangle\langle w_2,h_2,\pi_2\rangle$. Similarly for right distributivity. $\square$

Finally, we show correctness of our 3-tuple-based algorithm. Recall that the goal of this 3-tuple transformation is to retain that $\delta \approx \mathbf{d}_n = \mathbf{d}\mathbf{A}^N$, where addition and multiplication are defined as lmin and $+_{\text{rhs}}$, respectively. More accurately, we have $\mathbf{d}_n(v) = \langle \mathbf{\Delta}(s,v), |p|, u\rangle$, where $p = \langle s, \ldots, u, v\rangle$ is a smallest (size) path from s to v. Correctness is captured by the following lemma.

Lemma 5.3. *Consider a graph with designated vertex s. Let adjacency matrix $\mathbf{A}$ and initial distance vector $\mathbf{d}_0$ be set up according to equations (5.3) and (5.4), and let $\mathbf{d}_k = \mathbf{d}_{k-1}$ lmin.$+_{rhs}$ A for $k \geq 1$. If $h \leq k \in \mathbb{N}$ is the minimum size (number of hops) of a shortest ($\leq k$)-hop path from s to v, then there exists $u \in V$ such that $\mathbf{d}_k(v) = \langle \mathbf{\Delta}_k(s,v), h, u\rangle$, $u = \min\{v : v \in \Pi_k(s,v)\}$, and there exists an h-hop path $p = \langle s, \ldots, u, v\rangle$ with $w(p) = \mathbf{\Delta}_k(s,v)$. If no ($\leq k$)-hop path exists, then $\mathbf{d}_k(v) = \langle \infty, \infty, \infty\rangle$. For the source vertex, $\mathbf{d}_k(s) = \langle 0, 0, \text{NIL}\rangle$.*

Proof. By induction on k. The lemma trivially holds with $k = 0$.

Let $\langle w, h, u\rangle = \mathbf{d}_k(v)$ and suppose that a ($\leq k$)-hop path (for $k > 0$) from s to v exists.

First, we show that w and h are correct. For every $u_i \neq v \in V$, let $\langle w_i, h_i, \pi_i\rangle = \mathbf{d}_{k-1}(u_i)$. Let $\langle w', h', \pi'\rangle = \mathbf{d}_{k-1}(v)$. By definition, $\mathbf{d}_k(v) = \text{lmin}_{u' \in V} \mathbf{d}_{k-1}(u') +_{\text{rhs}} A(u', v) = \left(\text{lmin}_{u' : (u',v) \in E}\ \mathbf{d}_{k-1}(u') +_{\text{rhs}} A(u', v)\right) \text{lmin}\ \mathbf{d}_{k-1}(v)$. Thus, by nature of the lmin, we have $\langle w, h\rangle = \text{lmin}\{\langle w', h'\rangle, \text{lmin}_i\{\langle w_i + \mathbf{W}(u_i, v), h_i + 1\rangle\}\}$. The inductive assumption gives $w_i = \mathbf{\Delta}_{k-1}(s, u_i)$; we conclude that $w = \mathbf{\Delta}_k(s,v)$ since the shortest path using $\leq k$ hops must start with a shortest path using $\leq k-1$ hops. Similarly, h is correct since each h_i corresponds to the smallest number of hops necessary to achieve the shortest ($\leq k-1$)-hop path.

Next, we show that u is correct. Notice that the lmin across all neighbors u_i of v can be reduced to the lmin across all neighbors having $w_i + \mathbf{W}(u_i, v) = w = \mathbf{\Delta}_k(s,v)$ and $h_i + 1 = h$, or equivalently (from Lemma 5.1) the $u_i \in \Pi_k(s,v)$. The lmin is also affected by $\langle w', h', \pi'\rangle = \mathbf{d}_{k-1}(v)$ if $w' = w$ and $h' = h$. For any neighbor $u_i \in \Pi_k(s,v)$, we have $\mathbf{d}_{k-1}(u_i) +_{\text{rhs}} A(u_i, v) = \langle w, h, u_i\rangle$. For v itself, $\mathbf{d}_{k-1}(v) +_{\text{rhs}} A(v,v) = \mathbf{d}_{k-1}(v)$. By inductive assumption, $\pi' = \min\{u' : u' \in \Pi_{k-1}(s,v)\}$. Thus, since $w' = w$ implies $\Pi_k(s,v) = \Pi_{k-1}(s,v)$, we have that the result value u is affected by all and only the members of $\Pi_k(s,v)$. In other words, we have reduced our expression to $\mathbf{d}_k(v) = \text{lmin}_{\pi \in \Pi_k(s,v)} \langle w, h, \pi\rangle$. □

Lemmas 5.2 and 5.3 together imply that $\mathbf{d} = \mathbf{d}\mathbf{A}^N$ gives a result vector d that is consistent with the Bellman–Ford algorithm, regardless of the ordering of multiplications. As a result, either the iterative or repeated squaring algorithm still works.

Computing parent pointers at the end

In this approach, we use the one-extra-multiplication idea of equation (5.2), but we modify the scalars to make the equation work for 0-weight cycles. This approach still uses tuples, but they are slightly simpler.

We use 2-tuples of the $\langle w, h\rangle \in \mathbb{R}_\infty \times (\mathbb{N} \cup \{0, \infty\})$. These tuple values have the same meaning as in the 3-tuple approach, but now there is no mention of parent pointers in the tuple. Setting up A and $\mathbf{d}_0$ should be obvious given this new definition.

The Bellman–Ford algorithm using 2-tuples remains the same as that using 3-tuples, except now we use lmin for addition and the simpler + for multiplication. In other words, we have $\mathbf{d}_k = \mathbf{d}_{k-1}$ lmin.+ $\mathbf{A}$. Correctness follows from Lemma 5.3 since these operations are exactly the same as those used by the 3-tuple approach, except that the third tuple element is ignored. Note that + and lmin are virtually identical to + and min, so we now have a commutative semiring.

To compute the parent pointers, set $\pi = \mathbf{d}_n$ argmin.+ $(\mathbf{A} + \text{diag}(\langle\infty,\infty\rangle))$. In the following lemma, we claim that the values in $\pi(v)$ are correct for all $v \neq s$ that are reachable from s.

Lemma 5.4. *Consider a graph with designated vertex s and no negative-weight cycles. Let $\mathbf{d}_n(v) = \langle \mathbf{\Delta}(s,v), h_v\rangle$, where h_v is the smallest size of a shortest path from s to v. Let $\pi = \mathbf{d}_n$ argmin.+ $(\mathbf{A} + \text{diag}(\langle\infty,\infty\rangle))$. Then for all $v \in V - \{s\}$ with $\mathbf{\Delta}(s,v) < \infty$, we have $\pi(v) \in \Pi_n(s,v)$.*

Proof. Consider any vertex $v \in V - \{s\}$ that is reachable from s. The product $\pi = \mathbf{d}_n$ argmin.+ $(A + \text{diag}(\langle\infty,\infty\rangle))$ gives $\pi(v) = \text{argmin}_{u \neq v}\{\langle \mathbf{\Delta}(s,u) + \mathbf{W}(u,v), h_u + 1\rangle\}$. By Lemma 5.1, we have $u \in \Pi_n(s,v)$ if and only if $\mathbf{\Delta}(s,u) + \mathbf{W}(u,v) = \mathbf{\Delta}(s,v)$ and $h_u = h_v - 1$. Thus, $\pi(v) = \text{argmin}_{u \in \Pi_n(s,v)}\{\langle \mathbf{\Delta}(s,v), h_v\rangle\}$, giving us the result that $\pi(v) \in \Pi_n(s,v)$. □

For any vertex that is not reachable from s, or for s itself, this approach does not yield a correct answer, and we must do some additional cleanup to get π to contain correct values. Alternatively, we can modify argmin to take on a value of ∞ if the operands are all $\langle\infty,\infty\rangle$, and then the only cleanup necessary is setting $\pi(s) = \text{NIL}$.

5.1.3 Floyd–Warshall

The Floyd–Warshall algorithm solves the all-pairs shortest paths problem. Given a graph $G = (V, E)$ with edge weights w and no negative-weight cycles, this algorithm returns the shortest paths distances $\mathbf{\Delta}(u,v)$ for all $u, v \in V$.

Like Bellman–Ford, Floyd–Warshall is a dynamic programming solution. Without loss of generality, vertices are numbered $V = \{1, 2, \ldots, N\}$. This algorithm uses $\mathbf{D}_k(u,v)$ to represent the shortest path from u to v using only intermediate vertices in $\{1, 2, \ldots, k\} \subseteq V$. Thus, we have

$$\mathbf{D}_k(u,v) = \begin{cases} \mathbf{W}(u,v), & k = 0, \\ \min\{\mathbf{D}_{k-1}(u,v), \mathbf{D}_{k-1}(u,k) + \mathbf{D}_{k-1}(k,v)\}, & k \geq 1 \end{cases}$$

Algorithm 5.3 gives pseudocode for the Floyd–Warshall algorithm. The running time is $\Theta(N^3)$ because of the triply nested loops. Although we explicitly store every $\mathbf{D}_k$ in this version of the algorithm, it is only necessary to store $\mathbf{D}_k$ and $\mathbf{D}_{k-1}$ at any given time. Thus, the space usage is $\Theta(N^2)$.

Algorithm 5.3. Floyd–Warshall.
A standard implementation of the Floyd–Warshall algorithm [Cormen et al. 2001].

```
FLOYD–WARSHALL(V, E, w)
for u = 1 to N
    do for v = 1 to N
        do D_0(u, v) = W(u, v)
           if u = v
              then Π_0(u, v) = NIL
           if u ≠ v and W(u, v) < ∞
              then Π_0(u, v) = u
           ▷ else Π_0(u, v) is undefined
for k = 1 to N
    do for u = 1 to N
        do for v = 1 to N
            do if D_{k−1}(u, k) + D_{k−1}(k, v) ≤ D_{k−1}(u, v)
                 then D_k(u, v) = D_{k−1}(u, k) + D_{k−1}(k, v)
                      Π_k(u, v) = Π_{k−1}(k, v)
                 else D_k(u, v) = D_{k−1}(u, v)
                      Π_k = Π_{k−1}(u, v)
```

The table entries $\Pi_k(u, v)$ in this algorithm refer to the penultimate vertex on a shortest path from u to v using only vertices $\{1, 2, \ldots, k\}$. This value should not be confused with Π_k given in Section 5.1.2, which is the set of penultimate vertices on every shortest ($\leq k$)-hop path from u to v. Note that if we define $\pi_k(v) = \Pi_k(s, v)$, then π_k comprises a shortest path tree from the source s. The entire set of edges selected by Π_k does not necessarily form a tree.

Algebraic Floyd–Warshall

To represent Floyd–Warshall as matrix and vector operations, we use the (dense) $N \times N$ shortest path matrices $\mathbf{D}_k$ to store the path weights. We initialize $\mathbf{D}_0 = \mathbf{A}$, or $\mathbf{D}_0(u, v) = \mathbf{W}(u, v)$, with $\mathbf{D}_0(u, u) = 0$.

The recursive definition of $\mathbf{D}_k(u, v)$ simply considers two possibilities—either the optimal path from u to v using intermediate vertices $\{1, 2, \ldots, k\}$ uses vertex k, or it does not. For our algebraic Floyd–Warshall, we compute the weight of the shortest paths using vertex k for all u, v at once. In particular, the kth column of $\mathbf{D}_k$, denoted $\mathbf{D}_k(:, k)$, contains the shortest paths from any u to k. Similarly, the kth row $\mathbf{D}_k(k, :)$ contains the shortest paths from k to any u. Thus, we simply need to add every pair of these values to get the shortest path weights going through k, which we can do by taking $\mathbf{D}_k(:, k)$ min.+ $\mathbf{D}_k(k, :)$. Note that strictly speaking, there is no scalar addition occurring in this outer product, so we do not have to use min.

Thus the full value for $\mathbf{D}_k$ is

$$\mathbf{D}_k = \mathbf{D}_{k-1} \text{ .min } (\mathbf{D}_{k-1}(:, k) \text{ min.+ } \mathbf{D}_{k-1}(k, :))$$

This expression yields the algebraic Floyd–Warshall algorithm given in Algorithm 5.4. This algorithm performs N dense vector multiplications, giving complexity $\Theta(N^3)$.

Algorithm 5.4. Algebraic Floyd–Warshall.
An algebraic implementation of the Floyd–Warshall algorithm.

FLOYD–WARSHALL($\mathbf{A}$)

$\mathbf{D} = \mathbf{A}$
for $k = 1$ **to** N
    do $\mathbf{D} = \mathbf{D}$ .min $[\mathbf{D}(:, k)$ min.+ $\mathbf{D}(k, :)]$

To compute parent pointers, we can use the same tuples as in Bellman–Ford.

5.2 Minimum spanning tree

In the minimum-spanning-tree problem, we are given a weighted, undirected graph $G = (V, E)$ with edge weights $w : E \to \mathbb{R}_\infty$, where $\mathbf{W}(u, v) = \infty$ if $(u, v) \notin E$. For simplicity, $\mathbf{W}(v, v) = 0$ for all $v \in V$.

A *spanning tree* is a subset $T \subseteq E$ of edges forming a "tree" that connects the entire graph. A *tree* is a set of edges forming an acyclic graph. To be a spanning tree, it must be that $|T| = N - 1$. The *weight* of a spanning tree is the total weight $w(T) = \sum_{e \in T} \mathbf{W}(e)$. A spanning tree is a *minimum spanning tree* if for all other spanning trees T', we have $w(T) \leq w(T')$.

Minimum spanning trees are unique if all edge weights in the graph are unique, but otherwise there is no guarantee of uniqueness. Without loss of generality, we assume that all edge weights are unique (since they can be made to be unique by creating tuples including indices, or equivalently by appending unique lower-order bits to the ends of the weights).

This section considers the minimum-spanning-tree problem. A similar problem that is not any "harder" is the problem of finding a minimum spanning forest in an unconnected graph, which can be solved directly or by finding minimum spanning trees in each connected component of the graph. For the purposes of exposition, we assume the graphs to be connected.

We describe Prim's algorithm for solving the minimum-spanning-tree problem. The algebraic variant of Prim's algorithm does not achieve as efficient a bound as the standard algorithm that it replaces. However, a variant of Borůvka's algorithm (not presented here) does have matching complexity.

5.2.1 Prim's

Prim's algorithm solves the minimum-spanning-tree problem by growing a single set S of vertices belonging to the spanning tree. On each iteration, we add the "closest" vertex not in S to S. Specifically, we say that an edge (u, v) is a **lightest** edge leaving S if $u \in S$, $v \notin S$, and $\mathbf{W}(u, v) = \min\{(u', v') : u' \in S, v' \notin S\}$. Suppose that the edge (u, v) is a lightest edge leaving S, with $u \in S$. Then Prim's algorithm updates $S = S \cup \{v\}$, and $T = T \cup \{(u, v)\}$. This process is repeated $N - 1$ times, at which point T is a spanning tree.

Prim's is typically implemented using a priority queue. A priority queue Q is a data structure that maintains a dynamic set of keyed objects and supports the following operations:

- INSERT(Q, x): inserts a keyed object x into the queue (i.e., $Q = Q \cup \{x\}$).
- Q = BUILD-QUEUE($x_1, x_2, \ldots, x_n$): performs a batched insert of keyed objects $x_1, x_2, \ldots, x_n$ into an empty queue Q (i.e., $Q = \{x_1, x_2, \ldots, x_n\}$).
- EXTRACT-MIN(Q): removes and returns the element in Q with the smallest key (i.e., assuming unique keys, if $key(x) = \min_{y \in Q}\{key(y)\}$, then $Q = Q - \{x\}$).
- DECREASE-KEY(Q, x, k): decreases the value of x's key to k, assuming that $k \leq key(x)$. Here, x is a pointer to the object.

Several other operations may be supported, but they are not important to Prim's algorithm. A naive implementation of a priority queue is an unsorted list. The INSERT and DECREASE-KEY operations are trivially $\Theta(1)$. The EXTRACT-MIN operation is $\Theta(V)$ in the worst case to scan the entire list. Fredman and Tarjan's [Fredman & Tarjan 1987] fibonacci heap has $O(1)$ amortized cost for all of these operations except EXTRACT-MIN, which has $O(\log N)$ amortized cost.

A standard implementation of Prim's algorithm using a priority queue is given in Algorithm 5.5. This algorithm performs one BUILD-QUEUE, M DECREASE-KEY, and N EXTRACT-MIN operations, yielding a runtime of $O(M + N \log N)$ using a fibonacci heap.

Algorithm 5.5. Prim's.
A standard implementation of the Prim's algorithm [Cormen et al. 2001].

```
PRIMS(V, E, w)
foreach v ∈ V
      do key(v) = ∞
         π(v) = NIL
weight = 0
▷ choose some arbitrary vertex s ∈ V
key(s) = 0
Q = BUILD-QUEUE(V)
while Q ≠ ∅
      do u = EXTRACT-MIN(Q)
         weight = weight + W(π(u), u)
         foreach v with (u, v) ∈ E
               do if v ∈ Q and W(u, v) < key(v)
                     then DECREASE-KEY(Q, v, W(u, v))
                          π(v) = u
```

Algebraic Prim's

We use the (sparse) $N \times N$ adjacency matrix $\mathbf{A}$ to store edge weights, a $1 \times N$ vector $\mathbf{s}$ to indicate membership in the set S, and a $1 \times N$ vector $\mathbf{d}$ to store the weights of the edges leaving the set S. We maintain $\mathbf{s}$ and $\mathbf{d}$ according to

$$\mathbf{s}(v) = \begin{cases} \infty & \text{if } v \in S \\ 0 & \text{otherwise} \end{cases}$$

and

$$\mathbf{d}(v) = \min_{u \in S} \mathbf{W}(u, v)$$

If $v \notin S$, then $\mathbf{d}(v)$ gives the lightest edge connecting v to S. If $v \in S$, then $\mathbf{d}(v) = 0$.

Our algebraic version of Prim's is given in Algorithm 5.6. line 7 finds the vertex that is closest to S (without being in S). This step performs an argmin across an entire vector, which is essentially $\Theta(N)$ work per iteration. The vector addition in line 10 decreases the key of all neighbors of closest vertex u. This step may cause $\Theta(N)$ work in a single iteration, but in total we only update each edge once, for a total work of $O(M)$ across all iterations. The algorithm thus has complexity $\Theta(N^2)$, which is caused by the slow argmin of line 7. To understand this bound, the total work is $\Theta(N^2)$, whereas the conventional Prim's with a priority queue achieves $O(M + N \log N)$. This bound is significantly worse for sparse graphs but equivalent for very dense graphs.

Computing the tree. Thus far, we have ignored the spanning-tree edges as we do with shortest path trees. Keeping track of edges is relatively simple for Prim's algorithm using a similar tuple-based approach. Each entry in the adjacency matrix $\mathbf{A}$ (and vector $\mathbf{d}$) is a 2-tuple $\mathbf{A}(u, v) = \langle \mathbf{W}(u, v), u \rangle$. The only necessary change to the algorithm is to keep a $1 \times N$ vector π that stores the spanning tree for all vertices in S. In particular, we modify the algorithm according to Algorithm 5.7. We add line 12 to store the edge at the time it is added to the spanning tree (because this edge may be destroyed by the update to $\mathbf{d}$ in line 13).

Algorithm 5.6. Algebraic Prim's.
An algebraic implementation of Prim's algorithm.

PRIMS($\mathbf{A}$)

1 $\mathbf{s} = 0$
2 *weight* $= 0$
3 ▷ choose arbitrary vertex $\in V$ to start from
4 $\mathbf{s}(1) = \infty$
5 $\mathbf{d} = \mathbf{A}(1, :)$
6 **while** $\mathbf{s} \neq \infty$
7   **do** $u = \text{argmin}\{\mathbf{s} + \mathbf{d}\}$
8         $\mathbf{s}(u) = \infty$
9   *weight* $=$ *weight* $+ \mathbf{d}(u)$
10         $\mathbf{d} = \mathbf{d}$ .min $\mathbf{A}(u, :)$

Algorithm 5.7. Algebraic Prim's with tree.
An algebraic implementation of Prim's algorithm that also computes the tree.

```
PRIMS(A)
  ⋮
  π = NIL
  while s ≠ ∞
        do u = argmin{s + d}
           s(u) = ∞
           ⟨w, p⟩ = d(u)
           weight = weight + w
           π(u) = p
           d = d .min A(u, :)
```

Since the entries in **d** and **A** correspond exactly to particular edges, it should be obvious that this version of the algorithm correctly returns the spanning tree edges as well as the total weight. Note that storing edges here is strikingly simpler than in shortest paths of Section 5.1.2. Part of the reason for the simplicity is that the **d** vector here stores just edges, whereas in Section 5.1.2, we store *paths.* Another cause is the structure of the algorithm—we are not operating over a single semiring here, so iterations of the algorithm do not associate anyway, so there is no need to come up with fancy logic.

References

[Bellman 1958] R. Bellman. On a routing problem. *Quarterly of Applied Mathematics*, 16:87–90, 1958.

[Cormen et al. 2001] T. Cormen, C. Leiserson, and R. Rivest. *Introduction to Algorithms, Second Edition.* Cambridge, Mass.: MIT Press and McGraw–Hill, 2001.

[Ford & Fulkerson 1962] L.R. Ford, Jr. and D.R. Fulkerson. *Flows in Networks.* Princeton, N.J.: Princeton University Press, 1962.

[Fredman & Tarjan 1987] M.L. Fredman and R.E. Tarjan. Fibonacci heaps and their uses in improved network optimization algorithms. *Journal of the ACM*, 34:596–615, 1987.

Chapter 6
Complex Graph Algorithms

*Eric Robinson**

Abstract

This chapter discusses the representation of several complex graph algorithms as algebraic operations. Even though the underlying algorithms already exist, the algebraic representation allows for easily expressible efficient algorithms with appropriate matrix constructs. This chapter gives algorithms for clustering, vertex betweenness centrality, and edge betweenness centrality.

6.1 Graph clustering

Graph clustering is the problem of determining natural groups with high connectivity in a graph. This can be useful in fields such as machine learning, data mining, pattern recognition, image analysis, and bioinformatics. There are numerous methods for graph clustering, many of which involve performing random walks through the graph. Here, a peer pressure clustering technique [Reinhardt et al. 2006] is examined.

6.1.1 Peer pressure clustering

Peer pressure clustering capitalizes on the fact that given any reasonable cluster approximation for the graph, a vertex's cluster assignment will be the same as the

*MIT Lincoln Laboratory, 244 Wood Street, Lexington, MA 02420 (erobinson@ll.mit.edu).

This work is sponsored by the Department of the Air Force under Air Force Contract FA8721-05-C-0002. Opinions, interpretations, conclusions, and recommendations are those of the author and are not necessarily endorsed by the United States Government.

cluster assignment of the majority of its neighbors. Consider the graph in Figure 6.1, where each vertex except vertex 4 is in the proper cluster.

If the incoming edges to each vertex in this graph are examined to determine from which cluster they originate, this clustering can be considered suboptimal. As seen in Figure 6.2, all vertices except vertex 4 have the most incoming edges from the cluster they are in.

In addition, if the clustering is adjusted accordingly by placing vertex 4 in the red cluster, the result now shows the clustering to be optimal. Each vertex has a majority of incoming edges originating from its own cluster. This is shown in Figure 6.3.

Traditionally, peer pressure clustering is performed by iteratively refining a cluster approximation for the graph. Given a cluster approximation, each vertex in the graph first votes for all of its neighbors to be in its current cluster. These votes are then tallied and a new cluster approximation is formed by moving vertices to the cluster for which they obtained the most votes. The algorithm typically terminates when a fixed point is reached (when two successive cluster approximations are identical).

Algorithm 6.1 shows the recursive definition for peer pressure clustering. This algorithm can also be performed in a loop, keeping track of the current and previous cluster approximation and terminating when they agree.

Algorithm 6.1. Peer pressure. Recursive algorithm for clustering vertices.*

PEERPRESSURE$(G = (V, E), C_i)$
1 **for** $(u, v, w) \in E$
2   **do** $T(v)(C(u)) \leftarrow T(v)(C(u)) + w$
3 **for** $n \in V$
4   **do** $C_f(n) \leftarrow i : \forall j \in V : T(n)(j) \leq T(n)(i)$
5 **if** $C_i == C_f$
6   **then return** C_f
7   **else return** PEERPRESSURE(G, C_f)

In this algorithm, the loop at line 1 is responsible for the voting, and the loop at line 3 tallies those votes to form a new cluster approximation. It is assumed that the structure T is stored as an array of lists, which keeps track of, for each vertex, the number of votes that vertex gets for each cluster for which it receives votes.

Unfortunately, convergence is not guaranteed for unclustered graphs and pathological cases do exist where the algorithm must make $O(n)$ recursive calls before a repetition. In any well-clustered graph, however, this algorithm will terminate after a small number of recursive calls, typically on the order of five.

*V.B. Shah, *An Interactive System for Combinatorial Scientific Computing with an Emphasis on Programmer Productivity*, Ph.D. thesis, Computer Science, University of California, Santa Barbara, June 2007.

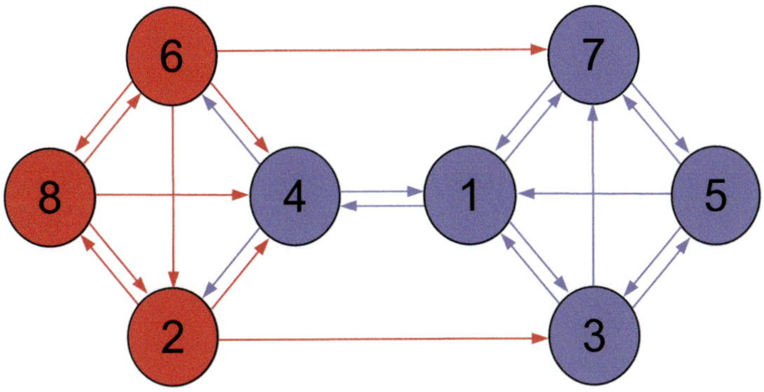

Figure 6.1. Sample graph with vertex 4 clustered improperly.

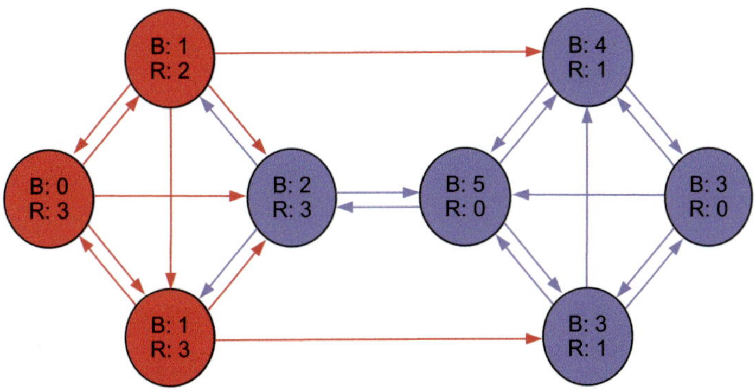

Figure 6.2. Sample graph with count of edges from each cluster to each vertex.

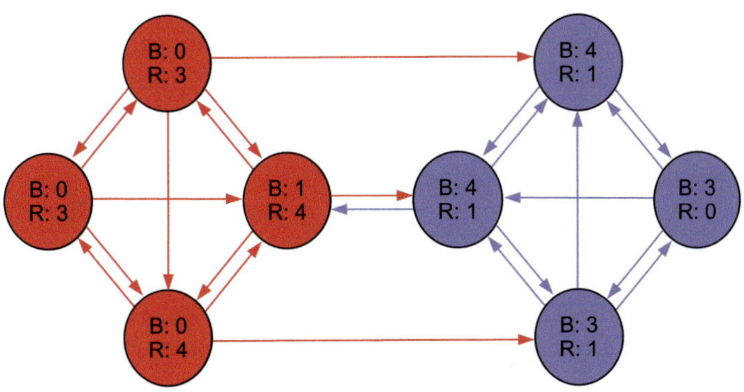

Figure 6.3. Sample graph with correct clustering and edge counts.

Starting approximation

In order to get things going with the peer pressure clustering algorithm, an initial cluster approximation must be chosen. One strategy for this is to simply perform a single round of Luby's maximal independent set algorithm [Luby 1986]. This will yield a reasonable cluster approximation assuming that the individual clusters in the graph are highly connected. However, this is completely unnecessary. For graphs that actually contain clusters, the solution arrived at by the peer pressure algorithm is highly independent from the initial cluster approximation. For this reason, a naive starting approximation as shown below, where each vertex is in a cluster by itself, suffices to start things off

for $v \in V$
  do $C_i(v) = v$

Assuring vertices have equal votes

In graphs where there is a large discrepancy between the out-degree of vertices, such as real-world power law graphs, vertices with a large out-degree will have a larger influence on the cluster approximations. These vertices will have more votes in each cluster refinement because they have more outgoing edges. This can be easily remedied by normalizing the votes of each vertex to one. This can be done by summing up the weights of the outgoing edges of each vertex, and then dividing those weights by that sum. The code for performing this is shown below

for $e = (u, v, w) \in E$
  do $S(u) = S(u) + w$
for $e = (u, v, w) \in E$
  do $w = w/S(u)$

Preserving small clusters and unclustered vertices

Depending upon the desired result, it may be advantageous to either group vertices with a single connection to a large cluster or to keep these vertices separate in clusters of their own. Typically, a one-vertex cluster makes little sense. However, one can view these as simply unclustered vertices. Smaller clusters and single vertices tend to get subsumed by larger ones because the vertices in larger clusters tend to have a very high-combined out-degree. The majority of these edges go to other vertices in the large cluster, but a few may go outward to smaller clusters. If a couple of vertices in a large cluster have these outward connections to the same small cluster, then that cluster may be pulled into the larger one.

To remedy this, each vertex's votes can be scaled according to the size of the cluster it is in. It is typically not desirable to divide the vertex's vote directly by the size of its cluster, as that will lead to a cluster with a single vertex having the same voting strength as all the vertices combined in a larger cluster. However, some scaling is clearly desired. To do this, rather than scaling directly by the number of vertices in the cluster, a strength $0 \leq p \leq 1$ is chosen, where 0 indicates no scaling

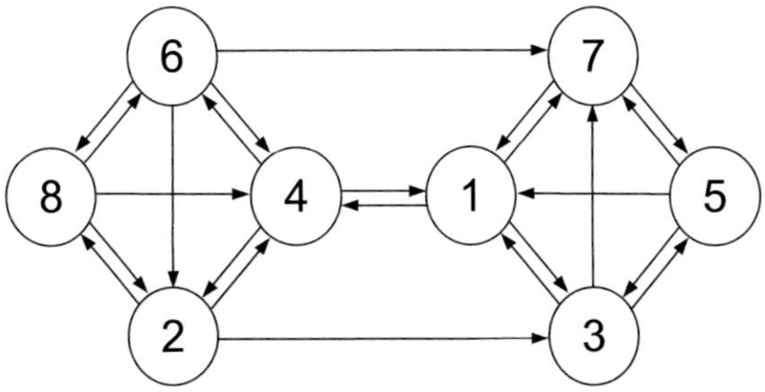

V	Out-degree
1	4
2	4
3	4
4	4
5	4
6	5
7	3
8	4

Figure 6.4. Sample graph.

and 1 indicates full scaling. The inverse of the scaling is then raised to this power to determine the actual scaling. This can be done as shown below

for $u \in V$
  do $s(C(u)) = s(C(u)) + 1$
for $e = (u, v, w) \in E$
  do $w = w/s(C(u))^p$

The value for p need not be static. One option is to compute p based on the number of current clusters. When there are many clusters, p should be close to 0 as scaling is most likely not necessary. As the number of clusters shrinks, p can grow to prevent the larger of those clusters from dominating.

Settling ties

Line 4 of Algorithm 6.1 does not specify what to do if two clusters tie for the maximum number of votes for a vertex. Currently, as the algorithm is shown, the vertex would not be put into a cluster. This, however, is not a desirable outcome. Instead, one of the clusters with the maximum number of votes for that vertex should be chosen. This can be done in a deterministic manner by selecting the cluster with the lowest number. This deterministic method also helps to speed the algorithm to convergence by having all vertices choose the same "leader" for their cluster early in the algorithm.

Sample calculation

A clustering example will be demonstrated on the graph shown in Figure 6.4, which is the same graph as that shown in Figure 6.3. The vertex numbers are shown here for the purposes of tie breakers (which will go to the vertex/cluster with the lowest number). The obvious clusters, as with the previous example, are between vertices 1, 3, 5, 7 and vertices 2, 4, 6, 8.

The initialization of the clustering algorithm requires computing weights for each vertex's outgoing edges based on the overall out-degree of the vertex. In addition, each vertex must be placed in a cluster by itself. Both of these steps are shown in Figure 6.5. Note that it is assumed that there is a self-edge to each vertex, and the initial cluster number for each vertex corresponds to its vertex number.

Consider a single pass through the peer pressure clustering algorithm for this graph. Vertices 1, 5, and 7 will be pulled into cluster 7, as its outgoing edges hold the most value (there are the fewest of them). From there, all of the rest of the votes will be ties outside of those from vertex 6, which will lose because it has more outgoing edges than the rest of the vertices. As a result, vertices 3 and 4 are pulled into cluster 1 (which no longer contains vertex 1), vertices 2 and 8 are pulled into cluster 2, and finally vertex 6 is pulled into cluster 4 (which no longer contains vertex 4). These results are shown in Figure 6.6.

After the first iteration through the peer pressure clustering algorithm, there are only four clusters remaining out of the initial eight. The second iteration reduces this number further. Vertices 1, 5, and 7 all have enough votes to remain in cluster 7. In addition, vertices 1 and 5 vote for vertex 8 to be in their cluster, resulting in a total vote of 0.5, enough to pull vertex 3 into cluster 7, as the only other votes it receives are from cluster 2 with a weight 0.25. Vertices 2 and 8 also have enough votes to remain in cluster 2. In addition, they both vote for vertex 4 to be in their cluster for a total weight of 0.5, beating out the votes from cluster 4 of 0.2 and cluster 7 of 0.25, and shifting vertex 4 into cluster 2. Finally, vertex 6 receives 0.25 votes from cluster 2, 0.25 votes from cluster 1, and 0.2 votes from cluster 4. The tie moves vertex 6 to the cluster with the lowest number, cluster 1, as shown in Figure 6.7.

One final iteration is required to arrive at the result and reduce the total number of clusters from three, as in the previous round, to two. In this iteration, vertices 2, 4, and 8 have enough votes from their cluster to remain in cluster 2. Vertices 1, 3, 5, and 7 do as well, and they remain in cluster 7. Vertex 6 receives 0.5 votes from cluster 2 and 0.2 votes from cluster 1 (itself), shifting it to cluster 2. This iteration is shown in Figure 6.8 and is the final clustered solution for this graph. Note that the next pass will keep the same cluster approximation, indicating that a fixed point has been reached.

Space complexity

The above algorithm has a space complexity of $O(M)$ due to the cluster tally T. T can be assigned at most one entry per edge in the graph because each edge represents a vote. Since it is possible that all of these edges represent votes for different vertex cluster pairs, T may contain up to M different vote combinations.

Time complexity

The voting loop at line 1 must process $O(M)$ votes, one for each edge in the graph. After this, the maximum for each vertex must be obtained by the tallying loop at line 3. Because T has a maximum of $O(M)$ entries, one for each vote made, this requires $O(M)$ operations as well. This leads to $O(M)$ operations per recursive call, and, where p is the number of passes before convergence, a total runtime of $O(p \times M)$. As noted earlier, p is typically a small constant for well-clustered graphs on the order of five. In this case, the total runtime is just $O(M)$.

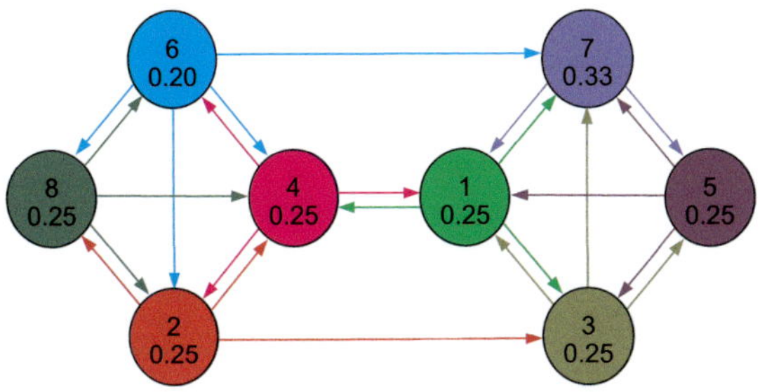

V	Color-Coded Votes
1	0.25 0.25 0.25 0.25 0.33
2	0.25 0.25 0.20 0.25
3	0.25 0.25 0.25
4	0.25 0.25 0.25 0.20 0.25
5	0.25 0.25 0.33
6	0.25 0.20 0.25
7	0.25 0.25 0.25 0.20 0.33
8	0.25 0.20 0.25

Figure 6.5. Initial clustering and weights.

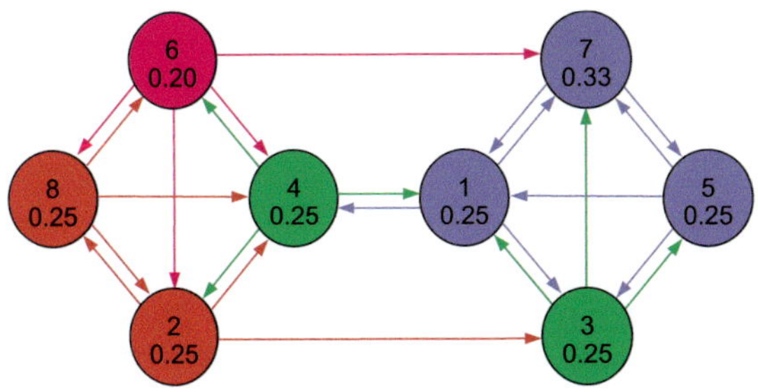

V	Color-Coded Votes
1	0.50 0.83
2	0.25 0.50
3	0.25 0.25 0.50
4	0.25 0.50 0.20 0.25
5	0.25 0.58
6	0.25 0.25 0.20
7	0.25 0.20 0.83
8	0.50 0.20

Figure 6.6. Clustering after first iteration.

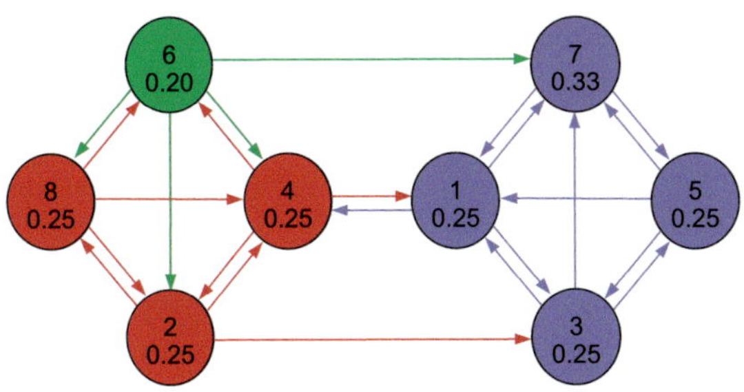

V	Color-Coded Votes
1	0.25 1.08
2	0.20 0.75
3	0.25 0.75
4	0.20 0.75 0.25
5	0.83
6	0.20 0.50
7	0.20 1.08
8	0.20 0.50

Figure 6.7. Clustering after second iteration.

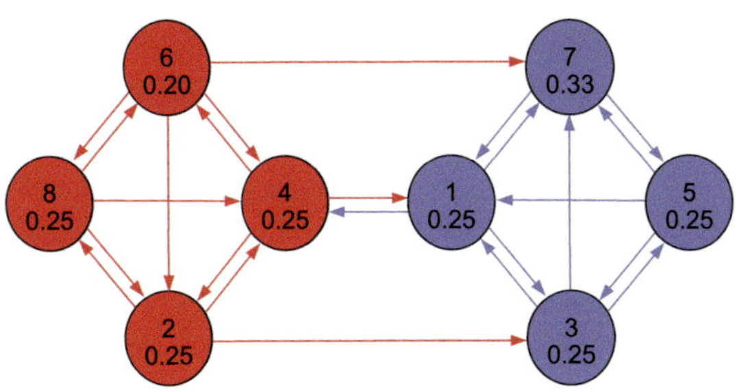

V	Color-Coded Votes
1	0.25 1.08
2	0.95
3	0.25 0.75
4	0.95 0.25
5	0.83
6	0.70
7	0.20 1.08
8	0.70

Figure 6.8. Final clustering.

6.1.2 Matrix formulation

Where the graph is represented as a weighted adjacency matrix, $G = \mathbf{A} : \mathbb{R}_+^{N \times N}$, the clustering algorithm can be performed in the same manner. Let $\mathbf{C} : \mathbb{B}^{N \times N}$ be the cluster approximation, where if $c_{ij} == 1$, then vertex j is in cluster i.

With this representation, voting can be expressed as a simple matrix matrix multiplication

$$\mathbf{T} = \mathbf{C}\mathbf{A}$$

Here $\mathbf{T}$ represents a tallying matrix where if $t_{ij} == k$, then there are k votes for vertex j to be in cluster i.

Once the votes have been performed, the new clusters need to be selected. This can be done with the following operations

$$\mathbf{m} = \mathbf{T} \;\; \text{max.}$$
$$\mathbf{C} = \mathbf{m} \;\; .\!== \;\; \mathbf{T}$$

Here the max vote for each vertex in each cluster is found, then the cluster approximation is set appropriately according to that value.

From this approximation, the peer pressure algorithm can be formed (see Algorithm 6.2). Line 2 performs the voting, and lines 3 and 4 tally those votes to form a new cluster approximation.

Algorithm 6.2. Peer pressure matrix formulation.

PeerPressure($G = \mathbf{A} : \mathbb{R}_+^{N \times N}, \mathbf{C_i} : \mathbb{B}^{N \times N}$)

1 $\mathbf{T} : \mathbb{R}_+^{N \times N}$ $\mathbf{C_f} : \mathbb{B}^{N \times N}$ $\mathbf{m} : \mathbb{R}_+^{N}$
2 $\mathbf{T} = \mathbf{C_i}\mathbf{A}$
3 $\mathbf{m} = \mathbf{T} \;\; \text{max.}$
4 $\mathbf{C_f} = \mathbf{m} \;\; .\!== \;\; \mathbf{T}$
5 **if** $\mathbf{C_i} == \mathbf{C_f}$
6   **then return** $\mathbf{C_f}$
7   **else return** PeerPressure($G, \mathbf{C_f}$)

Starting approximation

As before, an initial approximation must be selected. If each vertex is in a cluster by itself, with the cluster number being equal to the vertex number, then $\mathbf{C_i} = \mathbf{I}$, or the identity matrix.

Assuring vertices have equal votes

Normalizing the out-degrees of the vertices in the graph corresponds to normalizing the rows of the adjacency matrix. This can be done easily as shown below

$$\mathbf{w} = \mathbf{A} \;\; +.$$
$$\mathbf{A} = \mathbf{1}/\mathbf{w} \;\; .\!\times \;\; \mathbf{A}$$

Preserving small clusters and unclustered vertices

Applying weights to clusters requires computing the size of each cluster, which is a sum over the rows of the cluster approximation. It can be performed as follows

$$\mathbf{w} = \mathbf{C} \;\; +.$$
$$\mathbf{A} = (\mathbf{1}/\mathbf{w} \wedge p) \;\; .\times \;\; \mathbf{A}$$

Settling ties

Settling ties for votes in this clustering algorithm requires selecting the lowest numbered cluster with the highest number of votes. In many linear algebra packages, this simply corresponds to a call to max, finding the location of the maximum values in each column. Typically, the location corresponds to the first maximum value in that column, or the smallest cluster number among those who tie for maximums.

Space complexity

Algorithm 6.2 uses the following variables

Name	Type	Number of Elements
$\mathbf{A}$	$\mathbb{R}_+^{N \times N}$	$O(M)$
$\mathbf{T}$	$\mathbb{R}_+^{N \times N}$	$O(M)$
$\mathbf{C_i}$	$\mathbb{B}^{N \times N}$	$O(N)$
$\mathbf{C_f}$	$\mathbb{B}^{N \times N}$	$O(N)$
$\mathbf{m}$	$\mathbb{R}_+^{N}$	$O(N)$

$\mathbf{T}$ requires only M entries because each edge casts a vote. There cannot be more votes than there are edges in the graph. Given this, the overall space complexity of the matrix version of peer pressure clustering is $O(M)$.

Time complexity

Each edge casts a single vote in this algorithm, leading to the voting to require only $O(M)$ operations. Tallying those votes also only requires $O(M)$ operations. Where the algorithm requires p passes to complete, the overall time is $O(p \times M)$. As with the vertex/edge representation, assuming p is a small constant, the overall time is just $O(M)$.

6.1.3 Other approaches

Here the Markov clustering algorithm is examined and compared to the peer pressure clustering algorithm presented above.

Comparison to Markov clustering

Another popular clustering algorithm using an adjacency matrix representation known as Markov clustering is presented here in Algorithm 6.3.

Algorithm 6.3. Markov clustering. Recursive algorithm for clustering vertices.*

$\textsc{Markov}(\mathbf{C_i} : \mathbb{R}_+^{N \times N}, e, i)$

$\mathbf{C_f} = \mathbf{C_i}^i$
$\mathbf{C_f} = \mathbf{C_f} \ .\!\wedge\ r$
$\mathbf{w} = \mathbf{C_f}\ +.$
$\mathbf{C_f} = \mathbf{w}\ .\!\times\ \mathbf{C_f}$
if $\mathbf{C_i} == \mathbf{C_f}$
  then return $\mathbf{C_f}$
  else return $\textsc{Markov}(\mathbf{C_f}, e, i)$

Here, $\mathbf{C_i}$ is initialized to be $\mathbf{A}$. The parameters e and i are the expansion and inflation parameters, respectively. By tuning these parameters, the algorithm can theoretically be made to discover clusters of any coarseness unlike the peer pressure clustering algorithm, which only discovers fine-grained clusters. However, the performance differences between these two algorithms are substantial.

The tallying phase in peer pressure clustering flattens the cluster approximation, reducing the size of subsequent computations. For Markov clustering, there is nothing like this. Therefore, C can become very dense during the computation. The total space required by Markov clustering is $O(N^2)$. The total time required by the algorithm, assuming it takes a small constant number of passes p to converge, is $O(p \times N^3)$. Both of these are larger than the requirements for peer pressure clustering. In addition, while Markov clustering is guaranteed to converge, it typically does so at a much slower rate than peer pressure clustering. Markov clustering can require up to $p = 20$ or $p = 30$ passes before converging to a solution.

6.2 Vertex betweenness centrality

Centrality metrics for vertices in general seek to generate information about the importance of vertices in a graph. Betweenness centrality rates this importance on the basis of the number of shortest paths a vertex falls on between all other pairs of vertices. It is commonly used in fault tolerance and key component analysis.

6.2.1 History

The original vertex betweenness centrality algorithm was straightforward. It ran an all-pairs shortest paths search over the graph, keeping track of the distances between all pairs along with the number of shortest paths between those pairs. After this, the centrality for each vertex was computed by looking at all other pairs of vertices

*S. van Dongen, *Graph Clustering by Flow Stimulation.* Ph.D. thesis, University of Utrecht, May 2000.

and adding the appropriate value to the centrality if the vertex in question was on a shortest path between the other two vertices. The computation time for this method was dominated by the second metric, which had to examine $O(N^2)$ pairs for each of the $O(N)$ points, leading to a runtime of $O(N^3)$. In addition, the path lengths between all pairs of vertices corresponded to a dense matrix and required $O(N^2)$ storage.

In order to remedy this problem, a new algorithm [Brandes 2001] was developed that utilized cascading updates to betweenness centrality and performed single-source shortest paths searches rather than an all-pairs shortest path search. This algorithm reduced the computation time to $O(NM)$ for unweighted graphs and $O(NM + N^2 \log(N))$ for weighted graphs. It also allowed the computation space to remain only $O(N + M)$, making it especially preferred when looking at sparsely connected graphs.

6.2.2 Brandes' algorithm

Here the improved algorithm for vertex betweenness centrality in unweighted graphs (see [Brandes 2001]) is examined. This algorithm centers on performing a single-source shortest paths search for every vertex in the graph. During this search, the number of shortest paths to all other vertices is recorded, along with the order in which the vertices are seen and the shortest path predecessors for every vertex. After the shortest paths search has been performed for an individual vertex, the betweenness centrality metrics for each vertex are updated in the reverse order from which they were discovered during the search.

Traditional algorithm

Algorithm 6.4 shows Brandes' algorithm for betweenness centrality using a vertex/edge set representation.

The loop on line 10 performs the single-source shortest paths search (through a breadth-first search) from an individual vertex. This requires $O(M)$ time. The loop on line 25 updates the centrality metrics. Through all the iterations, it performs at most one update per each predecessor on the shortest path for a vertex. There cannot be more of these updates than there are edges in the graph, so this operation requires at most $O(M)$ time. Over all iterations, the algorithm requires $O(NM)$ time.

Also, the only additional storage requirement that can have a size larger than $O(N)$ is P. As stated before, there cannot be more predecessors than there are edges in the graph, so P's size is limited to $O(M)$, and the algorithm requires at most $O(N + M)$ additional storage.

Sample calculation

Consider the graph shown in Figure 6.9. A simple shortest paths calculation, as well as a centrality update, for paths starting at vertex 1 is presented here.

Algorithm 6.4. Betweenness centrality.

```
BetweennessCentrality(G = (V, E))
C_B[v] ← 0, ∀v ∈ V
for ∀s ∈ V
      do
         S ← empty stack
         P[w] ← empty list, ∀w ∈ V
         σ[t] ← 0, ∀t ∈ V, σ[s] ← 1
         d[t] ← −1, ∀t ∈ V, d[s] ← 0
         Q ← empty queue
         enqueue(Q, s)
         while ¬empty(Q)
               do
                  v ← dequeue(Q)
                  push(S, v)
                  for ∀w ∈ neighbors(v)
                        do
                           if d[w] < 0
                              then
                                    enqueue(Q, w)
                                    d[w] ← d[v] + 1
                           if d[w] = d[v] + 1
                              then
                                    σ[w] ← σ[w] + σ[v]
                                    append(P[w], v)
         δ[v] ← 0, ∀v ∈ V
         while ¬empty(S)
               do
                  w ← pop(S)
                  for v ∈ P[w]
                        do δ[v] ← δ[v] + (σ[v]/σ[w]) × (1 + δ[w])
                  if w ≠ s
                     then C_B[w] ← C_B[w] + δ[w]
```

The first step in computing the betweenness centrality updates induced by paths originating at vertex 1 is to perform a breadth-first search from that vertex, keeping track of the number of shortest paths to each vertex as the search progresses. It is easy to see that the number of shortest paths to a vertex v at depth d is simply the sum of the number of shortest paths to all vertices with edges going to v at depth $d-1$.

This progression is shown in Figure 6.10. The two interesting steps are seen in Figure 6.10(c) and Figure 6.10(e). Joins of multiple predecessors with shortest paths occur there. In Figure 6.10(c), three vertices, all with one shortest path, combine to form a shortest path count of three for their neighboring vertex. In Figure 6.10(e), two vertices, all with three shortest paths, combine to form a shortest path count of six for their neighboring vertex.

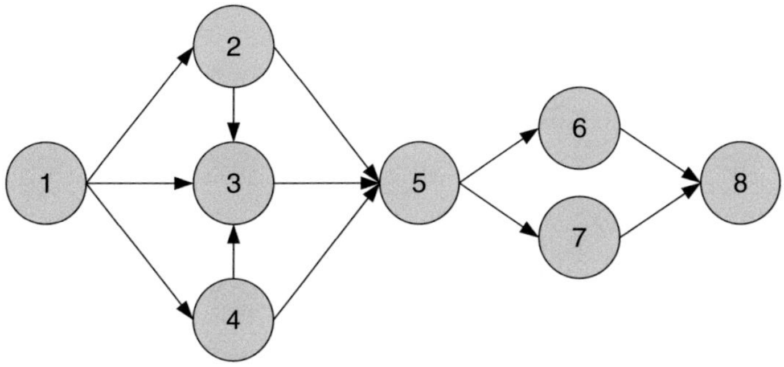

Figure 6.9. Sample graph.

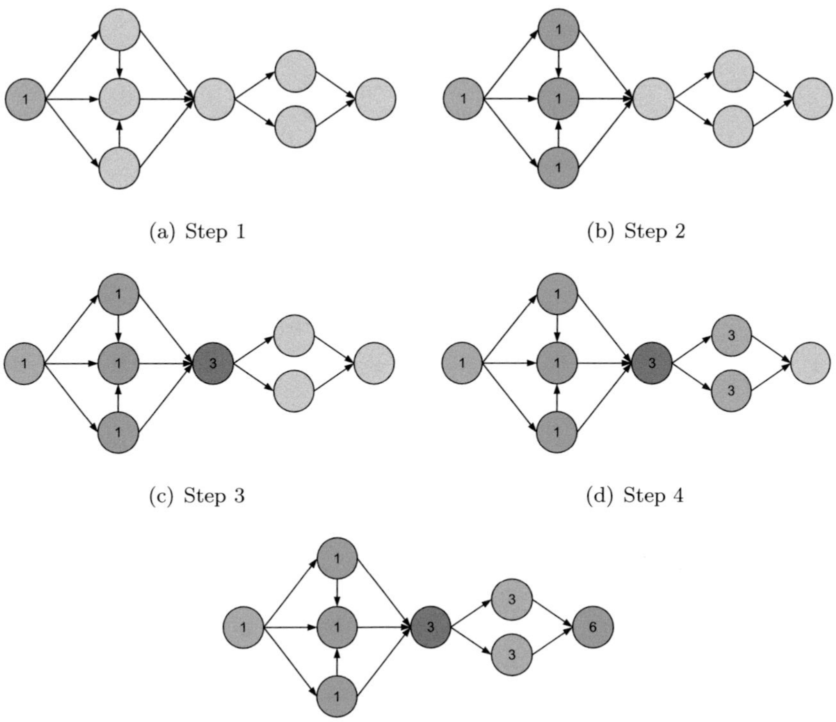

(a) Step 1

(b) Step 2

(c) Step 3

(d) Step 4

(e) Step 5

Figure 6.10. Shortest path steps.
Procedure for computing the shortest paths in the sample graph.

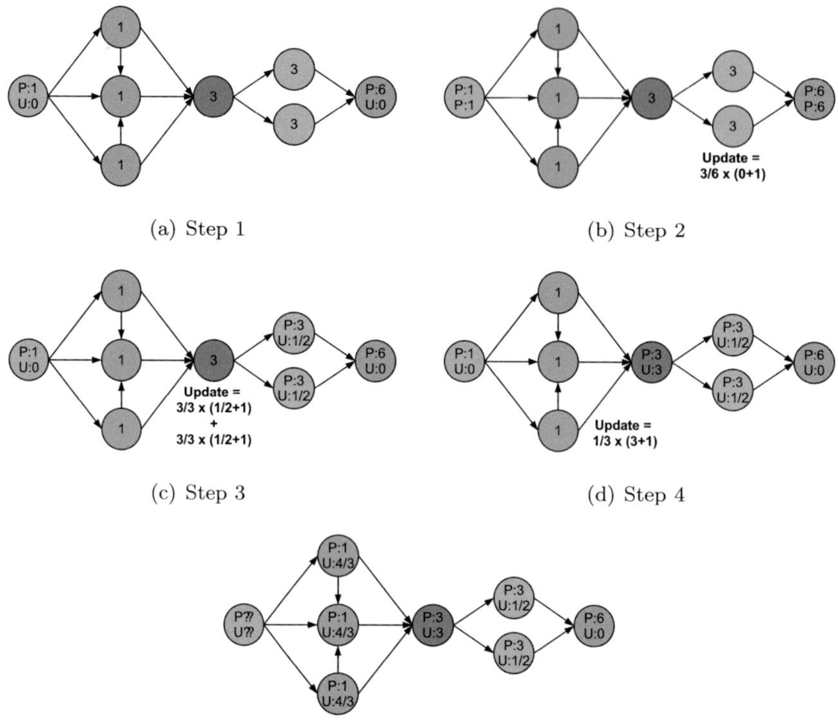

(a) Step 1 (b) Step 2

(c) Step 3 (d) Step 4

(e) Step 5

Figure 6.11. Betweenness centrality updates.
Steps for computing the betweenness centrality updates in the sample graph.

Once the shortest paths have been computed, the centrality updates can begin. The updates are computed in reverse depth order. The update for a vertex corresponds to the updates for all its shortest path successors, plus one for the successor itself, multiplied by the ratio of the shortest paths that the edge contributed to the successor, which is simply the number of shortest paths to the vertex itself divided by the number of shortest paths to its successor.

This cascading update is shown in Figure 6.11. The updates for the root and all end vertices are set to zero immediately. From this, the rest of the updates are computed. Figure 6.11(b) shows the first set of update computations. Because there are six paths to the final vertex, and only three paths to each of its predecessors, each predecessor receives only half of its centrality update plus one. Figure 6.11(c) joins the updates from its two successors, leading to a total centrality update of three for it. Figure 6.11(d) shows how its three predecessors split this score, leading to a centrality update of $\frac{4}{3}$ for each of them.

This completes the centrality updates induced by shortest paths originating from vertex 1. To verify this answer, the depths of each vertex in the graph from vertex 1 minus one can be summed up. This sum should be equivalent to the sum of the centrality updates. As can be seen, vertices 2, 3, and 4 each contribute $1-1=0$ to the depth sum. Vertex 5 contributes $2-1=1$ to the depth sum. Vertices 6 and 7 each contribute $3-1=2$ to the depth sum. Finally, vertex 8 contributes $4-1=3$ to the depth sum. This sum, $2+2\times 2+3=8$ is the same as the sum of the centrality updates, $\frac{4}{3}+\frac{4}{3}+\frac{4}{3}+3+\frac{1}{2}+\frac{1}{2}=8$.

Linear algebraic formulation

In order to take advantage of breadth-first search using matrix vector multiplication, it is necessary to be able to perform updates to the parent and path information a full depth in the search at a time. In addition, to retain the advantage of a linear algebra representation, the second loop, the betweenness centrality updates, should also be able to be performed a full depth at a time. Fortunately, both of these are possible.

In the first loop, the number of shortest paths is calculated in the natural way using a breadth-first search. In addition, rather than keeping track of shortest path parents, it suffices to remember the depth of each vertex in the breadth-first search. From this, the shortest path parents for a vertex v at breadth-first search depth d can be computed easily as $\forall u \in V : \mathit{depth}(u) = d-1$ and $A(u,v)=1$.

During the second loop, since the only dependence between the betweenness centrality updates are between parent and child, performing the updates a depth at a time causes no conflicts. The updates are done by selecting edges going to vertices at the current depth that are coming from vertices at the previous depth. These edges, which correspond to the betweenness centrality updates for their source vertices, are then weighted accordingly and summed up.

Algorithm 6.5 shows Brandes' algorithm for computing betweenness centrality by using linear algebra. The variables used in the algorithm are shown in the following table

Name	Type	Description
$\mathbf{S}$	$\mathbb{B}^{N\times N}$	the search, keeps track of the depth at which each vertex is seen
$\mathbf{p}$	$\mathbb{Z}^N$	the number of shortest paths to each vertex
$\mathbf{f}$	$\mathbb{Z}^N$	the fringe, the number of shortest paths to vertices at the current depth
$\mathbf{w}$	$\mathbb{R}^N$	the weights for the BC updates
$\mathbf{b}$	$\mathbb{R}^N$	the BC score for each vertex
$\mathbf{u}$	$\mathbb{R}^N$	the BC update for each vertex
r	$\mathbb{Z}$	the current root value, or starting vertex for which to compute BC updates
d	$\mathbb{Z}$	the current depth being examined

Algorithm 6.5. Betweenness centrality matrix formulation.

$b = \textsc{BetweennessCentrality}(G = A : \mathbb{B}^{N \times N})$

```
1   b = 0
2   for 1 ≤ r ≤ N
3       do
4           d = 0
5           S = 0
6           p = 0, p(r) = 1
7           f = A(r, :)
8           while f ≠ 0
9               do
10                  d = d + 1
11                  p = p + f
12                  S(d, :) = f
13                  f = fA × ¬p
14          while d ≥ 2
15              do
16                  w = S(d, :) × (1 + u) ÷ p
17                  w = Aw
18                  w = w × S(d − 1, :) × p
19                  u = u + w
20                  d = d − 1
21          b = b + u
```

Line 8 performs the breadth-first search. After updating the search based on the new fringe obtained for the previous level, it selects the outgoing edges of that fringe, weighting them by the number of shortest paths that go to their parents and summing them up. It then filters out those that go to vertices that have been seen previously, resulting in the new fringe for the next level. The loop terminates when no new values are seen on the fringe.

Line 14 performs the betweenness centrality updates. It processes the vertices in reverse depth order. First, it computes the weights corresponding to those derived from the children values (including filtering out edges that do not go to vertices at the current depth). It applies the computed weights to the adjacency matrix and sums up the row values. It then applies the weights corresponding to those derived from parent values (including filtering out vertices not at the previous depth) and adds in the new centrality updates. The loop updates centrality for all parents not including the root value.

Algorithm 6.5 has a small optimization. Lines 16–19 can be combined into a single parenthesized linear algebraic operation. This reduction allows the w variable to be optimized out during implementation.

Time complexity

The breadth-first search performed in the loop on line 8 requires the standard $O(N + M)$ operations because each vertex is seen at most once over the course of the search, which leads to each outgoing edge being selected at most once in the vector matrix multiplication. While the negation operation can take up to $O(N)$ time for each depth in the search, in reality, the operation is filtering out existing edges in the fringe and can be performed through subtraction or exclusive or to remedy this issue. The fringe over all passes of the search will contain each vertex at most once.

The centrality updates performed in the loop on line 14 start by computing the weight vector corresponding to the child weights. This is filtered by the vertices seen at the current depth, leading to at most $O(N)$ operations being performed through every loop. This also applies to the parent weights computation and the summation in the end. Here, the weights are filtered by the vertices seen at the previous depth. The matrix vector multiplication only requires $O(M)$ time through all loops because the vector has an entry for a vertex at most once through all loops. This leads to $O(N + M)$ for the second loop as well.

Since these two loops are performed $O(N)$ times, the total time required by the algorithm is $O(N^2 + NM)$.

Space complexity

The space required by each data structure in Algorithm 6.5 is listed below

Name	Space
$\mathbf{S}$	$O(N)$
$\mathbf{p}$	$O(N)$
$\mathbf{f}$	$O(N)$
$\mathbf{w}$	$O(N)$
$\mathbf{b}$	$O(N)$
$\mathbf{u}$	$O(N)$
r	$O(1)$
d	$O(1)$

While the $\mathbf{S}$ matrix has N^2 entries, its storage is only $O(N)$. This is because there are only N nonzero values in $\mathbf{S}$, one per column, where the row of the entry corresponds to the depth at which the vertex for that column was seen. Given this, the overall additional space used by the algorithm is only $O(N)$.

6.2.3 Batch algorithm

Algorithm 6.5 performs a single-source all-destinations breadth-first search. Rather than processing this for each root vertex in a loop, it can be modified to process all of the root vertices at once by using matrix matrix multiplications rather than matrix vector multiplications. This modification leads to similar performance as the original betweenness centrality algorithm, in that it requires space quadratic in N; $O(N^2)$, however, only requires $O(NM)$ time.

The Brandes algorithm and its modification are only two extremes of a parameterized algorithm for computing the betweenness centrality. This algorithm considers batches of vertices at a time rather than single vertices, as in the Brandes algorithm, or all vertices, as in the modified algorithm.

Linear algebraic formulation

Let P be a partitioning of the vertices V. For ease of exposition, it is assumed that P is selected in such a way that each partition $p \in P$ has the same size, $|p|$. Where the number of partitions is $|P|$, the following relation holds:

$$\forall p \in P, \frac{N}{|p|} = |P|$$

Algorithm 6.6 shows the vertex batching algorithm. Where $|P| = N$ and $|p| = 1$, it reduces to the Brandes algorithm. Where $|P| = 1$ and $|p| = N$, it reduces to the modified matrix algorithm described above. Algorithm 6.6 performs the same general operations as Algorithm 6.5. However, rather than performing operations on a single set of vertices (a vector), it operates on multiple sets of vertices (a matrix) simultaneously.

Time complexity

Algorithm 6.6 has the same time complexity as Algorithm 6.5, $O(N^2 + NM)$. It performs the exact same operations. However, it batches these operations in the form of a matrix matrix rather than matrix vector multiplication. While this does not improve the theoretical time, it can improve the actual runtime depending on the implementation.

Space complexity

The space required by each data structure in Algorithm 6.6 is listed below

Name	Space
$\mathbf{S}$	$O(\|p\| \times N)$
$\mathbf{p}$	$O(\|p\| \times N)$
$\mathbf{f}$	$O(\|p\| \times N)$
$\mathbf{w}$	$O(\|p\| \times N)$
$\mathbf{u}$	$O(\|p\| \times N)$
$\mathbf{b}$	$O(N)$
r	$O(1)$
d	$O(1)$

This leads to an overall space usage of $O(|p| \times N)$. While this is more than Algorithm 6.6 when the partition size is greater than one, if partitions are kept reasonably small, it may be a win. The increased performance gains from batch processing may be substantial enough to warrant the small increase in storage.

Algorithm 6.6. Betweenness centrality batch.

Var	Type	Description
$\mathbb{S}$	$\mathbb{B}^{N\times\lvert p\rvert\times N}$	the search, keeps track of the depth at which each vertex is seen for each starting vertex
$\mathbf{P}$	$\mathbb{Z}^{\lvert p\rvert\times N}$	the number of shortest paths to each vertex from each starting vertex
$\mathbf{F}$	$\mathbb{Z}^{\lvert p\rvert\times N}$	the fringe, the number of shortest paths to vertices at the current depth from each starting vertex
$\mathbf{W}$	$\mathbb{R}^{\lvert p\rvert\times N}$	the weights for the BC updates for each starting vertex
$\mathbf{B}$	$\mathbb{R}^{\lvert p\rvert\times N}$	the BC score for each vertex for each starting vertex
$\mathbf{U}$	$\mathbb{R}^{\lvert p\rvert\times N}$	the BC update for each vertex for each starting vertex
$\mathbf{r}$	$\mathbb{Z}^{\lvert p\rvert}$	the current root values, or starting vertices for which to compute BC updates
d	$\mathbb{Z}$	the current depth being examined

$b = \text{BetweennessCentrality}(G = A : \mathbb{B}^{N\times N}, P)$

```
b = 0
for r ∈ P
    do
        d = 0
        S = 0
        P = I(r, :)
        F = A(r, :)
        while F ≠ 0
            do
                d = d + 1
                P = P + F
                S(d, :, :) = F
                F = FA × ¬P
        while d ≥ 2
            do
                W = S(d, :, :) × (1 + U) ÷ P
                W = (AW')'
                W = W × S(d − 1, :, :) × P
                U = U + W
                d = d − 1
        b = b + ( +. U)
```

6.2.4 Algorithm for weighted graphs

While it would be possible to implement the version of this algorithm for weighted graphs, which can typically run in $O(N^2 \log(N))$ time, it would require $O(N^3)$ time using matrix and vector operations. This increase is caused by the fact that the rows of the matrix cannot be stored and operated on as a priority queue. The traditional algorithm relies on a priority queue storage scheme to reduce the breadth-first search cost from N^3 to $N^2 \log(N)$ in the runtime.

6.3 Edge betweenness centrality

Centrality metrics for edges in general seek to generate information about the importance of edges in a graph. Betweenness centrality rates this importance based on the number of shortest paths an edge falls on between all pairs of vertices. It is commonly used in fault tolerance and key component analysis. The algorithm is similar to the vertex betweenness centrality algorithm described in Section 6.2. Like that algorithm, an approach with cascading updates is generally preferred.

6.3.1 Brandes' algorithm

The Brandes' algorithm for edge betweenness centrality in unweighted graphs operates in much the same way as it did for the vertex formulation [Brandes 2001]. The first loop to determine the number of shortest paths to each vertex, in fact, is identical. The second loop to perform centrality updates must update edges rather than vertices, however.

Vertex/edge set formulation

Algorithm 6.7 shows Brandes' algorithm for betweenness centrality using a vertex/edge set representation. The loop on line 10 performs the single-source shortest paths search (through a breadth-first search) from an individual vertex. This requires $O(M)$ time. The loop on line 25 updates the centrality metrics. Through all the iterations, it performs at most one update per each edge in the graph, so this requires at most $O(M)$ time. Over all iterations, the algorithm requires $O(NM)$ time.

Also, there are no data structures requiring more than $O(M)$ storage, the storage required for the result. This leads to an overall storage requirement of $O(M)$.

Sample calculation

The same example as the one used in vertex betweenness centrality is considered here, using the graph shown in Figure 6.9. A simple shortest paths calculation, as well as a centrality update for paths starting at vertex 1, is presented here.

Algorithm 6.7. Edge betweenness centrality.

```
EdgeBetweennessCentrality(G = (V, E))
C_B[(v, w)] ← 0, ∀(v, w) ∈ E
for ∀s ∈ V
    do
        S ← empty stack
        P[w] ← empty list, ∀w ∈ V
        σ[t] ← 0, ∀t ∈ V, σ[s] ← 1
        d[t] ← −1, ∀t ∈ V, d[s] ← 0
        Q ← empty queue
        enqueue(Q, s)
        while ¬empty(Q)
            do
                v ← dequeue(Q)
                push(S, v)
                for ∀w ∈ neighbors(v)
                    do
                        if d[w] < 0
                          then
                              enqueue(Q, w)
                              d[w] ← d[v] + 1
                        if d[w] = d[v] + 1
                          then
                              σ[w] ← σ[w] + σ[v]
                              append(P[w], v)
        δ[v] ← 0, ∀v ∈ V
        while ¬empty(S)
            do
                w ← pop(S)
                for v ∈ P[w]
                    do
                        δ[v] ← δ[v] + σ[v] × (δ[w]/σ[w] + 1)
                        C_B[(v, w)] ← C_B[(v, w)] + σ[v] × (δ[w]/σ[w] + 1)
```

The first step in computing the betweenness centrality updates induced by paths originating at vertex 1 is to perform a breadth-first search from that vertex, keeping track of the number of shortest paths to each vertex as the search progresses. It is easy to see that the number of shortest paths to a vertex v at depth d is simply the sum of the number of shortest paths to all vertices with edges going to v at depth $d-1$. This progression is identical to the one for vertex betweenness centrality and is shown in Figure 6.10.

Once the shortest paths have been computed, the centrality updates can begin. The updates are computed in reverse depth order. As the updates proceed, each vertex also keeps track of the sum of all the updates from edges originating from it. The update for an edge corresponds to the number of shortest paths to the

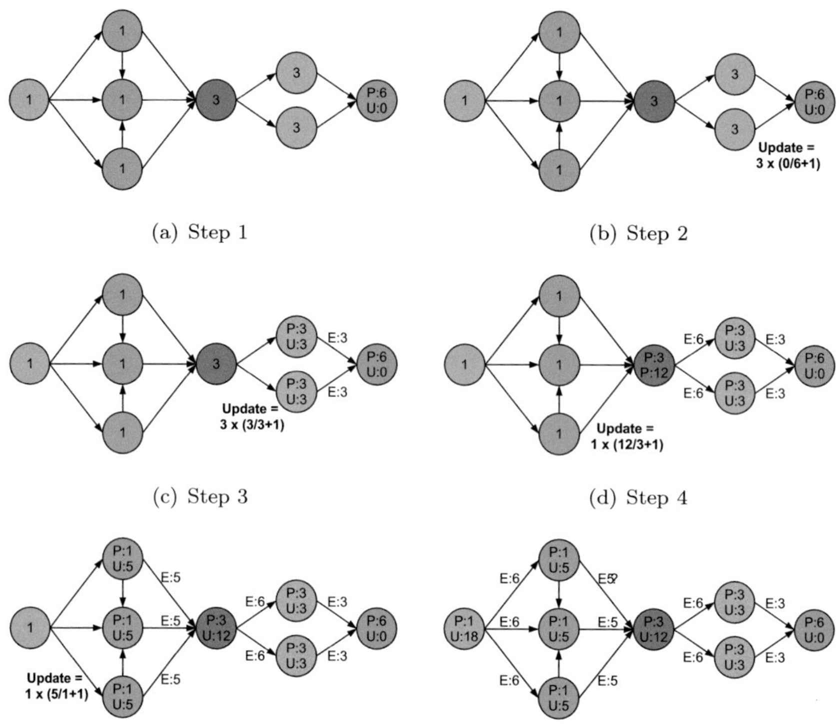

(a) Step 1 (b) Step 2

(c) Step 3 (d) Step 4

(e) Step 5 (f) Step 5

Figure 6.12. Edge centrality updates.
Steps for computing the edge centrality updates in sample graph.

originating vertex times the ratio of all the updates for all the edges originating from the edge's destination vertex over the number of shortest paths to destination vertex plus one.

This cascading update is shown in Figure 6.12. The update flow for all end vertices is set to zero immediately. From this, the rest of the updates and update flows are computed. Figure 6.12(b) shows the first set of update computations. Because there are six paths to the final vertex, and only three paths to each of its predecessors, each predecessor receives only half of those six paths. They each have a single predecessor so the flow through those predecessors is just the edge's centrality updates. In Figure 6.12(c), each of the two edges contribute three paths to their successor and the end vertex, making the update six for each edge. There is a single predecessor for those two edges, so that vertex's update flow is twelve. In Figure 6.12(d), the update of twelve is split between each of the three shortest path edges flowing in. A shortest path edge is then added in for the current vertex, giving each of the edges a score of five. They each have a single predecessor, so they each get a flow update of five as well. Finally, in Figure 6.12(e), each of the preceding edges receives one additional path for the subsequent vertex, giving each of those edges a score of six. The update flow through the root vertex is then eighteen.

This completes the centrality updates induced by shortest paths originating from vertex 1. Unlike vertex betweenness centrality, the sum of the shortest paths is not related to the sum of the edge centrality scores. This is due to the fact that the score is not scaled based on the percent of shortest paths the edge lies on, only the number of shortest paths.

Adjacency matrix formulation

Similar to vertex betweenness centrality, edge betweenness centrality is processed a full depth in the search at a time both during the breadth-first search and the centrality update loops. The first loop involves a breadth-first search just as in the vertex formulation of the problem, Algorithm 6.5. The second loop is modified to compute edge rather than vertex centrality updates.

Algorithm 6.8 shows Brandes' algorithm for computing edge betweenness centrality using linear algebra.

Line 10 performs the breadth-first search. After updating the search based on the new fringe obtained for the previous level, it selects the outgoing edges of that fringe, weighting them by the number of shortest paths that go to their parents, and summing them up. It then filters out those that go to vertices that have been seen previously. This becomes the new fringe for the next level. The loop terminates when no new vertices are on the fringe.

Line 16 performs the betweenness centrality updates. It processes the edges in reverse depth order. First, it computes the weights stemming from the children vertices (including filtering out edges that do not go to vertices at the current depth). It applies this to the columns of the adjacency matrix. It then applies the weights stemming from parent vertices (including filtering out edges that do not originate from the previous depth) to the rows of the matrix. Finally, it adds this update to the betweenness centrality scores and computes the vertex flow by summing over the rows of the current update. The loop updates centrality for all shortest path edges from the current root in the graph.

Algorithm 6.8 has a small optimization. Lines 18–21 can be combined into a single parenthesized linear algebraic operation. This reduction allows the w variable to be optimized out during implementation.

Time complexity

The breadth-first search performed in the loop on line 10 requires the standard $O(N + M)$ operations because each vertex is seen at most once over the course of the search, which leads to each outgoing edge being selected at most once in the vector matrix multiplication. While the negation operation can take up to $O(N)$ time for each depth in the search, in reality, the operation is filtering out existing edges in the fringe and can be performed through subtraction or exclusive or to remedy this issue. The fringe over all passes of the search will contain each vertex at most once.

Algorithm 6.8. Edge betweenness centrality matrix formulation.

Var	Type	Description
$\mathbf{S}$	$\mathbb{B}^{N\times N}$	the search, keeps track of the depth at which each vertex is seen
$\mathbf{p}$	$\mathbb{Z}^{N}$	the number of shortest paths to each vertex
$\mathbf{f}$	$\mathbb{Z}^{N}$	the fringe, the number of shortest paths to vertices at the current depth
$\mathbf{w}$	$\mathbb{R}^{N}$	the column/row weights for the BC updates
$\mathbf{B}$	$\mathbb{R}^{N\times N}$	the BC score for each edge
$\mathbf{U}$	$\mathbb{R}^{N\times N}$	the BC update for each edge
$\mathbf{v}$	$\mathbb{Z}^{N}$	the BC flow through the vertices
r	$\mathbb{Z}$	the current root value, or starting vertex for which to compute BC updates
d	$\mathbb{Z}$	the current depth being examined

$B = \text{EdgeBetweennessCentrality}(G = A : \mathbb{B}^{N\times N})$

```
B = 0
for 1 ≤ r ≤ N
    do
        d = 0
        S = 0
        p = 0, p(r) = 1
        U = 0
        v = 0
        f = A(r, :)
        while f ≠ 0
            do
                d = d + 1
                p = p + f
                S(d, :) = f
                f = fA × ¬p
        while d ≥ 2
            do
                w = S(d, :) ÷ p × v + S(d, :)
                U = A .× w
                w = S(d − 1, :) × p
                U = w .× U
                B = B + U
                v = U +.
                d = d − 1
```

The centrality updates performed in the loop on line 16 start by computing the weight vector corresponding to the child weights. This is filtered by the vertices seen at the current depth, leading to at most $O(N)$ operations being performed through every loop. This also applies to the parent weights computation and the summation in the end. Here, the weights are filtered by the vertices seen at the previous depth. The matrix vector scaling only requires $O(N+M)$ time through all loops because the vector has an entry for a vertex at most once through all loops. This leads to $O(N + M)$ for the second loop as well.

Since these two loops are performed $O(N)$ times, the total time required by the algorithm is $O(N^2 + NM)$.

Space complexity

The space required by each data structure in Algorithm 6.8 is listed below

Name	Space
$\mathbf{S}$	$O(N)$
$\mathbf{p}$	$O(N)$
$\mathbf{f}$	$O(N)$
$\mathbf{w}$	$O(N)$
$\mathbf{B}$	$O(M)$
$\mathbf{U}$	$O(M)$
$\mathbf{v}$	$O(N)$
r	$O(1)$
d	$O(1)$

While the $\mathbf{S}$ matrix has N^2 entries, its storage is only $O(N)$. This is because there are only N nonzero values in $\mathbf{S}$, one per column, where the row of the entry corresponds to the depth at which the vertex for that column was seen. In addition, both $\mathbf{B}$ and $\mathbf{U}$ have at most one entry (centrality score) per edge. Given this, the overall additional space used by the algorithm is only $O(M)$.

6.3.2 Block algorithm

Much like vertex betweenness centrality, edge betweenness centrality can also be performed using blocks of root vertices. Here, rather than matrix vector multiplication and matrix vector scaling, matrix matrix multiplication and matrix-tensor scaling are used. Since support for tensor operations is limited both in our linear algebra notation as well as in most mathematical software, an in-depth examination of this approach is not included. However, it performs similarly to the batched vertex version of the algorithm, requiring extra space $O(|p| \times M)$, where $|p|$ is the size of a block. It still requires the same $O(N^2 + NM)$ time, though the blocking may introduce constant speedups.

6.3.3 Algorithm for weighted graphs

While it would be possible to implement the version of this algorithm for weighted graphs, which can typically run in $O(N^2 \log(N))$ time, it would require $O(N^3)$ time using matrix and vector operations. This increase is caused by the fact that the rows of the matrix cannot be stored and operated on as a priority queue. The traditional algorithm relies on a priority queue storage scheme to reduce the breadth-first search cost from N^3 to $N^2 \log(N)$ in the run time.

References

[Brandes 2001] U. Brandes. A faster algorithm for betweenness centrality. *Journal of Mathematical Sociology*, 25:163–177, 2001.

[Luby 1986] M. Luby. A simple parallel algorithm for the maximal independent set problem. *SIAM Journal on Computing*, 15:1036–1053, 1986.

[Reinhardt et al. 2006] S. Reinhardt, J.R. Gilbert, and V. Shah. High performance graph algorithms from parallel sparse matrices. *Workshop on State of the Art in Scientific and Parallel Computing*, 2006.

Chapter 7

Multilinear Algebra for Analyzing Data with Multiple Linkages

Daniel M. Dunlavy, Tamara G. Kolda*†*, and W. Philip Kegelmeyer*†*

Abstract

Tensors are a useful tool for representing multi-link graphs, and tensor decompositions facilitate a type of link analysis that incorporates all link types simultaneously. An adjacency tensor is formed by stacking the adjacency matrix for each link type to form a three-way array. The CANDECOMP/PARAFAC (CP) tensor decomposition provides information about adjacency tensors of multi-link graphs analogous to that produced for adjacency matrices of single-link graphs using the singular value decomposition (SVD). The CP tensor decomposition generates feature vectors that incorporate all linkages simultaneously for each node in a multi-link graph. Feature vectors can be used to analyze bibliometric data in a variety of ways, for example, to analyze five years of publication data from journals published by the Society for Industrial and Applied Mathematics (SIAM). Experiments presented include analyzing a body of work, distinguishing between papers written by different authors with the same name, and predicting the journal in which a paper is published.

*Computer Science and Informatics Department, Sandia National Laboratories, Albuquerque, NM 87185–1318 (dmdunla@sandia.gov).

†Informatics and Decision Sciences Department, Sandia National Laboratories, Livermore, CA 94551-9159 (tgkolda@sandia.gov, wpk@sandia.gov).

Note: First appeared as Sandia National Laboratories Technical Report SAND2006-2079, Albuquerque, NM and Livermore, CA, April 2006.

7.1 Introduction

Multi-link graphs, i.e., graphs with multiple link types, are challenging to analyze, yet such data are ubiquitous. For example, Adamic and Adar [Adamic & Adar 2005] analyzed a social network where nodes are connected by organizational structure, i.e., each employee is connected to his or her boss, and also by direct email communication. Social networks clearly have many types of links—familial, communication (phone, email, etc.), organizational, geographical, etc.

Our overarching goals are to analyze data with multiple link types and to derive feature vectors for each individual node (or data object). As a motivating example, we use journal publication data—specifically considering several of the many ways that two papers may be linked. The analysis is applied to five years of journal publication data from eleven journals and a set of conference proceedings published by the Society for Industrial and Applied Mathematics (SIAM). The nodes represent published papers. Explicit, directed links exist whenever one paper cites another. Undirected similarity links are derived based on title, abstract, keyword, and authorship. Historically, bibliometric researchers have focused solely on citation analysis or text analysis, but not both simultaneously. Though this work focuses on the analysis of publication data, the techniques are applicable to a wide range of tasks, such as higher order web link graph analysis [Kolda & Bader 2006, Kolda et al. 2005].

Link analysis typically focuses on a single link type. For example, both PageRank [Brin & Page 1998] and HITS [Kleinberg 1999] consider the structure of the web and decompose the adjacency matrix of a graph representing the hyperlink structure. Instead of decomposing an adjacency matrix that represents a single matrix, our approach is to decompose an adjacency *tensor* that represents multiple link types.

A tensor is a multidimensional, or N-way, array. For multiple linkages, a three-way array can be used, where each two-dimensional *frontal slice* represents the adjacency matrix for a single link type. If there are N nodes and K link types, then the data can be represented as a three-way tensor of size $N \times N \times K$ where the (i, j, k) entry is nonzero if node i is connected to node j by link type k. In the example of Adamic and Adar [Adamic & Adar 2005] discussed above, there are two links types: organization connections versus email communication connections. For bibliometric data, the five different link types mentioned above correspond to (frontal) slices in the tensor; see Figure 7.1.

The CANDECOMP/PARAFAC (CP) tensor decomposition (see, for instance, [Carroll & Chang 1970, Harshman 1970]) is a higher order analog of the matrix singular value decomposition (SVD). The CP decomposition applied to the adjacency tensor of a multi-link graph leads to the following types of analysis.

- The CP decomposition reveals "communities" within the data and how they are connected. For example, a particular factor may be connected primarily by title similarity while another may depend mostly on citations.
- The CP decomposition also generates feature vectors for the nodes in the graph, which can be compared directly to get a similarity score that combines the multiple linkage types.

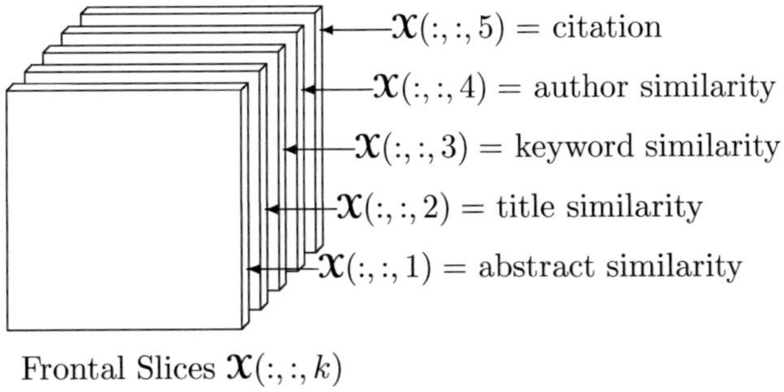

Figure 7.1. Tensor slices.
Slices of a third-order tensor representing a multi-link graph.

- The average of a set of feature vectors represents a *body of work*, e.g., by a given author, and can be used to find the most similar papers in the larger collection.
- The feature vectors can be used for disambiguation. In this case, the feature vectors associated with the body of work for two or more authors indicate whether they are the same authors or not. For example, is H. SIMON the same as H. S. SIMON?
- By inputting the feature vectors to a supervised learning method (decision trees and ensembles), the publication journal for each paper can be predicted.

This chapter is organized as follows. A description of the CP tensor decomposition and how to compute it is provided in Section 7.2. We discuss the properties of the data and how they are represented as a sparse tensor in Section 7.3. Numerical results are provided in Section 7.4. Related work is discussed in Section 7.5. Conclusions and ideas for future work are discussed in Section 7.6.

7.2 Tensors and the CANDECOMP/PARAFAC decomposition

This section provides a brief introduction to tensors and the CP tensor decomposition. For a survey of tensors and their decompositions, see [Kolda & Bader 2009].

7.2.1 Notation

Scalars are denoted by lowercase letters, e.g., c. Vectors are denoted by boldface lowercase letters, e.g., $\mathbf{v}$. The ith entry of $\mathbf{v}$ is denoted by $\mathbf{v}(i)$. Matrices are denoted by boldface capital letters, e.g., $\mathbf{A}$. The jth column of $\mathbf{A}$ is denoted by $\mathbf{A}(:,j)$ and element (i,j) by $\mathbf{A}(i,j)$. Tensors (i.e., N-way arrays) are denoted by boldface Euler script letters, e.g., $\mathcal{X}$. Element (i,j,k) of a third-order tensor $\mathcal{X}$ is denoted by $\mathcal{X}(i,j,k)$. The kth frontal slice of a three-way tensor is denoted by $\mathcal{X}(:,:,k)$; see Figure 7.1.

7.2.2 Vector and matrix preliminaries

The symbol $\otimes$ denotes the *Kronecker product of vectors*; for example

$$\mathbf{x} = \mathbf{a} \otimes \mathbf{b} \quad \Leftrightarrow \quad \mathbf{x}(\ell) = \mathbf{a}(i)\mathbf{b}(j)$$
$$\text{where } \ell = j + (i-1)(J) \text{ for all } 1 \leq i \leq I,\ 1 \leq j \leq J$$

This is a special case of the Kronecker product of matrices.

The symbol $.*$ denotes the *Hadamard matrix product.* This is the element-wise product of two matrices of the same size.

The symbol $\odot$ denotes the *Khatri–Rao product* (or column wise Kronecker product) of two matrices [Smilde et al. 2004]. For example, let $\mathbf{A} \in \mathbb{R}^{I \times K}$ and $\mathbf{B} \in \mathbb{R}^{J \times K}$. Then

$$\mathbf{A} \odot \mathbf{B} = \begin{bmatrix} \mathbf{A}(:,1) \otimes \mathbf{B}(:,1) & \mathbf{A}(:,2) \otimes \mathbf{B}(:,2) & \cdots & \mathbf{A}(:,K) \otimes \mathbf{B}(:,K) \end{bmatrix}$$

is a matrix of size $(IJ) \times K$.

7.2.3 Tensor preliminaries

The norm of a tensor is given by the square root of the sum of the squares of all its elements; i.e., for a tensor $\mathcal{X}$ of size $I \times J \times K$

$$\|\mathcal{X}\|^2 \equiv \sum_{i=1}^{I} \sum_{j=1}^{J} \sum_{k=1}^{K} \mathcal{X}(i,j,k)^2$$

This is the higher order analog of the Frobenius matrix norm.

The symbol $\circ$ denotes the *outer product of vectors.* For example, let $\mathbf{a} \in \mathbb{R}^I$, $\mathbf{b} \in \mathbb{R}^J$, $\mathbf{c} \in \mathbb{R}^K$. Then

$$\mathcal{X} = \mathbf{a} \circ \mathbf{b} \circ \mathbf{c} \quad \Leftrightarrow \quad \mathcal{X}(i,j,k) = \mathbf{a}(i)\mathbf{b}(j)\mathbf{c}(k)$$
$$\text{for all} \quad 1 \leq i \leq I,\ 1 \leq j \leq J,\ 1 \leq k \leq K$$

A *rank-one tensor* is a tensor that can be written as the outer product of vectors. For $\boldsymbol{\lambda} \in \mathbb{R}^R$, $\mathbf{A} \in \mathbb{R}^{I \times R}$, $\mathbf{B} \in \mathbb{R}^{J \times R}$, and $\mathbf{C} \in \mathbb{R}^{K \times R}$, the *Kruskal operator* [Kolda 2006] denotes a sum of rank-one tensors

$$[\![\boldsymbol{\lambda}; \mathbf{A}, \mathbf{B}, \mathbf{C}]\!] \equiv \sum_{r=1}^{R} \boldsymbol{\lambda}(r)\ \mathbf{A}(:,r) \circ \mathbf{B}(:,r) \circ \mathbf{C}(:,r) \in \mathbb{R}^{I \times J \times K}$$

If $\boldsymbol{\lambda}$ is a vector of ones, then $[\![\mathbf{A}, \mathbf{B}, \mathbf{C}]\!]$ is used as shorthand.

Matricization, also known as *unfolding* or *flattening*, is the process of reordering the elements of an N-way array into a matrix; in particular, the mode-n matricization of a tensor $\mathcal{X}$ is denoted by $\mathbf{X}_{(n)}$; see, e.g., [Kolda 2006]. For a three-way tensor $\mathcal{X} \in \mathbb{R}^{I \times J \times K}$, the mode-$n$ unfoldings are defined as follows

$$\mathbf{X}_{(1)}(i,p) = \mathcal{X}(i,j,k) \qquad \text{where } p = j + (k-1)(J) \tag{7.1}$$
$$\mathbf{X}_{(2)}(j,p) = \mathcal{X}(i,j,k) \qquad \text{where } p = i + (k-1)(I) \tag{7.2}$$
$$\mathbf{X}_{(3)}(k,p) = \mathcal{X}(i,j,k) \qquad \text{where } p = i + (j-1)(I) \tag{7.3}$$

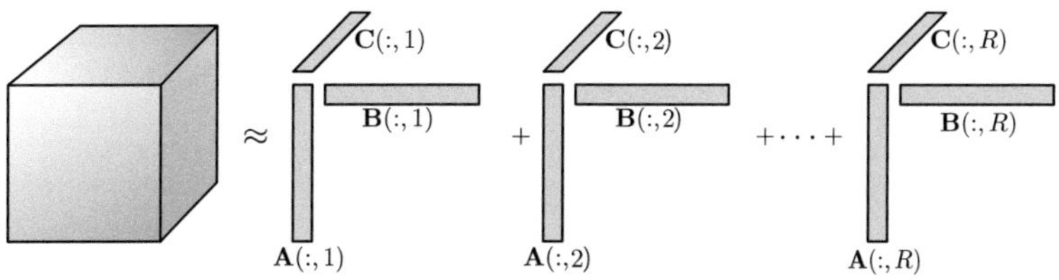

Figure 7.2. CP decomposition.
Approximates a tensor by a sum of rank-one factors.

7.2.4 The CP tensor decomposition

The CP decomposition, first proposed by Hitchcock [Hitchcock 1927] and later rediscovered simultaneously by Carroll and Chang [Carroll and Chang 1970] and Harshman [Harshman 1970], is a higher order analog of the matrix SVD. It should not be confused with the Tucker decomposition [Tucker 1966], a different higher order analog of the SVD.

CP decomposes a tensor into a sum of rank-one tensors. Let $\mathcal{X}$ be a tensor of size $I \times J \times K$. A CP decomposition with R factors approximates the tensor $\mathcal{X}$ as

$$\mathcal{X} \approx \sum_{r=1}^{R} \mathbf{A}(:,r) \circ \mathbf{B}(:,r) \circ \mathbf{C}(:,r) \equiv [\![\mathbf{A},\mathbf{B},\mathbf{C}]\!]$$

where $\mathbf{A} \in \mathbb{R}^{I\times R}$, $\mathbf{B} \in \mathbb{R}^{J\times R}$, and $\mathbf{C} \in \mathbb{R}^{K\times R}$. The matrices $\mathbf{A}$, $\mathbf{B}$, and $\mathbf{C}$ are called the *component matrices*. Figure 7.2 illustrates the decomposition.

It is useful to normalize the columns of the matrices $\mathbf{A}$, $\mathbf{B}$, and $\mathbf{C}$ to length one and rewrite the CP decomposition as

$$\mathcal{X} \approx \sum_{r=1}^{R} \boldsymbol{\lambda}(r)\ \mathbf{A}(:,r) \circ \mathbf{B}(:,r) \circ \mathbf{C}(:,r) \equiv [\![\boldsymbol{\lambda}\,;\mathbf{A},\mathbf{B},\mathbf{C}]\!]$$

where $\boldsymbol{\lambda} \in \mathbb{R}^R$. In contrast to the solution provided by the SVD, the factor matrices $\mathbf{A}$, $\mathbf{B}$, and $\mathbf{C}$ do not have orthonormal columns [Kolda 2001, Kolda & Bader 2009].

Each rank-one factor, $\boldsymbol{\lambda}(r)\ \mathbf{A}(:,r) \circ \mathbf{B}(:,r) \circ \mathbf{C}(:,r)$, represents a "community" within the data; see Section 7.4.1. The number of factors in the approximation, R, should loosely reflect the number of *communities* in the data. Often some experimentation is required to determine the most useful value of R.

7.2.5 CP-ALS algorithm

A common approach to fitting a CP decomposition is the ALS (alternating least squares) algorithm [Carroll & Chang 1970, Harshman 1970]; see also, [Tomasi 2006, Faber et al. 2003, Tomasi & Bro 2006]. At each inner iteration, the CP-ALS algorithm solves for one-component matrix while holding the others fixed. For example, it solves for the matrix $\mathbf{C}$ when $\mathbf{A}$ and $\mathbf{B}$ are fixed, i.e.,

$$\min_{\mathbf{C}} \| \mathcal{X} - [\![\mathbf{A}, \mathbf{B}, \mathbf{C}]\!] \| \tag{7.4}$$

In this case, $\boldsymbol{\lambda}$ is omitted because it will just be absorbed into the lengths of the columns of $\mathbf{C}$ when the computation is complete. Equation (7.4) can be rewritten as a matrix problem (see, e.g., [Smilde et al. 2004])

$$\min_{\mathbf{C}} \left\| \mathbf{X}_{(3)} - \mathbf{C} \left(\mathbf{B} \odot \mathbf{A} \right)^{\mathsf{T}} \right\| \tag{7.5}$$

Here $\mathbf{X}_{(3)}$ is the mode-3 matricization or unfolding from equation 7.3.

Solving this problem makes use of the pseudoinverse of a Khatri–Rao product, given by

$$(\mathbf{B} \odot \mathbf{A})^{\dagger} = \left((\mathbf{B}\mathbf{B}) \,.*\, (\mathbf{A}^{\mathsf{T}}\mathbf{A}) \right)^{\dagger} (\mathbf{B} \odot \mathbf{A})^{\mathsf{T}}$$

Note that only the pseudoinverse of an $R \times R$ matrix needs to be calculated rather than that of an $IJ \times R$ matrix [Smilde et al. 2004].

The optimal $\mathbf{C}$ is the least squares solution to equation (7.5)

$$\mathbf{C} = \mathbf{X}_{(3)} \left[(\mathbf{B} \odot \mathbf{A})^{\mathsf{T}} \right]^{\dagger} = \mathbf{X}_{(3)} (\mathbf{B} \odot \mathbf{A}) \left((\mathbf{B}^{\mathsf{T}}\mathbf{B}) \,.*\, (\mathbf{A}^{\mathsf{T}}\mathbf{A}) \right)^{\dagger}$$

which can be computed efficiently thanks to the properties of the Khatri–Rao product. The other component matrices can be computed in an analogous fashion using mode-1 and mode-2 matricizations of $\mathcal{X}$ in solving for $\mathbf{A}$ and $\mathbf{B}$, respectively.

It is generally efficient to initialize the ALS algorithm with the R leading eigenvectors of $\mathbf{X}_{(n)}\mathbf{X}_{(n)}^{\mathsf{T}}$ for the nth component matrix as long as the nth dimension of $\mathcal{X}$ is at least as big as R; see, e.g., [Kolda & Bader 2009]. Otherwise, random initialization can be used. Only two of the three initial matrices need to be computed since the other is solved for in the first step. The CP-ALS algorithm is presented in Algorithm 7.1.

Algorithm 7.1. CP-ALS.
CP decomposition via an alternating least squares. $\mathcal{X}$ is a tensor of size $I \times J \times K$, $R > 0$ is the desired number of factors in the decomposition, $M > 0$ is the maximum number of iterations to perform, and $\epsilon > 0$ is the stopping tolerance.

CP-ALS $(\mathfrak{X}, R, M, \epsilon)$

$m = 0$
$\mathbf{A} = R$ principal eigenvectors of $\mathbf{X}_{(1)}\mathbf{X}_{(1)}^{\mathsf{T}}$
$\mathbf{B} = R$ principal eigenvectors of $\mathbf{X}_{(2)}\mathbf{X}_{(2)}^{\mathsf{T}}$
repeat
    $m = m + 1$
    $\mathbf{C} = \mathbf{X}_{(3)}(\mathbf{B} \odot \mathbf{A})\left((\mathbf{B}^{\mathsf{T}}\mathbf{B}) \,.\!* \,(\mathbf{A}^{\mathsf{T}}\mathbf{A})\right)^{\dagger}$
    Normalize columns of $\mathbf{C}$ to length 1
    $\mathbf{B} = \mathbf{X}_{(2)}(\mathbf{C} \odot \mathbf{A})\left((\mathbf{C}^{\mathsf{T}}\mathbf{C}) \,.\!* \,(\mathbf{A}^{\mathsf{T}}\mathbf{A})\right)^{\dagger}$
    Normalize columns of $\mathbf{B}$ to length 1
    $\mathbf{A} = \mathbf{X}_{(1)}(\mathbf{C} \odot \mathbf{B})\left((\mathbf{C}^{\mathsf{T}}\mathbf{C}) \,.\!* \,(\mathbf{B}^{\mathsf{T}}\mathbf{B})\right)^{\dagger}$
    Store column norms of $\mathbf{A}$ in $\boldsymbol{\lambda}$ and normalize columns of $\mathbf{A}$ to length 1
until $m > M$ or $\| \mathfrak{X} - [\![\boldsymbol{\lambda} \,; \mathbf{A}, \mathbf{B}, \mathbf{C}]\!] \| < \epsilon$
return $\boldsymbol{\lambda} \in \mathbb{R}^{R}$; $\mathbf{A} \in \mathbb{R}^{I \times R}$; $\mathbf{B} \in \mathbb{R}^{J \times R}$; $\mathbf{C} \in \mathbb{R}^{K \times R}$
    such that $\mathfrak{X} \approx [\![\boldsymbol{\lambda} \,; \mathbf{A}, \mathbf{B}, \mathbf{C}]\!]$

In the discussion that follows, $\boldsymbol{\Lambda}$ denotes the $R \times R$ diagonal matrix whose diagonal is $\boldsymbol{\lambda}$.

All computations were performed using the Tensor Toolbox for MATLAB (see [Bader & Kolda 2006, Bader & Kolda 2007]), which was appropriate because of its ability to handle large-scale, sparse tensors.

7.3 Data

The data consist of publication metadata from eleven SIAM journals as well as SIAM proceedings for the period 1999–2004. There are 5022 articles; the number of articles per publication is shown in Table 7.1. The names of the journals used throughout this paper are their ISI abbreviations* and "SIAM PROC" is used to indicate the proceedings.

7.3.1 Data as a tensor

The data are represented as an $N \times N \times K$ tensor where $N = 5022$ is the number of documents and $K = 5$ is the number of link types. The five link types are described below; see also Figure 7.1.

(1) The first slice $(\mathfrak{X}(:,:,1))$ represents abstract similarity; i.e., $\mathfrak{X}(i,j,1)$ is the cosine similarity of the abstracts for documents i and j. The Text to Matrix Generator (TMG) v2.0 [Zeimpekis & Gallopoulos 2006] was used to generate a term-document matrix, $\mathbf{T}$. All words appearing on the default TMG stopword list as well as words starting with a number were removed. The matrix was weighted using term frequency and inverse document frequency local and global weightings

*http://www.isiknowledge.com/.

Table 7.1. SIAM publications.
Names of the SIAM publications along with the number of articles of each used as data for the experiments.

Journal Name	Articles
SIAM J APPL DYN SYST	32
SIAM J APPL MATH	548
SIAM J COMPUT	540
SIAM J CONTROL OPTIM	577
SIAM J DISCRETE MATH	260
SIAM J MATH ANAL	420
SIAM J MATRIX ANAL APPL	423
SIAM J NUMER ANAL	611
SIAM J OPTIM	344
SIAM J SCI COMPUT	656
SIAM PROC	469
SIAM REV	142

(tf.idf); this means that

$$\mathbf{T}(i,j) = f_{ij} \log_2(N/N_i)$$

where f_{ij} is the frequency of term i in document j and N_i is the number of documents that term i appears in. Each column of $\mathbf{T}$ is normalized to length one (for cosine scores). Finally

$$\mathcal{X}(:,:,1) = \mathbf{T}^{\mathsf{T}}\mathbf{T}$$

Because they are cosine scores, all are in the range $[0, 1]$. In order to sparsify the slice, only scores greater than 0.2 (chosen heuristically to reduce the total number of nonzeros in all three text similarity slices to approximately 250,000) are retained.

(2) The second slice ($\mathcal{X}(:,:,2)$) represents title similarity; i.e., $\mathcal{X}(i,j,2)$ is the cosine similarity of the titles for documents i and j. It is computed in the same manner as the abstract similarity slice.

(3) The third slice ($\mathcal{X}(:,:,3)$) represents author-supplied keyword similarity; i.e., $\mathcal{X}(i,j,3)$ is the cosine similarity of the keywords for documents i and j. It is computed in the same manner as the abstract similarity slice.

(4) The fourth slice ($\mathcal{X}(:,:,4)$) represents author similarity; i.e., $\mathcal{X}(i,j,4)$ is the similarity of the authors for documents i and j. It is computed as follows. Let $\mathbf{W}$ be the author-document matrix such that

$$\mathbf{W}(i,j) = \begin{cases} 1/\sqrt{M_j} & \text{if author } i \text{ wrote document } j, \\ 0 & \text{otherwise} \end{cases}$$

where M_j is the number of authors for document j. Then

$$\mathcal{X}(:,:,4) = \mathbf{W}^{\mathsf{T}}\mathbf{W}$$

(5) The fifth slice ($\mathcal{X}(:,:,5)$) represents citation information; i.e.,

$$\mathcal{X}(i,j,5) = \begin{cases} 2 & \text{if document } i \text{ cites document } j, \\ 0 & \text{otherwise} \end{cases}$$

For this document collection, a weight of 2 was chosen heuristically so that the overall *slice weight* (i.e., the sum of all the entries in $\mathcal{X}(:,:,k)$, see Table 7.3) would not be too small relative to the other slices. The interpretation is that there are relatively few connections in this slice, but each citation connection indicates a strong connection. In future work, we would like to consider less ad hoc ways of determining the value for citation links.

Each slice is an adjacency matrix of a particular graph. The first four slices are symmetric and correspond to undirected graphs; the fifth slice is asymmetric and corresponds to a directed graph. These graphs can be combined into a multi-link graph and a corresponding tensor representation since they are all on the same set of nodes.

These choices for link types are examples of what can be done—many other choices are possible. For instance, asymmetric similarity weights are an option; e.g., if document i is a subset of document j, the measure might say that document i is very similar to document j, but document j is not so similar to document i. Other symmetric measures include co-citation or co-publication in the same journal.

7.3.2 Quantitative measurements on the data

Table 7.2 shows overall statistics on the data set. Note that some of the documents in this data set have empty titles, abstracts, or keywords; the averages shown in the table are not adjusted for the lack of data for those documents. Recall that Table 7.1 shows the number of articles per journal. In Table 7.2, the citations are counted only when both articles are in the data set and reflect the number of citations *from* each article. The maximum citations *to* a single article is 15.

Table 7.3 shows the number of nonzero entries and the sums of the entries for each slice. The text similarity slices ($k = 1, 2, 3$) have large numbers of nonzeros but low average values, the author similarity slice has few nonzeros but a higher average value, and the citation slice has the fewest nonzeros but all values are equal to 2.

7.4 Numerical results

The results use a CP decomposition of the data tensor $\mathcal{X} \in \mathbb{R}^{N \times N \times K}$

$$\mathcal{X} \approx [\![\boldsymbol{\lambda} \, ; \mathbf{A}, \mathbf{B}, \mathbf{C}]\!]$$

where $\boldsymbol{\lambda} \in \mathbb{R}^{R}$, $\mathbf{A}, \mathbf{B} \in \mathbb{R}^{N \times R}$, and $\mathbf{C} \in \mathbb{R}^{K \times R}$. Using $R = 30$ factors worked well for the experiments and is the default value unless otherwise noted.

Table 7.2. SIAM journal characteristics.
Characteristics of the SIAM journal and proceedings data (5022 documents in total).

	Total in Collection	Per Document	
		Average	Maximum
Unique terms	16617	148.32	831
abstracts	15752	128.06	802
titles	5164	10.16	33
keywords	5248	10.10	40
Authors	6891	2.19	13
Citations (within collection)	2659	0.53	12

Table 7.3. SIAM journal tensors.
Characteristics of the tensor representation of the SIAM journal and proceedings data.

Slice (k)	Description	Nonzeros	$\sum_i \sum_j \mathcal{X}(i,j,k)$
1	Abstract Similarity	28476	7695.28
2	Title Similarity	120236	33285.79
3	Keyword Similarity	115412	16201.85
4	Author Similarity	16460	8027.46
5	Citation	2659	5318.00

7.4.1 Community identification

The rank-one CP factors (see Figure 7.2) reveal communities within the data. The largest entries for the vectors in each factor

$$(\mathbf{A}(:,r), \mathbf{B}(:,r), \mathbf{C}(:,r))$$

correspond to interlinked entries in the data. For the rth factor, high-scoring nodes in $\mathbf{A}(:,r)$ are connected to high-scoring nodes in $\mathbf{B}(:,r)$ with the high-scoring link types in $\mathbf{C}(:,r)$. Recall that the fifth link type, representing citations, is asymmetric; when that link type scores high in $\mathbf{C}(:,r)$, then the highest-scoring nodes in $\mathbf{A}(:,r)$ can be thought of as papers that cite the highest-scoring nodes in $\mathbf{B}(:,r)$.

For example, consider the first factor ($r = 1$). The link scores from $\mathbf{C}(:,1)$ are shown in Table 7.4. Title and keyword similarities are strongest. In fact, the top three link types are based on text similarity and so are symmetric. Therefore, it is no surprise that the highest-scoring nodes in $\mathbf{A}(:,1)$ and $\mathbf{B}(:,1)$, also shown in Table 7.4, are nearly identical. This community is related primarily by text similarity and is about the topic "conservation laws."

On the other hand, the tenth factor ($r = 10$) has citation as the dominant link type; see Table 7.5. Citation links are asymmetric, so the highest-scoring nodes in $\mathbf{A}(:,10)$ and $\mathbf{B}(:,10)$ are not the same. This is a community that is linked primarily

Table 7.4. First community in CP decomposition. Community corresponding to the first factor ($r = 1$) of the CP tensor decomposition with $R = 30$ factors.

Link scores in $\mathbf{C}(:, 1)$	
Score	**Link Type**
0.95	Title Similarity
0.28	Keyword Similarity
0.07	Abstract Similarity
0.06	Citation
0.06	Author Similarity

Paper node scores in $\mathbf{A}(:, 1)$ (top 10)

Score	Title
0.18	On the boundary control of systems of conservation laws
0.17	On stability of conservation laws
0.16	Two a posteriori error estimates for 1D scalar conservation laws
0.16	A free boundary problem for scalar conservation laws
0.15	Convergence of SPH method for scalar nonlinear conservation laws
0.15	Adaptive discontinuous Galerkin finite element methods for nonlinear hyperbolic ...
0.15	High-order central schemes for hyperbolic systems of conservation laws
0.15	Adaptive mesh methods for one- and two-dimensional hyperbolic conservation laws

Paper node scores in $\mathbf{B}(:, 1)$ (top 10)

Score	Title
0.18	On the boundary control of systems of conservation laws
0.18	On stability of conservation laws
0.16	Two a posteriori error estimates for one-dimensional scalar conservation laws
0.16	A free boundary problem for scalar conservation laws
0.16	Adaptive discontinuous Galerkin finite element methods for nonlinear hyperbolic ...
0.16	Convergence of SPH method for scalar nonlinear conservation laws
0.15	Adaptive mesh methods for one- and two-dimensional hyperbolic conservation laws
0.14	High-order central schemes for hyperbolic systems of conservation laws

because the high-scoring papers in $\mathbf{A}(:, 10)$ cite the high-scoring papers in $\mathbf{B}(:, 10)$. The topic of this community is "preconditioning," though the third paper in $\mathbf{B}(:, 10)$ is not about preconditioning directly but rather a graph technique that can be used by preconditioners—that is why it is on the "cited" side.

The choice to have symmetric or asymmetric connections affects the interpretation of the CP model. In this case, the tensor has four symmetric slices and one asymmetric slice. If all of the slices were symmetric, then this would be a special case of the CP decomposition called the INDSCAL decomposition [Carroll & Chang 1970] where $\mathbf{A} = \mathbf{B}$. In related work, Selee et al. [Selee et al. 2007] have investigated this situation.

7.4.2 Latent document similarity

The CP component matrices $\mathbf{A}$ and $\mathbf{B}$ provide latent representations (i.e., feature vectors) for each document node. These feature vectors can, in turn, be used to

Table 7.5. Tenth community in CP decomposition. Community corresponding to the tenth factor ($r = 10$) of the CP tensor decomposition with $R = 30$ factors.

Link scores in $\mathbf{C}(:,10)$	
Score	**Link Type**
0.96	Citation
0.19	AuthorSim
0.16	TitleSim
0.10	KeywordSim
0.06	AbstractSim

Paper node scores in $\mathbf{A}(:,10)$ (top 10)

Score	Title
0.36	Multiresolution approximate inverse preconditioners
0.20	Preconditioning highly indefinite and nonsymmetric matrices
0.16	A factored approximate inverse preconditioner with pivoting
0.16	On two variants of an algebraic wavelet preconditioner
0.14	A robust and efficient ILU that incorporates the growth of the inverse triangular factors
0.11	An algebraic multilevel multigraph algorithm
0.11	On algorithms for permuting large entries to the diagonal of a sparse matrix
0.11	Preconditioning sparse nonsymmetric linear systems with the Sherman–Morrison formula

Paper node scores in $\mathbf{B}(:,10)$ (top 10)

Score	Title
0.27	Ordering anisotropy and factored sparse approximate inverses
0.25	Robust approximate inverse preconditioning for the conjugate gradient method
0.23	A fast and high-quality multilevel scheme for partitioning irregular graphs
0.20	Orderings for factorized sparse approximate inverse preconditioners
0.19	The design and use of algorithms for permuting large entries to the diagonal of ...
0.17	BILUM: Block versions of multielimination and multilevel ILU preconditioner ...
0.16	Orderings for incomplete factorization preconditioning of nonsymmetric problems
0.15	Preconditioning highly indefinite and nonsymmetric matrices

compute document similarity scores inclusive of text, authorship, and citations. Since there are two applicable component matrices, $\mathbf{A}$, $\mathbf{B}$, or some combination can be used. For example

$$\mathbf{S} = \frac{1}{2}\mathbf{A}\mathbf{A}^{\mathsf{T}} + \frac{1}{2}\mathbf{B}\mathbf{B}^{\mathsf{T}} \tag{7.6}$$

Here $\mathbf{S}$ is an $N \times N$ similarity matrix where the similarity for documents i and j is given by $\mathbf{S}(i,j)$.

It may also be desirable to incorporate $\mathbf{\Lambda}$, e.g.,

$$\mathbf{S} = \frac{1}{2}\mathbf{A}\mathbf{\Lambda}\mathbf{A}^{\mathsf{T}} + \frac{1}{2}\mathbf{B}\mathbf{\Lambda}\mathbf{B}^{\mathsf{T}}$$

This issue is reminiscent of the choice facing users of latent semantic indexing (LSI) [Dumais et al. 1988] which uses the SVD of a term-document matrix, producing term and document matrices. In LSI, there is a choice of how to use the diagonal scaling for the queries and comparisons [Berry et al. 1995].

Table 7.6. Articles similar to *Link Analysis* .... Comparison of most similar articles to *Link Analysis: Hubs and Authorities on the World Wide Web* using different numbers of factors in the CP decomposition.

R = 10

Score	Title
0.000079	Ordering anisotropy and factored sparse approximate inverses
0.000079	Robust approximate inverse preconditioning for the conjugate gradient method
0.000077	An interior point algorithm for large-scale nonlinear programming
0.000073	Primal-dual interior-point methods for semidefinite programming in finite precision
0.000068	Some new search directions for primal-dual interior point methods in semidefinite ...
0.000068	A fast and high-quality multilevel scheme for partitioning irregular graphs
0.000067	Reoptimization with the primal-dual interior point method
0.000065	Superlinear convergence of primal-dual interior point algorithms for nonlinear ...
0.000064	A robust primal-dual interior-point algorithm for nonlinear programs
0.000063	Orderings for factorized sparse approximate inverse preconditioners

R = 30

Score	Title
0.000563	Skip graphs
0.000356	Random lifts of graphs
0.000354	A fast and high-quality multilevel scheme for partitioning irregular graphs
0.000322	The minimum all-ones problem for trees
0.000306	Rankings of directed graphs
0.000295	Squarish k-d trees
0.000284	Finding the k-shortest paths
0.000276	On floor-plan of plane graphs
0.000275	1-Hyperbolic graphs
0.000269	Median graphs and triangle-free graphs

As an example of how these similarity measures can be used, consider the paper *Link analysis: Hubs and authorities on the World Wide Web* by Ding et al., which presents an analysis of an algorithm for web graph link analysis. Table 7.6 shows the most similar articles to this paper based on equation (7.6) for *two different* CP decompositions with $R = 10$ and $R = 30$ factors. In the $R = 10$ case, the results are not very good because the "most similar" papers include several papers on interior point methods that are not related. The results for $R = 30$ are all focused on graphs and are therefore related. Observe that there is also a big difference in the magnitude of the similarity scores in the two different cases. This example illustrates that, just as with LSI, choosing the number of factors of the approximation (R) is heuristic and affects the similarity scores.

In the next section, feature vectors from the CP factors are combined to represent a body of work.

7.4.3 Analyzing a body of work via centroids

Finding documents similar to a body of work may be useful in a literature search or in finding other authors working in a given area. This subsection and the next discuss two sets of experiments using centroids, corresponding to a term or an author, respectively, to analyze a body of work.

Consider finding collections of articles containing a particular term (or phrase). All articles containing the term in either the title, abstract, or keywords are identified and then the centroids $\mathbf{g}_A$ and $\mathbf{g}_B$ are computed using the columns of the matrices $\mathbf{A}$ and $\mathbf{B}$, respectively, for the identified articles. The similarity scores for all documents to the body of work are then computed as

$$\mathbf{s} = \frac{1}{2}\mathbf{A}\mathbf{g}_A + \frac{1}{2}\mathbf{B}\mathbf{g}_B \tag{7.7}$$

Consequently, $\mathbf{s}(i)$ is the similarity of the ith document to the centroid.

Table 7.7 shows the results of a search on the term "GMRES," which is an iterative method for solving linear systems. The table lists the top-scoring documents using a combination of matrices $\mathbf{A}$ and $\mathbf{B}$. In order not to overemphasize the papers that cite many of the papers about GMRES (i.e., using only the components from $\mathbf{A}$) or those which are most cited (i.e., using only the components from $\mathbf{B}$), combining the two sets of scores takes into account the content of the papers (i.e., abstracts, titles, and keywords) as an average of these two extremes. Thus, the average scores result in a more balanced look at papers about GMRES.

Similarly, centroids were used to analyze a body of work associated with a particular author. All of the articles written by an author were used to generate a centroid and similarity score vector as above. Table 7.8 shows the most similar papers to the articles written by V. KUMAR, a researcher who focuses on several research areas, including graph analysis. In these ten articles in the table, only three papers (including the two authored by V. KUMAR) are explicitly linked to V. KUMAR by coauthorship or citations. Furthermore, several papers that are closely related to those written by V. KUMAR focused on graph analysis, while some are not so obviously linked. Table 7.8 lists the authors as well to illustrate that such results could be used as a starting point for finding authors related to V. KUMAR that are not necessarily linked by coauthorship or citation. In this case, the author W. P. TANG appears to be linked to V. KUMAR.

Analysis of centroids derived from tensor decompositions can be useful in understanding small collections of documents. For example, such analysis could be useful for matching referees to papers. In this case, program committee chairs could create a centroid for each member on a program committee, and work assignments could be expedited by automatically matching articles to the appropriate experts.

As a segue to the next section, note that finding a set of documents associated with a particular author is not always straightforward. In fact, in the example above, there is also an author named V. S. A. KUMAR, and it is not clear from article titles alone that this author is not the same one as V. KUMAR. The next section discusses the use of the feature vectors produced by tensor decompositions for solving this problem of author disambiguation.

7.4.4 Author disambiguation

A challenging problem in working with publication data is determining whether two authors are in fact a single author using multiple aliases. Such problems are often caused by incomplete or incorrect data or varying naming conventions for authors

Table 7.7. Articles similar to *GMRES*.
Articles similar to the centroid of articles containing the term *GMRES* using the component matrices of a CP tensor decomposition to compute similarity scores.

Highest-scoring nodes using $\frac{1}{2}\mathbf{A}\mathbf{g}_A + \frac{1}{2}\mathbf{B}\mathbf{g}_B$

Score	Title
0.0134	FQMR: A flexible quasi-minimal residual method with inexact ...
0.0130	Flexible inner-outer Krylov subspace methods
0.0114	Adaptively preconditioned GMRES algorithms
0.0112	Truncation strategies for optimal Krylov subspace methods
0.0093	Theory of inexact Krylov subspace methods and applications to ...
0.0086	Inexact preconditioned conjugate gradient method with inner-outer iteration
0.0085	Flexible conjugate gradients
0.0078	GMRES with deflated restarting
0.0065	A case for a biorthogonal Jacobi–Davidson method: Restarting and ...
0.0062	On the convergence of restarted Krylov subspace methods

Highest-scoring nodes using $\mathbf{A}\mathbf{g}_A$

Score		
$\mathbf{A}\mathbf{g}_A$	$\mathbf{B}\mathbf{g}_B$	Title
0.0240	0.0019	Flexible inner-outer Krylov subspace methods
0.0185	0.0082	FQMR: A flexible quasi-minimal residual method with inexact ...
0.0169	0.0017	Theory of inexact Krylov subspace methods and applications to ...
0.0132	0.0024	GMRES with deflated restarting
0.0127	0.0003	A case for a biorthogonal Jacobi–Davidson method: Restarting and ...
0.0107	0.0010	A class of spectral two-level preconditioners
0.0076	0.0011	An augmented conjugate gradient method for solving consecutive ...

Highest-scoring nodes using $\mathbf{B}\mathbf{g}_B$

Score		
$\mathbf{B}\mathbf{g}_B$	$\mathbf{A}\mathbf{g}_A$	Title
0.0217	0.0011	Adaptively preconditioned GMRES algorithms
0.0158	0.0014	Inexact preconditioned conjugate gradient method with inner-outer iteration
0.0149	0.0074	Truncation strategies for optimal Krylov subspace methods
0.0113	0.0056	Flexible conjugate gradients
0.0082	0.0185	FQMR: A flexible quasi-minimal residual method with inexact ...
0.0080	0.0007	Linear algebra methods in a mixed approximation of magnetostatic problems
0.0063	0.0060	On the convergence of restarted Krylov subspace methods

used by different publications (e.g., J. R. SMITH versus J. SMITH). In the SIAM articles, there are many instances where two or more authors share the same last name and at least the same first initial, e.g., V. TORCZON and V. J. TORCZON. In these cases, the goal is to determine which names refer to the same person.

The procedure for solving this author disambiguation problem works as follows. For each author name of interest, we extract all the columns from the matrix $\mathbf{B}$ corresponding to the articles written by that author name. Recall that the matrix $\mathbf{B}$ comes from the $R = 30$ CP decomposition. Because of the directional citation links in $\mathcal{X}(:,:,5)$, using the matrix $\mathbf{B}$ slightly favors author names that are co-cited (i.e., their papers are cited together in papers), whereas using $\mathbf{A}$ would have slightly

Table 7.8. Similarity to V. Kumar.
Papers similar to those by V. KUMAR using a rank $R = 30$ CP tensor decomposition.

Score	Authors	Title
0.0645	Karypis G, Kumar V	A fast and high-quality multilevel scheme for partitioning ...
0.0192	Bank RE, Smith RK	The incomplete factorization multigraph algorithm
0.0149	Tang WP, Wan WL	Sparse approximate inverse smoother for multigrid
0.0115	Chan TF, Smith B, Wan WL	An energy-minimizing interpolation for robust methods ...
0.0114	Henson VE, Vassilevski PS	Element-free AMGe: General algorithms
0.0108	Hendrickson B, Rothberg E	Improving the run-time and quality of nested dissection ...
0.0092	Karypis G, Kumar V	Parallel multilevel k-way partitioning scheme for irregular ...
0.0091	Tang WP	Toward an effective sparse approximate inverse preconditioner
0.0085	Saad Y, Zhang J	BILUM: Block versions of multielimination and multilevel ...
0.0080	Bridson B, Tang WP	A structural diagnosis of some IC orderings

favored author names that co-cite (i.e., their papers cite the same papers). The centroid of those columns from **B** is used to represent the author name. Two author names are compared by computing the cosine similarity of their two centroids, resulting in a value between -1 (least similar) and 1 (most similar). In the example above, the similarity score of the centroids for V. TORCZON and V. J. TORCZON is 0.98, and thus there is a high confidence that these names both refer to the same person (verified by manual inspection of the articles).

As an example use of author disambiguation, the following experiment was performed. (i) The top 40 author names of papers in the data set were selected, i.e., those with the most papers. (ii) For each author name in the top 40, all papers in the full document collection with any name sharing the same first initial and last name were retrieved. (iii) Next the centroids for each author name as in Section 7.4.3 were computed. (iv) The combined similarity scores using the formula in (7.7) were calculated for all papers of author names sharing the same first initial and last name. (v) Finally, the resulting scores were compared to manually performed checks to see which matches are correct.

According to the above criteria, there are a total of 15 pairs of names to disambiguate. Table 7.9 shows all the pairs and whether or not each is a correct match, which was determined manually.

Figure 7.3 presents plots of the similarity scores for these 15 pairs of author names using CP decompositions with $R = 15, 20, 25, 30$. The scores denoted by $+$ in the figure are those name pairs that refer to the same person, whereas those pairs denoted by $\circ$ refer to different people. Ideally, there will be a distinct cutoff between correct and incorrect matches. The figure shows that, in general, most correct matches have higher scores than the incorrect ones. However, there are several instances where there is not a clear separation between pairs in the two

Table 7.9. Author disambiguation.
Author name pairs to be disambiguated.

Pair	Name 1	Name 2	Same Person?
1	T. CHAN	T. F. CHAN	Yes
2	T. CHAN	T. M. CHAN	No
3	T. F. CHAN	T. M. CHAN	No
4	T. MANTEUFFEL	T. A. MANTEUFFEL	Yes
5	S. MCCORMICK	S. F. MCCORMICK	Yes
6	G. GOLUB	G. H. GOLUB	Yes
7	X. L. ZHOU	X. Y. ZHOU	No
8	R. EWING	R. E. EWING	Yes
9	S. KIM	S. C. KIM	No
10	S. KIM	S. D. KIM	Yes
11	S. KIM	S. J. KIM	No
12	S. C. KIM	S. D. KIM	No
13	S. C. KIM	S. J. KIM	No
14	S. D. KIM	S. J. KIM	No
15	J. SHEN	J. H. SHEN	Yes

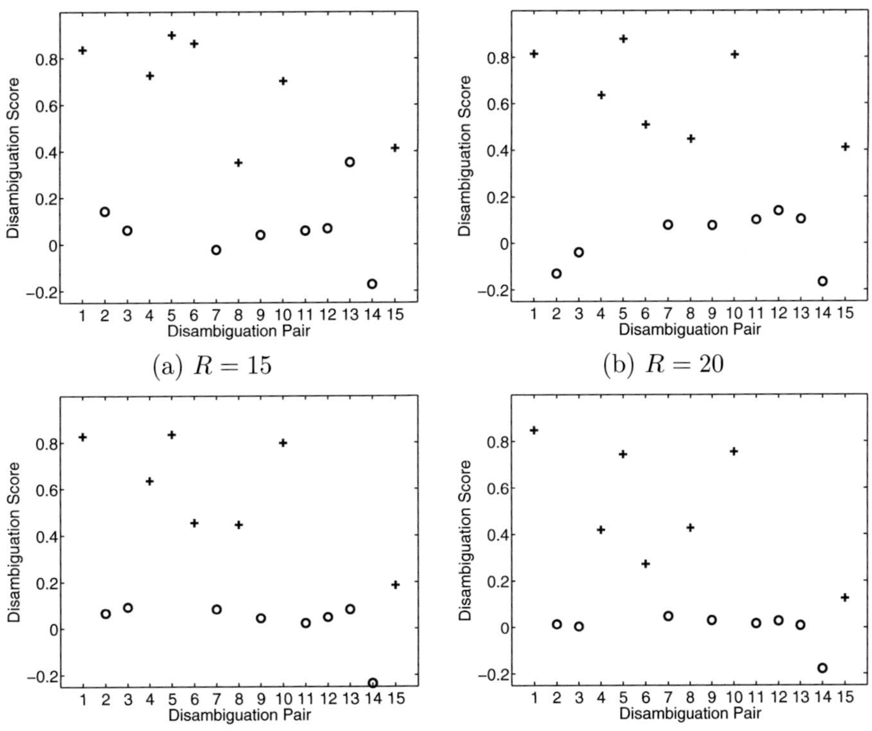

(a) $R = 15$ (b) $R = 20$ (c) $R = 25$ (d) $R = 30$

Figure 7.3. Disambiguation scores.
Author disambiguation scores for various CP tensor decompositions ($+$ = correct; $\circ$ = incorrect).

Table 7.10. Disambiguation before and after.
Authors with most papers before and after disambiguation.

Before Disambuguation		After Disambuguation	
Papers	Author	Papers	Author
17	Q. Du	17	Q. Du
15	K. Kunisch	16	T. F. Chan
15	U. Zwick	16	T. A. Manteuffel
14	T. F. Chan	16	S. F. McCormick
13	A. Klar	15	K. Kunisch
13	T. A. Manteuffel	15	U. Zwick
13	S. F. McCormick	13	A. Klar
13	R. Motwani	13	R. Motwani
12	G. H. Golub	13	G. H. Golub
12	M. Y. Kao	12	M. Y. Kao
12	S. Muthukrishnan	12	S. Muthukrishnan
12	D. Peleg	12	D. Peleg
11	H. Ammari	12	S. D. Kim
11	N. J. Higham	11	H. Ammari
11	K. Ito	11	N. J. Higham
11	H. Kaplan	11	K. Ito
11	L. Q. Qi	11	H. Kaplan
11	A. Srinivasan	11	L. Q. Qi
11	X. Y. Zhou	11	A. Srinivasan
10	N. Alon	11	X. Y. Zhou

sets—e.g., pairs 8, 13, and 15 in Figure 7.3(a). The CP decomposition with $R = 20$ clearly separates the correct and incorrect matches. Future work in this area will focus on determining if there is an optimal value for R for the task of predicting cutoff values for separating correct and incorrect matches.

Table 7.10 shows how correctly disambiguating authors can make a difference in publication counts. The left column shows the top 20 authors before disambiguation, and the right column shows the result afterward. There are several author names—T. F. Chan, T. A. Manteuffel, S. F. McCormick, G. H. Golub, and S. D. Kim—that move up (some significantly) in the list when the ambiguous names are resolved correctly.

One complication that has not yet been addressed is that two different people may be associated with the same author name. This is particularly likely in the case that the name has only a single initial and a common last name. Consider the name Z. Wu—there are two papers in the collection with this author name and five others with author names with the same first initial and a different second initial. Table 7.11 lists the papers by these authors along with the full first name of the author, which was determined by manual inspection.

Two approaches for solving this name resolution problem are considered: treating Z. Wu as a single author and taking the centroid of the two papers and treating each paper as separate. In Table 7.12(a), Z. Wu, as the author of two papers, appears most similar to author 3. Separating the articles of Z. Wu and recomputing the scores provides much stronger evidence that authors 1b and 3 are the same author, and that author 1a is most likely not an alias for one of the other authors; see Table 7.12(b).

Table 7.11. Data used in disambiguating the author Z. Wu.

ID	Author	Title(s)
1a	Wu Z (Zhen)	Fully coupled forward-backward stochastic differential equations and ...
1b	Wu Z (Zili)	Sufficient conditions for error bounds
2	Wu ZJ (Zhijun)	A fast Newton algorithm for entropy maximization in phase determination
3	Wu ZL (Zili)	First order and second order conditions for error bounds
3	Wu ZL (Zili)	Weak sharp solutions of variational inequalities in Hilbert spaces
4	Wu ZN (Zi-Niu)	Steady and unsteady shock waves on overlapping grids
4	Wu ZN (Zi-Niu)	Efficient parallel algorithms for parabolic problems

Table 7.12. Disambiguation of author Z. Wu.

(a) Combination of all ambiguous authors.

	1	2	3	4
1	1.00	0.18	0.79	0.03
2		1.00	0.06	0.06
3			1.00	0.01
4				1.00

(b) Separation of all ambiguous authors.

	1a	1b	2	3	4
1a	1.00	0.01	0.21	0.03	0.07
1b		1.00	0.09	**0.90**	0.00
2			1.00	0.06	0.06
3				1.00	0.01
4					1.00

Manual inspection of all the articles by this group of authors indicates that authors 1b and 3 are in fact the same person, Zili Wu, and that author 1a is not an alias of any other author in this group. The verified full name of each author is listed in parentheses in Table 7.11.

The experiments and results presented in this section suggest several ways that tensor decompositions can be used for resolving ambiguity in author names. In particular, the use of centroids for characterizing a body of work associated with an author shows promise for solving this problem. In the next set of experiments, though, it can be observed that the utility of centroids may be limited to small, cohesive collections, as they fail to produce useful results for the problem of predicting which journal an article may appear in.

7.4.5 Journal prediction via ensembles of tree classifiers

Another analysis approach, supervised machine learning with the feature vectors obtained in Section 7.4.2, may be used to predict the journal that a given paper is published in.

Table 7.13. Summary journal prediction results.

ID	Journal Name	Size	Correct	Mislabeled as	
1	SIAM J APPL DYN SYST	1%	0%	2	(44%)
2	SIAM J APPL MATH	11%	58%	6	(10%)
3	SIAM J COMPUT	11%	56%	11	(20%)
4	SIAM J CONTROL OPTIM	11%	60%	2	(10%)
5	SIAM J DISCRETE MATH	5%	15%	3	(47%)
6	SIAM J MATH ANAL	8%	26%	2	(29%)
7	SIAM J MATRIX ANAL APPL	8%	56%	10	(19%)
8	SIAM J NUMER ANAL	12%	50%	10	(16%)
9	SIAM J OPTIM	7%	66%	4	(16%)
10	SIAM J SCI COMPUT	13%	36%	8	(21%)
11	SIAM PROC	9%	32%	3	(38%)
12	SIAM REV	3%	5%	2	(34%)

The approach from Section 7.4.3 of considering the centroid of a body of work does not yield useful results in the case of journals because the centroids are not sufficiently distinct. Therefore, classifiers trained on subsets of the data are used to predict the journals in which the articles not included in those training sets are published. The feature vectors were based on the matrix $\mathbf{A}$ from a CP decomposition with $R = 30$ components. Thus, each document is represented by a length-30 feature vector, and the journal in which it is published is used as the label value, i.e., the classification. The 5022 labeled feature vectors were split into ten disjoint partitions, stratified so that the relative proportion of each journal's papers remained constant across the partitions. Ten-fold cross validation was used, meaning that each one of the ten partitions (10% of the data) was used once as testing data and the remaining nine partitions (90% of the data) were used to train the classifier. This computation was done using OpenDT [Banfield et al. 2004] to create bagged ensembles [Dietterich 2000] of C4.5 decision trees. The ensemble size was 100; larger ensembles did not improve performance.

Table 7.13 provides an overview of the results giving, for each journal, its identification number, its size relative to the entire collection, the percentage of its articles that were correctly classified, and the journal that it was most often mislabeled as and how often that occurred. For instance, articles in journal 2, make up 11% of the total collection, are correctly identified 58% of the time, and are confused with journal 6 most often (10% of the time). The overall "confusion matrix" is given in Table 7.14; this matrix is obtained by combining the confusion matrices generated for each of the ten folds.

Figure 7.4 shows a graphical representation of the confusion matrix. Each journal is represented as a node, and the size of the node corresponds to the percentage of its articles that were correctly labeled (0–66%). There is a directed edge from journal i to journal j if journal i's articles were mislabeled as article j. A Barnes–Hut forced directed method (using the weighted edges) was used to determine the positions of the nodes [Beyer 2007]. Only those edges corresponding to mislabeling percentages of 5% or higher are actually shown in the image (though all were used for the layout); the thicker the edge, the greater the proportion of mislabeled articles.

Table 7.14. Predictions of publication.
Confusion matrix of predictions of publication of articles in the different SIAM publications. A classifier based on bagging and using decision trees as weak learners was used in this experiment. The bold entries are correct predictions.

	Predicted Journal											
	1	*2*	*3*	*4*	*5*	*6*	*7*	*8*	*9*	*10*	*11*	*12*
1	**0**	14	4	1	1	4	0	3	1	1	2	1
2	1	**318**	19	46	3	54	13	31	7	41	12	3
3	0	29	**303**	24	29	5	15	8	7	10	109	1
4	0	57	21	**346**	2	34	20	12	51	22	11	1
5	0	12	122	9	**40**	4	15	2	1	2	53	0
6	0	120	19	56	1	**108**	15	58	3	34	5	1
7	0	23	11	22	5	8	**235**	18	18	81	2	0
8	0	56	13	47	0	37	37	**304**	13	98	5	1
9	0	10	19	55	1	4	10	5	**228**	1	10	1
10	0	77	7	32	0	36	98	135	23	**237**	7	4
11	0	37	176	21	34	12	9	8	7	13	**149**	3
12	1	48	13	12	2	13	16	6	6	10	8	**7**

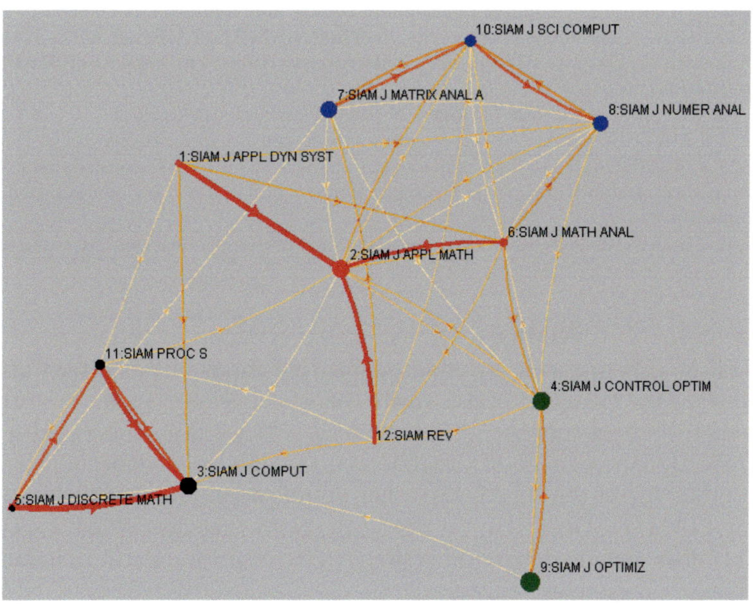

Figure 7.4. Journals linked by mislabeling.

The automatic layout generated by the Barnes–Hut algorithm visually yields four clusters, and the nodes in Figure 7.4 are color-coded according to their cluster labels. These journals along with their descriptions are presented in Table 7.15, and they are clearly clustered by overlap in topics. Observe that, for example,

Table 7.15. Journal clusters.
Journals grouped by how they are generally confused, with descriptions.

ID	Topic
	Red-Colored Nodes: Dynamical Systems
2	SIAM J APPL MATH: scientific problems using methods that are of mathematical interest such as asymptotic methods, bifurcation theory, dynamical systems theory, and probabilistic and statistical methods
6	SIAM J MATH ANAL: partial differential equations, the calculus of variations, functional analysis, approximation theory, harmonic or wavelet analysis, or dynamical systems; applications to natural phenomena
1	SIAM J APPL DYN SYST: mathematical analysis and modeling of dynamical systems and its application to the physical, engineering, life, and social sciences
12	SIAM REV: articles of broad interest
	Green-Colored Nodes: Optimization
4	SIAM J CONTROL OPTIM: mathematics and applications of control theory and on those parts of optimization theory concerned with the dynamical systems
9	SIAM J OPTIM: theory and practice of optimization
	Purple-Colored Nodes: Discrete Math and Computer Science
3	SIAM J COMPUT: mathematical and formal aspects of computer science and nonnumerical computing
5	SIAM J DISCRETE MATH: combinatorics and graph theory, discrete optimization and operations research, theoretical computer science, and coding and communication theory
11	SIAM PROC: Conference proceedings including SIAM Data Mining, ACM-SIAM Symposium on Discrete Algorithms, Conference on Numerical Aspects of Wave Propagation, etc.
	Cyan-Colored Nodes: Numerical Analysis
7	SIAM J MATRIX ANAL APPL: matrix analysis and its applications
8	SIAM J NUMER ANAL: development and analysis of numerical methods including convergence of algorithms, their accuracy, their stability, and their computational complexity
10	SIAM J SCI COMPUT: numerical methods and techniques for scientific computation

the scope of SIAM J COMPUT (3) includes everything in the scope of SIAM J DISCRETE MATH (5), so it is not surprising that many of the latter's articles are misidentified as the former. In cases where there is little overlap in the stated scope, there seems to be less confusion. For instance, articles from the SIAM J OPTIM (9) are correctly labeled 66% of the time and the only other journal it is confused with more than 5% of the time is the other optimization journal represented in the collection: SIAM J CONTROL OPTIM (4). Note that the SIAM J CONTROL OPTIM (4) does include dynamical systems in its description and is, in fact, linked to the "dynamical systems" cluster.

7.5 Related work

7.5.1 Analysis of publication data

Researchers look at publication data to understand the impact of individual authors and who is collaborating with whom, to understand the type of information being published and by which venues, and to extract "hot topics" and understand trends [Boyack 2004].

As an example of the interest in this problem, the 2003 KDD Cup challenge brought together 57 research teams from around the world to focus on the analysis of publication data for citation prediction (i.e., implicit link detection in a citation graph), citation graph creation, and usage estimation (downloads from a server of preprint articles) [Gehrke et al. 2003]. The data were from the high-energy physics community (a portion of the arXiv preprint server collection*). For this challenge, McGovern et al. looked at a number of questions related to the analysis of publication data [McGovern et al. 2003]. Of particular relevance to this paper, they found that clustering papers based only on text similarity did not yield useful clusters. Instead, they applied spectral-based clustering to a citation graph where the edges were weighted by the cosine similarity of the paper abstracts—combining citation and text information into one graph. Additionally, for predicting in which journal an article will be published, they used relational probability trees (see Section 7.5.3).

In other work, Barábasi et al. [Barábasi et al. 2002] considered the social network of scientific collaborations based on publication data, particularly the properties of the entire network and its evolution over time. In their case, the data were from publications in mathematics and neuroscience. The nodes correspond to authors and the links to coauthorship.

Hill and Provost [Hill & Provost 2003] used only citation information to predict authorship with an accuracy of 45%. They created a profile on each author based on his/her citation history (weighting older citations less). This profile can then be used to predict the authorship of a paper where only the citation information is known but not the authors. They did not use any text-based matching but observe that using such methods may improve accuracy.

Jo, Lagoze, and Giles [Jo et al. 2007] used citation graphs to determine topics in a large-scale document collection. For each term, the documents (nodes in the citation graph) were down-selected to those containing a particular term. The interconnectivity of those nodes within the "term" subgraph was used to determine whether or not it comprises a topic. The intuition of their approach was that, if a term represents a topic, the documents containing that term will be highly interconnected; otherwise, the links should be random. They applied their method to citation data from the arXiv (papers in physics) and Citeseer† papers in computer science) preprint server collections.

7.5.2 Higher order analysis in data mining

Tensor decompositions such as CP [Carroll & Chang 1970, Harshman 1970] and Tucker [Tucker 1966] (including HO-SVD [De Lathauwer et al. 2000] as a special case) have been in use for several decades in psychometrics and chemometrics and have recently become popular in signal processing, numerical analysis, neuroscience,

*http://www.arXiv.org/.
†http://citeseer.ist.psu.edu/.

computer vision, and data mining. See [Kolda & Bader 2009] for a comprehensive survey.

Recently, tensor decompositions have been applied to data-centric problems including analysis of click-through data [Sun et al. 2005] and chatroom analysis [Acar et al. 2005, Acar et al. 2006]. Liu et al. [Liu et al. 2005] presented a tensor space model which outperforms the classical vector space model for the problem of classification of Internet newsgroups. In the area of web hyperlink analysis, the CP decomposition has been used to extend the well-known HITS method to incorporate anchor text information [Kolda et al. 2005, Kolda & Bader 2006]. Bader et al. [Bader et al. 2007a, Bader et al. 2007b] used tensors to analyze the communications in the Enron e-mail data set. Sun et al. [Sun et al. 2006a, Sun et al. 2006b] dynamically updated Tucker models for detecting anomalies in network data. Tensors have also been used for multiway clustering, a method for clustering entities of different types based on both entity attributes as well as the connections between the different types of entities [Banerjee et al. 2007].

7.5.3 Other related work

Cohn and Hofmann [Cohn & Hofmann 2001] developed a joint probability model that combines text and links, with an application to categorizing web pages. Relational probability trees (RPTs) [Getoor et al. 2003, Getoor & Diehl 2005] offer a technique for analyzing graphs with different link and node types, with the goal of predicting node or link attributes.

For the problem of author disambiguation, addressed in this paper, Bekkerman and McCallum [Bekkerman & McCallum 2005] have developed an approach called multiway distributional clustering (MDC) that clusters data of several types (e.g., documents, words, and authors) based on interactions between the types. They used an instance of this method for disambiguation of individuals appearing in pages on the web.

7.6 Conclusions and future work

Multiple similarities between documents in a collection are represented as a three-way tensor ($N \times N \times K$), the tensor is decomposed using the CP-ALS algorithm, and relationships between the documents are analyzed using the CP component matrices. How to best choose the weights of the entries of the tensor is an open topic of research—the ones used here were chosen heuristically.

Different factors from the CP decomposition are shown to emphasize different link types; see Section 7.4.1. Moreover, the highest-scoring components in each factor denote an interrelated community. The component matrices ($\mathbf{A}$ and $\mathbf{B}$) of the CP decomposition can be used to derive feature vectors for latent similarity scores. However, the number of components (R) of the CP decomposition can strongly influence the quality of the matches; see Section 7.4.2. The choice of the number of components (R) and exactly how to use the component matrices are open questions, including how to combine these matrices, how to weight or normalize the features, and whether or not to incorporate the factor weightings, i.e., $\boldsymbol{\lambda}$.

This brings us to two disadvantages of the CP model. First, the factor matrices are not orthogonal, in contrast to the matrix SVD. A possible remedy for this is to instead consider the Tucker decomposition [Tucker 1966], which produces orthogonal component matrices and, moreover, can have a different number of columns for each component matrix; unfortunately, the Tucker decomposition is not unique and does not produce rank-one components like CP. Second, the best decomposition with R components is not the same as the first R factors of the optimal decomposition with $S > R$ components, again in contrast to the SVD [Kolda 2001]. This means that we cannot determine the optimal R by trial-and-error without great expense.

The centroids of feature vectors from the component matrices of the CP decomposition can be used to represent a small body of work (e.g., all the papers with the phrase "GMRES") in order to find related works. As expected, the feature vectors from the different component matrices produce noticeably different answers, either one of which may be more or less useful in different contexts; see Section 7.4.3. Combining these scores can be used to provide a ranked list of relevant work, taking into account the most relevant items from each of the component matrices.

A promising application of the similarity analysis is author disambiguation, where centroids are compared to predict which authors with similar names are actually the same. The technique is applied to the subset of authors with the most papers authored in the entire data set and affects the counts for the most published authors; see Section 7.4.4. In future work, we will consider the appropriate choice of the number of components (R) for disambiguation, identify how to choose the disambiguation similarity threshold, and perform a comparison to other approaches.

Using the feature vectors, it is possible to predict which journal each article was published in; see Section 7.4.5. Though the accuracy was relatively low, closer inspection of the data yielded clues as to why. For example, two of the publications were not focused publications. Overall, the results revealed similarities between the different journals. In future work, we will compare the results of using ensembles of decision trees to other learning methods (e.g., k-nearest neighbors, perceptrons, and random forests).

We also plan to revisit the representation of the data on two fronts. First, we wish to add authors as nodes. Hendrickson [Hendrickson 2007] observes that term-by-document matrices can be expanded to be (term *plus* document)-by-(term *plus* document) matrices so that term-term and document-document connections can be additionally encoded. Therefore, we intend to use a (document *plus* author) dimension so that we can explicitly capture connections between documents and authors as well as the implicit connections between authors, such as colleagues, conference co-organizers, etc. Second, in order to make predictions or analyze trends over time, we intend to incorporate temporal information using an additional dimension for time.

Though the CP decomposition has indications of the importance of each link in the communities it identifies (see Section 7.4.1), we do not exploit this information in reporting or computing similarities. As noted in [Ramakrishnan et al. 2005], understanding *how* two entities are related is an important issue and a topic for future work.

The reasons that the spectral properties of adjacency matrices aid in clustering are beginning to be better understood; see, e.g., [Brand & Huang 2003]. Similar analyses to explain the utility of the CP model for higher order data are needed.

7.7 Acknowledgments

Data used in this paper were extracted from the Science Citation Index Expanded, Thomson ISI, Philadelphia, PA, USA. We gratefully acknowledge Brett Bader for his work on the MATLAB tensor toolbox [Bader & Kolda 2006, Bader & Kolda 2007], which was used in our computations, and for providing the image used in Figure 7.2; the TMG Toolbox creators for providing a helpful tool for generating term-document matrices in MATLAB [Zeimpekis & Gallopoulos 2006]; Ann Yoshimura for TaMALE, which was used to create Figure 7.4; and Dirk Beyer for providing the code CCVisu* to generate the Barnes–Hut layouts.

References

[Acar et al. 2005] E. Acar, S.A. Çamtepe, M.S. Krishnamoorthy, and B. Yener. Modeling and multiway analysis of chatroom tensors. In *ISI 2005: Proceedings of the IEEE International Conference on Intelligence and Security Informatics*, vol. 3495 of *Lecture Notes in Computer Science*, 256–268, New York: Springer, 2005.

[Acar et al. 2006] E. Acar, S.A. Çamtepe, and B. Yener. Collective sampling and analysis of high order tensors for chatroom communications. In *ISI 2006: Proceedings of the IEEE International Conference on Intelligence and Security Informatics*, vol. 3975 of *Lecture Notes in Computer Science*, 213–224, New York: Springer, 2006.

[Adamic & Adar 2005] L. Adamic and E. Adar. How to search a social network. *Social Networks*, 27:187–203, 2005.

[Bader et al. 2007a] B.W. Bader, M.W. Berry, and M. Browne. Discussion tracking in Enron email using PARAFAC. In M.W. Berry and M. Castellanos, eds., *Survey of Text Mining: Clustering, Classification, and Retrieval, Second Edition*, 147–162, New York: Springer, 2007.

[Bader et al. 2007b] B.W. Bader, R.A. Harshman, and T.G. Kolda. Temporal analysis of semantic graphs using ASALSAN. In *ICDM 2007: Proceedings of the 7th IEEE International Conference on Data Mining*, 33–42, 2007.

[Bader & Kolda 2006] B.W. Bader and T.G. Kolda. Algorithm 862: MATLAB tensor classes for fast algorithm prototyping. *ACM Transactions on Mathematical Software*, 32:635–653, 2006.

*http://www.sosy-lab.org/~dbeyer.

[Bader & Kolda 2007] B.W. Bader and T.G. Kolda. Efficient MATLAB computations with sparse and factored tensors. *SIAM Journal on Scientific Computing*, 30:205–231, 2007.

[Banerjee et al. 2007] A. Banerjee, S. Basu, and S. Merugu. Multi-way clustering on relation graphs. In *SDM07: Proceedings of the 2007 SIAM International Conference on Data Mining*, http://www.siam.org/proceedings/datamining/2007/dm07.php, 145–156, 2007.

[Banfield et al. 2004] R. Banfield et al. OpenDT Web Page. http://opendt.sourceforge.net/, 2004.

[Barábasi et al. 2002] A.L. Barábasi, H. Jeong, Z. Néda, E. Ravasz, A. Schubert, and T. Vicsek. Evolution of the social network of scientific collaborations. *Physica A*, 311:590–614, 2002.

[Bekkerman & McCallum 2005] R. Bekkerman and A. McCallum. Disambiguating web appearances of people in a social network. In *WWW 2005: Proceedings of the 14th International Conference on World Wide Web*, 463–470, ACM Press, 2005.

[Berry et al. 1995] M.W. Berry, S.T. Dumais, and G.W. O'Brien. Using linear algebra for intelligent information retrieval. *SIAM Review*, 37:573–595, 1995.

[Beyer 2007] D. Beyer. CCVisu: A tool for co-change visualization and general force-directed graph layout, version 1.0. http://www.sosy-lab.org/~dbeyer, 2007.

[Boyack 2004] K.W. Boyack. Mapping knowledge domains: Characterizing PNAS. *Proceedings of the National Academy of Sciences*, 101:5192–5199, 2004.

[Brand & Huang 2003] M. Brand and K. Huang. A unifying theorem for spectral embedding and clustering. In *Proceedings of the Ninth International Workshop on Artificial Intelligence and Statistics*, 2003.

[Brin & Page 1998] S. Brin and L. Page. The anatomy of a large-scale hypertextual Web search engine. In *WWW7: Proceedings of the Seventh International World Wide Web Conference*, 107–117, Elsevier, 1998.

[Carroll & Chang 1970] J.D. Carroll and J.J. Chang. Analysis of individual differences in multidimensional scaling via an N-way generalization of 'Eckart-Young' decomposition. *Psychometrika*, 35:283–319, 1970.

[Cohn & Hofmann 2001] D. Cohn and T. Hofmann. The missing link—a probabilistic model of document content and hypertext connectivity. In *NIPS 2000: Advances in Neural Information Processing Systems*, 13:460–436, 2001.

[De Lathauwer et al. 2000] L. De Lathauwer, B. De Moor, and J. Vandewalle. A multilinear singular value decomposition. *SIAM Journal on Matrix Analysis and Applications*, 21:1253–1278, 2000.

[Dietterich 2000] T.G. Dietterich. Ensemble methods in machine learning. In Josef Kittler and Fabio Roli, eds., *First International Workshop on Multiple Classifier Systems*, no. 1857 in *Lecture Notes in Computer Science*, 1–15. New York: Springer, 2000.

[Dumais et al. 1988] S.T. Dumais, G.W. Furnas, T.K. Landauer, S. Deerwester, and R. Harshman. Using latent semantic analysis to improve access to textual information. In *CHI '88: Proceedings of the SIGCHI Conference on Fuman Factors in Computing Systems*, 281–285, ACM Press, 1988.

[Faber et al. 2003] N.M. Faber, R. Bro, and P.K. Hopke. Recent developments in CANDECOMP/PARAFAC algorithms: A critical review. *Chemometrics and Intelligent Laboratory Systems*, 65:119–137, 2003.

[Gehrke et al. 2003] J. Gehrke, P. Ginsparg, and J. Kleinberg. Overview of the 2003 KDD cup. *ACM SIGKDD Explorations Newsletter*, 5:149–151, 2003.

[Getoor et al. 2003] L. Getoor, N. Friedman, D. Koller, and B. Taskar. Learning probabilistic models of link structure. *Journal of Machine Learning Research*, 3:679–707, 2003.

[Getoor & Diehl 2005] L. Getoor and C.P. Diehl. Link mining: A survey. *ACM SIGKDD Explorations Newsletter*, 7:3–12, 2005.

[Harshman 1970] R.A. Harshman. Foundations of the PARAFAC procedure: Models and conditions for an "explanatory" multimodal factor analysis. *UCLA Working Papers in Phonetics*, 16:1–84, 1970. Available at http://www.psychology.uwo.ca/faculty/harshman/wpppfac0.pdf.

[Hendrickson 2007] B. Hendrickson. Latent semantic analysis and Fiedler retrieval. *Linear Algebra and Its Applications*, 421:345–355, 2007.

[Hill & Provost 2003] S. Hill and F. Provost. The myth of the double-blind review?: Author identification using only citations. *ACM SIGKDD Explorations Newsletter*, 5:179–184, 2003.

[Hitchcock 1927] F.L. Hitchcock. The expression of a tensor or a polyadic as a sum of products. *Journal of Mathematics and Physics*, 6:164–189, 1927.

[Jo et al. 2007] Y. Jo, C. Lagoze, and C.L. Giles. Detecting research topics via the correlation between graphs and texts. In *KDD '07: Proceedings of the 13th ACM SIGKDD International Conference on Knowledge Discovery and Data Mining*, 370–379, ACM Press, 2007.

[Kleinberg 1999] J.M. Kleinberg. Authoritative sources in a hyperlinked environment. *Journal of the ACM*, 46:604–632, 1999.

[Kolda 2001] T.G. Kolda. Orthogonal tensor decompositions. *SIAM Journal on Matrix Analysis and Applications*, 23:243–255, 2001.

[Kolda 2006] T.G. Kolda. Multilinear operators for higher-order decompositions. Technical Report SAND2006-2081, Sandia National Laboratories, Albuquerque, New Mexico and Livermore, Calif., April 2006.

[Kolda & Bader 2006] T.G. Kolda and B.W. Bader. The TOPHITS model for higher-order web link analysis. In *Workshop on Link Analysis, Counterterrorism and Security*, 2006.

[Kolda & Bader 2009] T.G. Kolda and B.W. Bader. Tensor decompositions and applications. *SIAM Review*, 51:455–500, 2009.

[Kolda et al. 2005] T.G. Kolda, B.W. Bader, and J.P. Kenny. Higher-order web link analysis using multilinear algebra. In *ICDM 2005: Proceedings of the 5th IEEE International Conference on Data Mining*, 242–249, IEEE Computer Society, 2005.

[Liu et al. 2005] N. Liu, B. Zhang, J. Yan, Z. Chen, W. Liu, F. Bai, and L. Chien. Text representation: From vector to tensor. In *ICDM 2005: Proceedings of the 5th IEEE International Conference on Data Mining*, 725–728, IEEE Computer Society, 2005.

[McGovern et al. 2003] A. McGovern, L. Friedland, M. Hay, B. Gallagher, A. Fast, J. Neville, and D. Jensen. Exploiting relational structure to understand publication patterns in high-energy physics. *ACM SIGKDD Explorations Newsletter*, 5:165–172, 2003.

[Ramakrishnan et al. 2005] C. Ramakrishnan, W.H. Milnor, M. Perry, and A.P. Sheth. Discovering informative connection subgraphs in multi-relational graphs. *ACM SIGKDD Explorations Newsletter*, 7:56–63, 2005.

[Selee et al. 2007] T.M. Selee, T.G. Kolda, W.P. Kegelmeyer, and J.D. Griffin. Extracting clusters from large datasets with multiple similarity measures using IMSCAND. In *CSRI Summer Proceedings 2007*, M. L. Parks and S. S. Collis, eds., Technical Report SAND2007-7977, Sandia National Laboratories, Albuquerque, NM and Livermore, CA, 87–103, December 2007.

[Smilde et al. 2004] A. Smilde, R. Bro, and P. Geladi. *Multi-Way Analysis: Applications in the Chemical Sciences.* West Sussex, England: Wiley, 2004.

[Sun et al. 2005] J.-T. Sun, H.-J. Zeng, H. Liu, Y. Lu, and Z. Chen. CubeSVD: A novel approach to personalized web search. In *WWW 2005: Proceedings of the 14th International Conference on World Wide Web*, 382–390, ACM Press, 2005.

[Sun et al. 2006a] J. Sun, S. Papadimitriou, and P.S. Yu. Window-based tensor analysis on high-dimensional and multi-aspect streams. In *ICDM 2006: Proceedings of the 6th IEEE Conference on Data Mining*, 1076–1080, IEEE Computer Society, 2006.

[Sun et al. 2006b] J. Sun, D.Tao, and C. Faloutsos. Beyond streams and graphs: Dynamic tensor analysis. In *KDD '06: Proceedings of the 12th ACM SIGKDD International Conference on Knowledge Discovery and Data Mining*, 374–383, ACM Press, 2006.

[Tomasi 2006] G. Tomasi. *Practical and computational aspects in chemometric data analysis.* PhD thesis, Department of Food Science, The Royal Veterinary and Agricultural University, Frederiksberg, Denmark, 2006. Available at http://www.models.life.ku.dk/theses/.

[Tomasi & Bro 2006] G. Tomasi and R. Bro. A comparison of algorithms for fitting the PARAFAC model. *Computational Statistics & Data Analysis*, 50:1700–1734, 2006.

[Tucker 1966] L.R. Tucker. Some mathematical notes on three-mode factor analysis. *Psychometrika*, 31:279–311, 1966.

[Zeimpekis & Gallopoulos 2006] D. Zeimpekis and E. Gallopoulos. TMG: A MATLAB toolbox for generating term-document matrices from text collections. In Jacob Kogan, Charles Nicholas, and Marc Teboulle, eds., *Grouping Multidimensional Data: Recent Advances in Clustering*, 187–210, New York: Springer, 2006.

Chapter 8

Subgraph Detection

Jeremy Kepner[*]

Abstract

Detecting subgraphs of interest in larger graphs is the goal of many graph analysis techniques. The basis of detection theory is computing the probability of a "foreground" with respect to a model of the "background" data. Hidden Markov Models represent one possible foreground model for patterns of interaction in a graph. Likewise, Kronecker graphs are one possible model for power law background graphs. Combining these models allows estimates of the signal to noise ratio, probability of detection, and probability of false alarm for different classes of vertices in the foreground. These estimates can then be used to construct filters for computing the probability that a background graph contains a particular foreground graph. This approach is applied to the problem of detecting a partially labeled tree graph in a power law background graph. One feature of this method is the ability to a priori estimate the number of vertices that will be detected via the filter.

8.1 Graph model

Graphs come in many shapes and sizes: directed, undirected, multi-edge, and hypergraphs, to name just a few. In addition, graphs are far more than just a set of

[*]MIT Lincoln Laboratory, 244 Wood Street, Lexington, MA 02420 (kepner@ll.mit.edu).

This work is sponsored by the Department of the Air Force under Air Force Contract FA8721-05-C-0002. Opinions, interpretations, conclusions, and recommendations are those of the author and are not necessarily endorsed by the United States Government.

connections. They also contain metadata about the vertices and edges. The first step in developing detection theory for graphs is to define a high dimensional model of a graph that can handle this diversity information. If there is an edge from vertex i to vertex j at time t of type l, then let the N vertex graph adjacency tensor have

$$\mathcal{A}(i, j, t) = l$$

where $\mathcal{A} : \mathbb{Z}^{N \times N \times N_t}$. Likewise, if there is no edge, then $\mathcal{A}(i, j, t) = 0$.

Other convenient forms of the same notation include

$$\mathcal{A}(i, j, k(t), l) = 1$$

where $\mathcal{A} : \mathbb{B}^{N \times N \times N_k \times N_l}$ and $k(t)$ represents the binning of t into discrete time windows. If e_{ijk} is a pointer to an edge, then another form is

$$\mathcal{A}(i, j, k) = e_{ijk}$$

where $\mathcal{A} : \mathbb{Z}^{N \times N \times N_k}$. In this form, each edge can also have metadata in the form of an arbitrary number of additional parameters such as $t(e_{ijk})$ and $l(e_{ijk})$. Likewise, vertices can also have similar parameters $t(v_i)$ and $l(v_i)$. Finally, the number of edges in the graph is computed by counting the number of nonzero entries in $\mathcal{A}$:

$$M = |\mathcal{A}| = \sum_{i,j,k,l} \{\mathcal{A}(i, j, k, l) \neq 0\}$$

8.1.1 Vertex/edge schema

The above multidimensional adjacency matrix description of a graph allows a rich variety of data to be directly incorporated into the graph model. However, the model says nothing about what is a vertex and an edge. In some instances, the vertices and edges are obvious. For example, in a webgraph (i.e., a graph of links between web pages) the vertices are webpages and $\mathbf{A}(i, j) = 1$ implies that webpage i has a link pointing to webpage j. Similarly, in a citation graph, $\mathbf{A}(i, j) = 1$ implies that paper i cites paper j. In both of these graphs, all the vertices are of the same type (i.e., all webpages or all documents) and that type happens to be the same type as the items in the corpus (webpages or documents). Such graphs can be referred to as having monotype vertices.

Another common type of graph associates entities in the items of the corpus with those items. For example, a graph showing all the words in a corpus of webpages or documents might have a graph in which $\mathbf{A}(i, j) = 1$ implies that webpage/paper i contains word j. Such graphs can be referred to as having dual-type vertices. These dual-type vertex graphs can be converted to monotype vertex graphs by squaring the adjacency matrix with itself. A document $\times$ word adjacency matrix can be converted into a word $\times$ word adjacency matrix via the inner square product

$$\mathbf{W} = \mathbf{A}^T \mathbf{A}$$

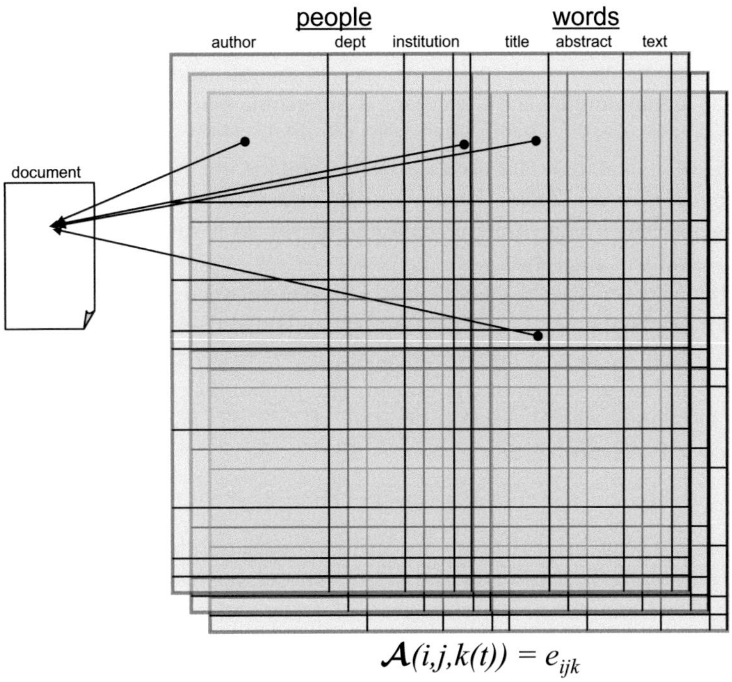

Figure 8.1. Multitype vertex/edge schema.
Edges point to documents. All pairs of items in a document create an edge.

where $\mathbf{W}(i,j)$ contains the number of documents that contain both words i and j. Likewise, a document $\times$ document matrix can be constructed via the outer square product

$$\mathbf{D} = \mathbf{A}\mathbf{A}^T$$

where $\mathbf{W}(i,j)$ contains the number of words that are in both documents i and j.

The utility of a graph for answering particular questions is usually dictated by the vertex/edge schema. The more specific the schema, the better it will be at addressing a specific question. When the questions to be answered are still being formulated, it may not be apparent how to organize the graph or even what is a vertex or an edge. In this situation, a multitype vertex schema is often useful as it can be applied to a range of data. In a multitype vertex graph, the items in the corpus are the edges and the vertices are many types: documents, people, words, etc. Each document generates a clique of edges all pointing to that document. In a multitype vertex graph, all interrelationships between all entities are stored (see Figure 8.1). Clearly, the size of the multitype vertex graph can be significantly larger than a monotype or dual-type vertex graph. Once the data is understood and the kinds of questions to be answered are defined, then the graph can be simplified to the appropriate monotype vertex or dual-type vertex graph.

8.2 Foreground: Hidden Markov model

The first component of a detection theory is a mathematical model for the object that is being sought, in this case, a subgraph embedded in a larger graph. There are many ways to formulate such a model. Here a hidden Markov model (HMM) is used to model the underlying evolving process that generates the subgraph [Weinstein et al. 2009]. It may be the case that more is known about these underlying processes (e.g., how a vertex is added to a subgraph) than about the specific subgraphs themselves.

An HMM describes the evolution of a system between states s. The probability of transitioning from state s to s' is given by the HMM transition probability matrix $\mathbf{S}_h(s,s')$. Within each s, a number of possible edges e_{ijl} that are characteristic of that state could be observed in the graph. The probability of observing an edge e_{ijl} in state s is given by the observation matrix $\mathbf{B}_h(s,e_{ijl})$. For a given state $h(s)$, this can be recoded as a probability graph

$$\mathcal{A}_{h(s)}(i,j,l) = \mathbf{B}_h(s,e_{ijl})$$

If the states are sequential (i.e., no revisiting of earlier states), then the probability graph for each state is stationary (i.e., does not change over time). Thus, s can be substituted for t and the probability graph for the entire HMM is

$$\mathcal{A}_h(i,j,s,l) = \mathbf{B}_h(s,e_{ijl})$$

8.2.1 Path moments

A model of the underlying process used for generating a subgraph is only practical if there is some way to connect the model to real observations (e.g., an observed path in a graph). This connection can be achieved by computing the a priori probability of a particular path from the model and then comparing it with the observations. Likewise, the observations of particular paths can be used to compute the parameters in the model.

Consider any path of length $r-1$ consisting of vertices $v_1 v_2 v_3 \ldots v_r$ and edges $e_{12} e_{23} e_{(r-1)r}$, where the vertices are not necessarily unique. Such a path will generate a specific sequence of vertex and edge parameters (or types):

$$l(v_1)l(e_{12})l(v_2)l(e_{23})l(v_3)\ldots l(v_{r-1})l(e_{(r-1)r})l(v_r)$$

or

$$l^v_1 l^e_{12} l^v_2 l^e_{23} l^v_3 \cdots l^v_{r-1} l^e_{(r-1)r} l^v_r$$

The distribution of all paths $< r$ is denoted by P_r where

$$P_r(l^v_1, l^e_{12}, l^v_2 l^e_{23}, l^v_3, \ldots, l^v_{r-1}, l^e_{(r-1)r}, l^v_r) = 1$$

Typically, distributions will be limited to smaller values of r. $P(l^v)$ is the distribution of all vertex types. $P(l^e)$ is the distribution of all edge types. $P(l^v, l^e)$ is the joint distribution of vertex and edge types and is the probability that a particular vertex type will have an edge of a particular edge type. $P(l^v_1, l^e_{12}, l^v_2)$ is the joint distribution of all vertex, edge, vertex types and is the probability that two vertices of given vertex types will have an edge of a particular edge type.

Special cases

If the number of vertex types is equal to the number of vertices $N_{l^v} = N$ and there is only one edge type $N_{l^e} = 1$, then the $P(l^v)$ is simply the degree distribution of the graph

$$P(l^v) = P(l(v_i)) = |\mathbf{\mathcal{A}}(i,:,:)| = \sum_{j,k} \mathbf{\mathcal{A}}(i,j,k)$$

or, equivalently,

$$P(l^v) = P(l(v_j)) = |\mathbf{\mathcal{A}}(:,j,:)| = \sum_{i,k} \mathbf{\mathcal{A}}(i,j,k)$$

The joint distribution $P(l_1^v, l_1^v)$ is simply the $N \times N$ adjacency matrix of the graph

$$P(l_1^v, l_2^v) = P(l(v_i), l(v_j)) = \mathbf{A}(i,j) = |\mathbf{\mathcal{A}}(i,j,:)| = \sum_{k} \mathbf{\mathcal{A}}(i,j,k)$$

If the number of vertex types is one ($N_{l^v} = 1$) and the number of edge types equals the number of edges ($N_{l^e} = M$), then $P(l^e)$ is the edge distribution of the graph

$$P(l^e) = |\mathbf{\mathcal{A}} = l^e| = \sum_{i,j,k} \{G(i,j,k) = l^e\}$$

If the number of vertex types is equal to the number of vertices $N_{l^v} = N$ and the edge types equal the number of edges $N_{l^e} = M$, then $P(l_1^v, l_2^v, l_{12}^e)$ recovers the original graph

$$P(P(l_1^v, l_2^v, l_{12}^e)) = P(l(v_i), l(v_j), l(e_{ij})) = \mathbf{\mathcal{A}}(i,j,k)$$

Expected paths

If the vertex and edge types are uncorrelated, then the joint probability is the product of the probabilities of vertex and edge types

$$P(l(v_i), l(v_j), l(e_{ij})) = P(l(v_i))P(l(v_j))P(l(e_{ij}))$$

Comparing the ratio of the left side to the actual counts in the data can be used to determine if any higher order correlations exist within data. Note: if $\mathbf{\mathcal{A}}(i,j,k)$ has strong correlations (e.g., in a power law distributions), then the above formula may need to be modified to take into account Poisson sampling effects (i.e., if one vertex has many more edges, then this will create "artificial" correlations with that vertex type).

For a given $\mathbf{\mathcal{A}}_{h(s)}$, the expected path-type distributions can be computed by integrating over all paths $P_{h(s)}$. In fact, this distribution can be used to define the $h(s)$. In other words, two HMM states can be viewed as path equivalent if their expected path type distributions are the same.

Computing P_r from the graph and comparing with all $P_{h(s)}$ allow the probability that a particular vertex is part of a particular $h(s)$ to be computed.

8.3 Background model: Kronecker graphs

The second component of a detection theory is a mathematical model for the background. There are many ways to formulate such a model. Here a Kronecker product approach is used because it naturally generates a wide range of graphs. The Kronecker graph generation algorithm (see [Leskovec 2005, Chakrabarti 2004]) is quite elegant and can be described as follows. First, let $\mathbf{A} \in \mathbb{R}^{M_B M_C \times N_B N_C}$, $\mathbf{B} \in \mathbb{R}^{M_B \times N_B}$, and $\mathbf{C} \in \mathbb{R}^{M_C \times N_C}$. Then the Kronecker product is defined as follows (see [Van Loan 2000])

$$\mathbf{A} = \mathbf{B} \otimes \mathbf{C} = \begin{pmatrix} \mathbf{B}(1,1)\mathbf{C} & \mathbf{B}(1,2)\mathbf{C} & \cdots & \mathbf{B}(1,M_B)\mathbf{C} \\ \mathbf{B}(2,1)\mathbf{C} & \mathbf{B}(2,2)\mathbf{C} & \cdots & \mathbf{B}(2,M_B)\mathbf{C} \\ \vdots & \vdots & & \vdots \\ \mathbf{B}(N_B,1)\mathbf{C} & \mathbf{B}(N_B,2)\mathbf{C} & \cdots & \mathbf{B}(N_B,M_B)\mathbf{C} \end{pmatrix}$$

Now let $\mathbf{A} \in \mathbb{R}^{N \times N}$ be an adjacency matrix. The Kronecker exponent to the power k is as follows

$$\mathbf{A}^{\otimes k} = \mathbf{A}^{\otimes k-1} \otimes \mathbf{A}$$

which generates an $N^k \times N^k$ adjacency matrix. This simple model naturally produces self-similar graphs, yet even a small $\mathbf{A}$ matrix provides ample parameters to fine-tune the generator to a particular real-world data set. The model also lends itself to the analytic computation of a wide range of graph measures.

The physical intuition as to why Kronecker products naturally generate power law graphs is as follows. Given a simple single state HMM

$$\mathcal{A}_{h(s)}(i,j,l) = \mathbf{A}_h(i,j) = \mathbf{B}_h(i,j)$$

The product $\mathbf{A}_h(i,j)\mathbf{A}_h(i',j')$ is the probability of a vertex being reached via the HMM process $e_{ij}e_{i'j'}$. Assuming that all such edges in a graph are the result of such a process, then $\mathbf{A}_h^{\otimes k}$ contains the probability of all possible processes of length k. Thus each edge in the resulting adjacency matrix indicates the precise set of HMM process steps used to generate that edge.

8.4 Example: Tree finding

The aforementioned detection theory concepts are illustrated by applying them to the simple example of finding a small tree embedded in a larger power law graph. This problem is related to the minimal spanning tree problem (see, for instance, [Furer & Raghavachari 1992] and the references therein).

8.4.1 Background: Power law

Consider a directed Kronecker graph with M edges and $N = 2^k$ vertices with boolean adjacency matrix $\mathbf{A} : \mathbb{B}^{N \times N}$ given by

$$\mathbf{A} \stackrel{M}{\leftarrow} \mathbf{A}_h^{\otimes k}$$

where the generator $\mathbf{A}_h : \mathbb{R}^{2\times 2}$ has values such that the resulting graph is a power law graph. Typical values of $\mathbf{A}_h$ are (see http://www.GraphAnalysis.org/benchmark)

$$\mathbf{A}_h = \begin{pmatrix} 0.55 & 0.1 \\ 0.1 & 0.25 \end{pmatrix}$$

The above matrix describes a stationary HMM process whereby the probability of staying in the same vertex is higher than transitioning to the other vertex.

8.4.2 Foreground: Tree

Consider a tree consisting of k_T levels and n_T branches per node. An HMM process for generating such a tree consists of k_T products beginning with the branch process given by the row vector of ones $\mathbf{1}_{1\times n_T}$. Next, k_T-1 applications of the identity matrix $\mathbf{I}_{n_T\times n_T}$ replicates each branch n_T times. Finally, the multiplication by the column vector creates the leaf nodes

$$(\ 1\ \overbrace{0\ldots0}^{n_T-1}\ )'$$

where $'$ denotes matrix transpose. Using this process, the adjacency matrix $\mathbf{A}_{T\times T} : \mathbb{B}^{N_T\times N_T}$ describing a directed tree with $N_T = n_T^{k_T}$ nodes can be constructed by

$$\mathbf{A}_{T\times T} = (\ 1\ \overbrace{0\ldots0}^{n_T-1}\ )' \otimes \mathbf{I}_{n_T\times n_T}^{\otimes k_T-1} \otimes \mathbf{1}_{1\times n_T}$$

The above tree will have k_T levels and n_T branches per node. Each branch node will have a degree of $n_T + 1$ and each leaf will have a degree of 1. See Figure 8.2 for the case in which $n_T = 2$ and $k_T = 4$.

8.4.3 Detection problem

Let T be a set of N_T random vertices drawn from $[1, N]$. The tree is inserted into the Kronecker graph via the assignment operation $\mathbf{A}(T,T) = \mathbf{A}_{T\times T}$. The detection problem is to find T given $\mathbf{A}$. Given this information, the only means for identifying T would be topological. The primary topological tool that can be used to find T is its degree distribution. For a nominal power law graph where $M/N \sim 8$, and a nominal binary tree ($n_T = 2$) with $M_T/N_T \sim 2$, the degree distribution can be used to identify vertices that are more likely in T. Unfortunately, if N and M are large compared to N_T and M_T, this problem is extremely challenging (see SNR calculation in next section) and likely does not have a unique solution because $\mathbf{A}$ will contain many trees that are topologically identical to $\mathbf{A}_{T\times T}$.

A common variant of the above problem is the case in which some of the vertices in the tree are known and some are not. Thus, the problem can be made more tractable if additional metadata on the vertices or edges are provided that allow a subset V of T to be identified such as $V = \{v : l(v) = l^T\}$ or $V = \{v : l(e_{vv'}) = l^T\}$. T is then split into two sets such that $T = V \cup U^\star$ (see Figure 8.3), where $f = N_V/N_T$ and $1 - f = N_{U^\star}/N_T$. The problem is then to find $U^\star$ given V.

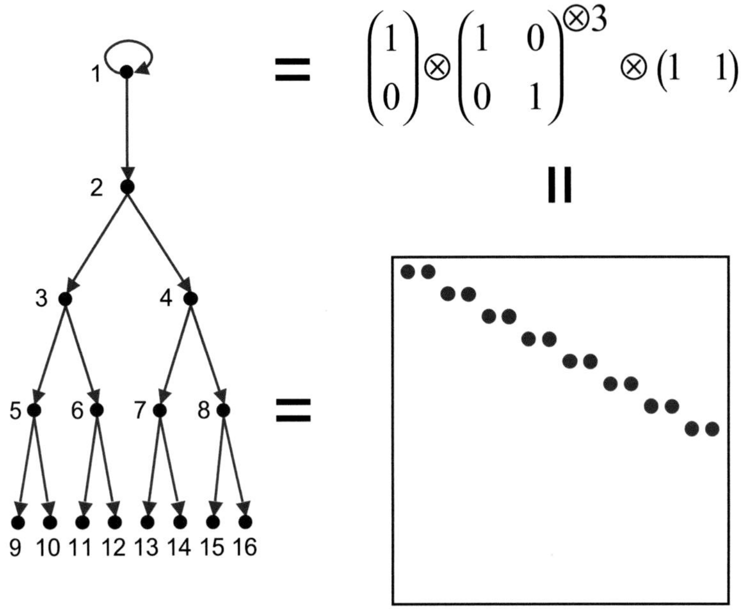

Figure 8.2. Tree adjacency matrix.
Graphical and algebraic depiction of the adjacency matrix of a tree with $k_T = 4$ levels and $n_T = 2$ branches per node.

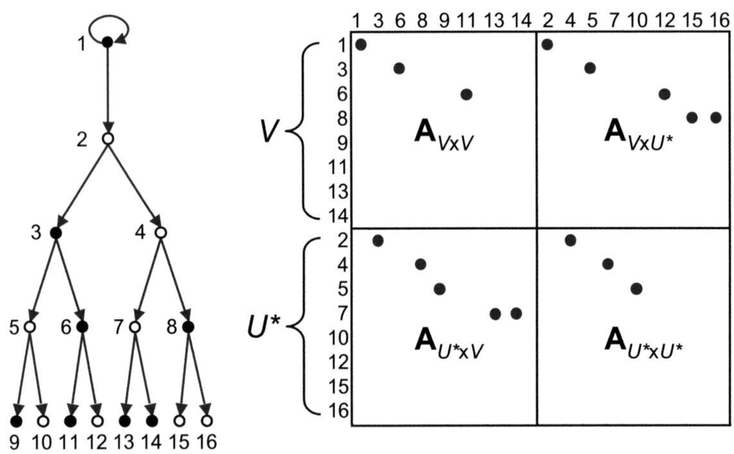

Figure 8.3. Tree sets.
Adjacency matrix of a tree split into two sets $T = V \cup U^\star$.

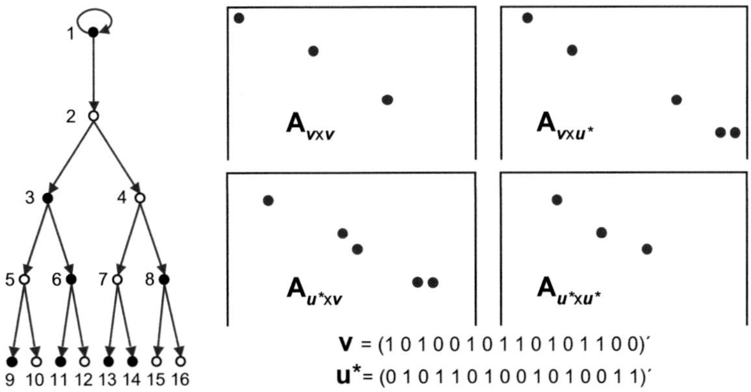

Figure 8.4. Tree vectors.
Adjacency matrices of a tree split into two vectors $\mathbf{t} = \mathbf{v} + \mathbf{u}^\star$.

8.4.4 Degree distribution

The primary topological tool that can be used to find T is the degree distribution. The degree of each vertex in the tree can be computed as follows

$$\mathbf{d}_{T\times T} = (\mathbf{A}_{T\times T} + \mathbf{A}'_{T\times T})\ \mathbf{1}_{N_T\times 1}$$

For a tree, half the vertices will have the max degree $d^{max}_{T\times T} = n_T + 1$, and the other half will have a degree of 1.

The above type of calculation can be simplified by introducing the following set/vector notation. Let set T also be represented by the boolean column vector $\mathbf{t} : \mathbb{B}^{N\times 1}$ where $\mathbf{t}(T) = 1$ and zero otherwise (see Figure 8.4). Furthermore, let $|\mathbf{t}| = N_T$ denote the number of nonzero elements in $\mathbf{t}$. The advantages of using $\mathbf{t}$ instead of T are that it preserves the overall context of the graph and allows creating subsets with less bookkeeping. Let $\mathbf{I}_\mathbf{t} = \text{diag}(\mathbf{t})$ be the identity matrix with the vector $\mathbf{t}$ along the diagonal. The subgraph consisting of T can now be written as

$$\mathbf{A}_{\mathbf{t}\times\mathbf{t}} = \mathbf{A}_{\mathbf{t}^2} = \mathbf{I}_\mathbf{t}\ \mathbf{A}\ \mathbf{I}_\mathbf{t}$$

From $\mathbf{A}_{\mathbf{t}^2} : \mathbb{B}^{N\times N}$, the vertex degrees can be computed as follows

$$\mathbf{d}_{\mathbf{t}^2} = (\mathbf{A}_{\mathbf{t}^2} + \mathbf{A}'_{\mathbf{t}^2})\ \mathbf{1}_{N\times 1}$$

The nonzero values of $\mathbf{d}_{\mathbf{t}^2}$ will be the same as $\mathbf{d}_{T\times T}$.

Often, it will be necessary to compute the degree of a set of starting vertices $\mathbf{v}$ into a set of ending vertices $\mathbf{u}$. This begins by computing the corresponding adjacency matrices

$$\mathbf{A}_{\mathbf{v}\times\mathbf{u}} = \mathbf{I}_\mathbf{v}\ \mathbf{A}\ \mathbf{I}_\mathbf{u}$$

and

$$\mathbf{A}_{\mathbf{u}\times\mathbf{v}} = \mathbf{I}_\mathbf{u}\ \mathbf{A}\ \mathbf{I}_\mathbf{v}$$

The resulting degree distribution is then

$$\mathbf{d}_{\mathbf{v}\times\mathbf{u}} = (\mathbf{A}_{\mathbf{v}\times\mathbf{u}} + \mathbf{A}'_{\mathbf{u}\times\mathbf{v}})\ \mathbf{1}_{N\times 1}$$

8.5 SNR, PD, and PFA

An essential aspect of detection is estimating the difficulty of the problem. The signal to noise ratio (SNR) is a standard tool for computing this difficulty. In the context of subgraph detection, SNR will be defined as follows. Given a set of vertices selected containing some number of foreground vertices, the SNR is

$$\begin{aligned}\text{SNR}(\{\text{foreground vertices selected}\} \subset \{\text{vertices selected}\}) \\ = \frac{\#\text{foreground vertices selected}}{\#\text{background vertices selected}}\end{aligned}$$

At first glance, the overall SNR of any subgraph detection problem can be lower bounded by selecting all the vertices in the graph

$$\text{SNR}(\mathbf{u}^\star \subset \mathbf{a}) = |\mathbf{u}^\star|/|\mathbf{a} - \mathbf{u}^\star| \approx |\mathbf{u}^\star|/N$$

where $|\mathbf{u}^\star|$ is the number of nonzero elements in $\mathbf{u}^\star$ and $\mathbf{a} = \mathbf{1}_{N\times 1}$ is the vector representing all the vertices in $\mathbf{A}$. For a nominal power law graph ($N = 2^{20}$) and tree ($N_T = 2^7$, $f = 1/2$), the $\text{SNR}(\mathbf{u}^\star \subset \mathbf{a}) = 2^{-14}$.

Another important tool for assessing detection is the probability of detection (PD) and can be defined as

$$\begin{aligned}\text{PD}(\{\text{foreground items selected}\} \subset \{\text{foreground items}\}) \\ = \frac{\#\text{foreground items selected}}{\#\text{foreground items}}\end{aligned}$$

By definition, if all the items in the graph are selected, then PD = 1.

Probability of detection is usually computed in concert with the probability of false alarm (PFA), which is defined as

$$\begin{aligned}\text{PFA}(\{\text{background items selected}\} \subset \{\text{items selected}\}) \\ = \frac{\#\text{background items selected}}{\#\text{items selected}}\end{aligned}$$

If all the items in the graph are selected, then

$$\text{PFA}(\mathbf{a} - \mathbf{u}^\star \subset \mathbf{a}) = \frac{N - |\mathbf{u}^\star|}{N} \approx 1 - \text{SNR}(\mathbf{u}^\star \subset \mathbf{a}) = 1 - 2^{-14}$$

Combined together, the SNR, PD, and PFA are useful tools for assessing the difficulty of a detection problem. For example, problems with a low SNR are more difficult than problems with a high SNR. By computing the SNR and then identifying ways to increase the SNR, a simple filtering algorithm will emerge for the tree detection example problem.

8.5.1 First and second neighbors

The first step toward getting a higher SNR for $\mathbf{u}^\star$ is to use the metadata that identified $\mathbf{v} \subset \mathbf{t}$ and recognizing that $\mathbf{u}^\star$ will most likely be found in the first and second nearest neighbors of $\mathbf{v}$.

The number of first and second nearest neighbors to $\mathbf{v}$ can be approximated by

$$|\mathbf{v}_{\rm I}| \approx 2(M/N)|\mathbf{v}|$$

and

$$|\mathbf{v}_{\rm II}| \approx 2(M/N)|\mathbf{v}_{\rm I}| = 4(M/N)^2|\mathbf{v}|$$

where the factors of two come from not differentiating between incoming and outgoing edges.

Assuming that $\mathbf{u}^\star \subset \mathbf{v}_{\rm I} + \mathbf{v}_{\rm II}$, then the SNR can be computed as

$$\begin{aligned}\mathrm{SNR}(\mathbf{u}^\star \subset \mathbf{v}_{\rm I} + \mathbf{v}_{\rm II}) &= \frac{|\mathbf{u}^\star|}{|\mathbf{v}_{\rm I} + \mathbf{v}_{\rm II} - \mathbf{u}^\star|} \approx \frac{|\mathbf{u}^\star|}{|\mathbf{v}_{\rm II}|} \\ &= \frac{(1-f)|\mathbf{v}|/f}{4(M/N)^2|\mathbf{v}|} \\ &= [(1-f)/f]/[4(M/N)^2]\end{aligned}$$

For a nominal power law graph where $M/N = 8$ and a nominal binary tree ($n_T = 2$) with $f = 0.5$, $\mathrm{SNR}(\mathbf{u}^\star \subset \mathbf{v}_{\rm I} + \mathbf{v}_{\rm II}) \approx 2^{-8}$. Likewise, by assumption PD ≈ 1 and PFA $\approx 1 - $ SNR.

8.5.2 Second neighbors

The SNR for the tree detection problem can be refined by recognizing that nearly all vertices in $\mathbf{u}^\star$ can be categorized into two classes with very different SNRs

$$\mathbf{u}^\star \approx \mathbf{u}^\star_{\rm I} + \mathbf{u}^\star_{\rm II}$$

where $\mathbf{u}^\star_{\rm I}$ are in the set of $\mathbf{v}$'s first neighbors $\mathbf{v}_{\rm I}$, and $\mathbf{u}^\star_{\rm II}$ are in the set of $\mathbf{v}$'s second neighbors $\mathbf{v}_{\rm II}$.

Now suppose $|\mathbf{u}^\star_{\rm II}|$ can be approximated by the number of "fallen leaves" and "fallen branches" on the tree. A fallen leaf is a leaf node in $\mathbf{u}^\star$ whose branch node back to the tree is also in $\mathbf{u}^\star$. The number of leaf nodes in a tree is approximately $(1-n_T^{-1})|\mathbf{t}|$ and the number of leaf nodes in $\mathbf{u}^\star$ is approximately $(1-f)(1-n_T^{-1})|\mathbf{t}|$. The probability that a leaf node's branch node is in $\mathbf{u}^\star$ is $1-f$; thus, the number of fallen leaves is $(1-f)^2(1-n_T^{-1})|\mathbf{t}|$. A fallen branch occurs when a branch node and all its neighbors are in $\mathbf{u}^\star$. The number of branch nodes in the tree is approximately $|\mathbf{t}|/n_t$ with approximately $(1-f)|\mathbf{t}|/n_t$ in $\mathbf{u}^\star$. The probability that all the neighbors of a branch are also in $\mathbf{u}^\star$ is $(1-f)^{n_T+1}$; thus, the total number of fallen leaves and branches is

$$\begin{aligned}|\mathbf{u}^\star_{\rm II}| &\approx (1-f)^2(1-n_T^{-1})|\mathbf{t}| + (1-f)^{n_T+2}|\mathbf{t}|/n_t \\ &= (1-f)^2[1 - n_T^{-1} + n_T^{-1}(1-f)^{n_T}]|\mathbf{t}|\end{aligned}$$

The corresponding SNR is

$$\begin{aligned}\text{SNR}(\mathbf{u}^\star_{\text{II}} \subset \mathbf{v}_{\text{II}}) &= \frac{|\mathbf{u}^\star_{\text{II}}|}{|\mathbf{v}_{\text{II}} - \mathbf{u}^\star_{\text{II}}|} \approx \frac{|\mathbf{u}^\star_{\text{II}}|}{|\mathbf{v}_{\text{II}}|} \\ &= \frac{(1-f)^2[1 - n_T^{-1} + n_T^{-1}(1-f)^{n_T}]|\mathbf{t}|}{4(M/N)^2|\mathbf{v}|} \\ &= \frac{(1-f)^2[1 - n_T^{-1} + n_T^{-1}(1-f)^{n_T}]}{4(M/N)^2 f}\end{aligned}$$

For a nominal power law graph and tree, the $\text{SNR}(\mathbf{u}^\star_{\text{II}} \subset \mathbf{v}_{\text{II}}) = 5/2^{12}$. Likewise, the PD is

$$\text{PD}(\mathbf{u}^\star_{\text{II}} \subset \mathbf{u}^\star) = |\mathbf{u}^\star_{\text{II}}|/|\mathbf{u}^\star| \approx 5/2^4$$

Similarly, the PFA is

$$\text{PFA}(\mathbf{v}_{\text{II}} - \mathbf{u}^\star_{\text{II}} \subset \mathbf{v}_{\text{II}}) = \frac{|\mathbf{v}_{\text{II}} - \mathbf{u}^\star_{\text{II}}|}{|\mathbf{v}_{\text{II}}|} \approx 1 - \text{SNR}(\mathbf{u}^\star_{\text{II}} \subset \mathbf{v}_{\text{II}}) = 1 - 5/2^{12}$$

Note that $\text{SNR}(\mathbf{u}^\star_{\text{II}} \subset \mathbf{v}_{\text{II}})$ is much lower than $\text{SNR}(\mathbf{u}^\star \subset \mathbf{v}_{\text{I}} + \mathbf{v}_{\text{II}})$. Thus, by not including second neighbors and only targeting first neighbors, it should be possible to increase the SNR.

8.5.3 First neighbors

By restricting the selection to the first neighbors $\mathbf{v}_{\text{I}}$, the SNR can be increased

$$\begin{aligned}\text{SNR}(\mathbf{u}^\star_{\text{I}} \subset \mathbf{v}_{\text{I}}) &= \frac{|\mathbf{u}^\star_{\text{I}}|}{|\mathbf{v}_{\text{I}} - \mathbf{u}^\star_{\text{I}}|} \approx \frac{|\mathbf{u}^\star - \mathbf{u}^\star_{\text{II}}|}{|\mathbf{v}_{\text{I}}|} \\ &= \frac{1 - |\mathbf{u}^\star_{\text{II}}|/|\mathbf{u}^\star|}{|\mathbf{v}_{\text{I}}|/|\mathbf{u}^\star|} \\ &= \frac{1 - \text{PD}(\mathbf{u}^\star_{\text{II}} \subset \mathbf{u}^\star)}{2(M/N)(1-f)/f}\end{aligned}$$

For a nominal power law graph and tree, the $\text{SNR}(\mathbf{u}^\star_{\text{I}} \subset \mathbf{v}_{\text{I}}) = 11/2^8$. Likewise, the PD is

$$\text{PD}(\mathbf{u}^\star_{\text{I}} \subset \mathbf{u}^\star) = |\mathbf{u}^\star_{\text{I}}|/|\mathbf{u}^\star| = |\mathbf{u}^\star - \mathbf{u}^\star_{\text{II}}|/|\mathbf{u}^\star| = 1 - \text{PD}(\mathbf{u}^\star_{\text{II}} \subset \mathbf{u}^\star) = 1 - 5/2^4$$

Similarly, the PFA is

$$\text{PFA}(\mathbf{v}_{\text{I}} - \mathbf{u}^\star_{\text{I}} \subset \mathbf{v}_{\text{I}}) = 1 - \text{SNR}(\mathbf{u}^\star_{\text{I}} \subset \mathbf{v}_{\text{I}}) = 1 - 11/2^8$$

8.5.4 First neighbor leaves

Clearly, the SNR of the first neighbors will be higher than the SNR of the second neighbors or of the first and second neighbors combined. Likewise, the first neighbors can also be divided into two classes of vertices with higher and lower SNR. Let $\mathbf{u}_{\text{I}}$ be made up of the branch and leaf vertices $\mathbf{u}^\star_{\text{B}}$ and $\mathbf{u}^\star_{\text{L}}$ such that

$$\mathbf{u}^\star_{\text{I}} = \mathbf{u}^\star_{\text{B}} + \mathbf{u}^\star_{\text{L}}$$

$\mathbf{u}^\star_{\mathrm{B}}$ are branch vertices of $\mathbf{t}$ that are first nearest neighbors of $\mathbf{v}$. $\mathbf{u}^\star_{\mathrm{L}}$ are leaf vertices of $\mathbf{t}$ that are also first nearest neighbors of $\mathbf{v}$, so that $\mathbf{u}^\star_{\mathrm{B}} + \mathbf{u}^\star_{\mathrm{L}}$ is the set all the first neighbors of $\mathbf{v}$ in $\mathbf{u}^\star$.

The number of vertices $|\mathbf{u}^\star_{\mathrm{L}}|$ can be approximated by adding the number of leaves and leaflike branches in $\mathbf{u}^\star$. The probability that a leaf node's branch node is in $\mathbf{v}$ is f, and so the number of leaf nodes in $\mathbf{u}^\star$ with branches in $\mathbf{v}$ is approximately $(1-f)f(1-n_T^{-1})|\mathbf{t}|$. A leaflike branch occurs when a branch node has only one of its branches $\mathbf{u}^\star$, and so appears to look like a leaf node. In other words, branches in $\mathbf{u}^\star$ will become leaflike when all but one of their neighbors is in $\mathbf{u}^\star$, which occurs with a probability of $(n_T+1)(1-f)^{n_T}f$. Thus, the total number of leaflike nodes in $\mathbf{u}^\star$ is

$$\begin{aligned}|\mathbf{u}^\star_{\mathrm{L}}| &\approx (1-f)f(1-n_T^{-1})|\mathbf{t}| + (n_T+1)f^2(1-f)^{n_T}|\mathbf{t}|/n_t \\ &\approx [(1-f)(1-n_T^{-1}) + (1-f)^{n_T}f(1+n_T^{-1})]f|\mathbf{t}|\end{aligned}$$

and

$$|\mathbf{u}^\star_{\mathrm{L}}|/|\mathbf{v}_{\mathrm{I}}| = \frac{(1-f)(1-n_T^{-1}) + (1-f)^{n_T}f(1+n_T^{-1})}{2(M/N)}$$

For a nominal power law graph and tree, $|\mathbf{u}^\star_{\mathrm{L}}|/|\mathbf{v}_{\mathrm{I}}| \approx 7/2^8$. The corresponding SNR is

$$\mathrm{SNR}(\mathbf{u}^\star_{\mathrm{L}} \subset \mathbf{v}_{\mathrm{I}}) = |\mathbf{u}^\star_{\mathrm{L}}|/|\mathbf{v}_{\mathrm{I}} - \mathbf{u}^\star_{\mathrm{L}}| = (|\mathbf{u}^\star|/|\mathbf{u}^\star_{\mathrm{L}}| - 1)^{-1} \approx |\mathbf{u}^\star_{\mathrm{L}}|/|\mathbf{v}_{\mathrm{I}}| = 7/2^8$$

The PD is

$$\begin{aligned}\mathrm{PD}(\mathbf{u}^\star_{\mathrm{L}} \subset \mathbf{u}^\star) &= |\mathbf{u}^\star_{\mathrm{L}}|/|\mathbf{u}^\star| \\ &= \frac{[(1-f)(1-n_T^{-1}) + (1-f)^{n_T}f(1+n_T^{-1})]f|\mathbf{t}|}{(1-f)|\mathbf{t}|} \\ &= [(1-n_T^{-1}) + (1-f)^{n_T-1}f(1+n_T^{-1})]f = 5/2^4\end{aligned}$$

Likewise, the PFA is

$$\mathrm{PFA}(\mathbf{v}_{\mathrm{I}} - \mathbf{u}^\star_{\mathrm{L}} \subset \mathbf{v}_{\mathrm{I}}) = |\mathbf{v}_{\mathrm{I}} - \mathbf{u}^\star_{\mathrm{L}}|/|\mathbf{v}_{\mathrm{I}}| = 1 - |\mathbf{u}^\star_{\mathrm{L}}|/|\mathbf{v}_{\mathrm{I}}| = 1 - 7/2^8$$

8.5.5 First neighbor branches

Probably the easiest nodes to identify are tree branches that share neighbors in $\mathbf{v}$. The number of branchlike nodes that are neighbors of $\mathbf{v}$ is not large. The probability that two branch neighbors happen to be the same node by chance is

$$|\mathbf{v}_{\mathrm{B}}| \approx |\mathbf{v}_{\mathrm{I}}|^2/N = 4(M/N)^2|\mathbf{v}|^2/N = 4(M/N)^2|\mathbf{t}|^2/N$$

The number of branch nodes in a tree is approximately $|\mathbf{t}|/n_T$ and the number of branch nodes in $\mathbf{u}^\star$ is $(1-f)|\mathbf{t}|/n_T$. A branch node will appear as a branch if two or more of its neighbors are in $\mathbf{v}$, which has a probability of $1 - (f^{n_T+1} + (n_T + 1)(1-f)^{n_T}f)$. Thus the number of branch nodes that will appear branchlike is

$$|\mathbf{u}^\star_{\mathrm{B}}| \approx [1 - (f^{n_T+1} + (n_T+1)(1-f)^{n_T}f)](1-f)|\mathbf{t}|/n_T$$

The corresponding SNR is

$$\begin{aligned}\mathrm{SNR}(\mathbf{u}^\star_\mathrm{B} \subset \mathbf{u}^\star_\mathrm{B} + \mathbf{v}_\mathrm{B}) &= |\mathbf{u}^\star_\mathrm{B}|/|\mathbf{v}_\mathrm{B}| \\ &= \frac{[1-(f^{n_T+1}+(n_T+1)(1-f)^{n_T}f)](1-f)|\mathbf{t}|/n_T}{4(M/N)^2|\mathbf{t}|^2/N} \\ &= \frac{[1-(f^{n_T+1}+(n_T+1)(1-f)^{n_T}f)](1-f)(N/|\mathbf{t}|)}{4(M/N)^2 f^2 n_T}\end{aligned}$$

For a nominal power law graph and tree, $\mathrm{SNR}(\mathbf{u}^\star_\mathrm{B} \subset \mathbf{u}^\star_\mathrm{B}| + \mathbf{v}_\mathrm{B}) = (N/|\mathbf{t}|)/2^9 = 2^4$.
The PD is

$$\begin{aligned}\mathrm{PD}(\mathbf{u}^\star_\mathrm{B} \subset \mathbf{u}^\star) &= |\mathbf{u}^\star_\mathrm{B}|/|\mathbf{u}^\star| \\ &= \frac{[1-(f^{n_T+1}+(n_T+1)(1-f)^{n_T}f)](1-f)|\mathbf{t}|/n_T}{(1-f)|\mathbf{t}|} \\ &= [1-(f^{n_T+1}+(n_T+1)(1-f)^{n_T}f)](N/n_T) = 2^{-2}\end{aligned}$$

Likewise, the PFA is

$$\begin{aligned}\mathrm{PFA}(\mathbf{v}_\mathrm{B} \subset \mathbf{u}^\star + \mathbf{v}_\mathrm{B}) &= |\mathbf{v}_\mathrm{B}|/|\mathbf{u}^\star + \mathbf{v}_\mathrm{B}| = [1 + |\mathbf{u}^\star|//\mathbf{v}_\mathrm{B}|]^{-1} \\ &= [1 + \mathrm{SNR}(\mathbf{u}^\star_\mathrm{B} \subset \mathbf{u}^\star_\mathrm{B} + \mathbf{v}_\mathrm{B})]^{-1} \\ &\approx 1/\mathrm{SNR}(\mathbf{u}^\star_\mathrm{B} \subset \mathbf{u}^\star_\mathrm{B} + \mathbf{v}_\mathrm{B}) = 2^{-4}\end{aligned}$$

8.5.6 SNR hierarchy

The analysis of the previous sections indicates that the vertices in $\mathbf{u}^\star$ can be grouped into several classes with very different SNRs (see Figure 8.5). These different classes indicate that certain vertices in $\mathbf{u}^\star$ should be easy to detect while others will be quite difficult. The filtering approach described subsequently targets the higher SNR groups $\mathbf{u}^\star_\mathrm{L}$ and $\mathbf{u}^\star_\mathrm{B}$.

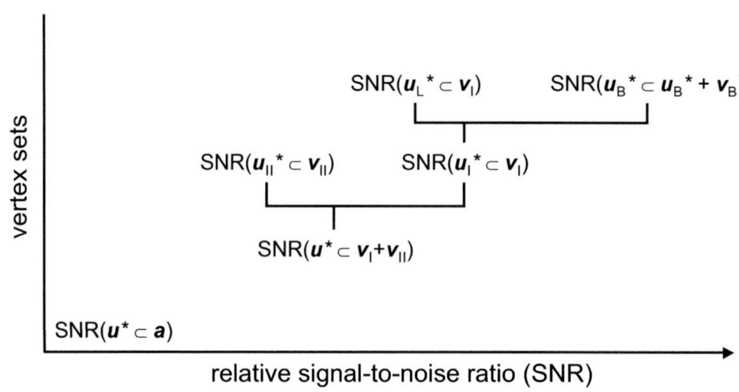

Figure 8.5. SNR hierarchy.
The SNR of different sets of unknown tree vertices $\mathbf{u}^\star$ as they are selected from the background.

8.6 Linear filter

This section presents a simple linear filtering approach for finding $\mathbf{u}^\star$ and follows the steps shown in the previous section. At each step, the goal is to increase the SNR until the SNR is sufficiently high that the probability of finding vertices in $\mathbf{u}^\star$ is good.

The algorithm relies on the fact that the maximum degree of a node in the tree is $n_T + 1$. This approach is chosen—as opposed to an approach using more tree-specific attributes—because in real applications a tree is most likely an approximation to the subgraph of interest. Thus, the algorithm is applicable to the class of subgraphs that have a maximum vertex degree.

The SNR analysis relied on a linear model of the background and assumed that all vertices had on average M/N edges. In reality, the degree distribution of the background is a power law. At several stages in the algorithm, adjustments will be made to eliminate the occasional high degree vertex.

The boundary between linear and nonlinear detection can be a little fuzzy. In this case, linear detection is characterized by not using nonlinear operations (e.g., $\mathbf{A}^2$) or recursive global optimization techniques to find the tree.

8.6.1 Find nearest neighbors

To begin, recall $\mathbf{t} = \mathbf{v} + \mathbf{u}^\star$. Step 0 is to find all nearest neighbors $\mathbf{u}_0$ of $\mathbf{t}$ by computing the adjacency matrices

$$\mathbf{A}_{\mathbf{v}\times\mathbf{u}_0} = \mathbf{I}_\mathbf{v}\ \mathbf{A} - \mathbf{A}_{\mathbf{v}^2}, \qquad \mathbf{A}_{\mathbf{u}_0\times\mathbf{v}} = \mathbf{A}\ \mathbf{I}_\mathbf{v} - \mathbf{A}_{\mathbf{v}^2}$$

The degree of the neighboring vertices with edges into $\mathbf{v}$ is then

$$\mathbf{d}_{\mathbf{u}_0\times\mathbf{v}} = (\mathbf{A}_{\mathbf{u}_0\times\mathbf{v}} + \mathbf{A}'_{\mathbf{v}\times\mathbf{u}_0})\ \mathbf{1}_{N\times 1}$$

$\mathbf{u}_0$ can now be found by finding all nonzeros in $\mathbf{d}_{\mathbf{u}_0\times\mathbf{v}}$

$$\mathbf{u}_0 = (\mathbf{d}_{\mathbf{u}_0\times\mathbf{v}} > 0)$$

This step selects the first nearest neighbors of V (see Figure 8.6).

8.6.2 Eliminate high degree nodes

One property of a power law graph is the existence of very high degree nodes that cannot be accounted for in linear theory. Likewise, a property of trees is that the maximum vertex degree is small and can be approximated by $d^{max}_{\mathbf{t}^2} \approx d^{max}_{\mathbf{v}^2}$. Step 0 is used to exploit this constraint. In general, step 0 is highly nonlinear in that it mostly eliminates a very few candidate vertices, but occasionally eliminates a large number.

Step 1 eliminates from $\mathbf{u}_0$ those vertices whose number of edges with $\mathbf{v}$ is greater than $d^{max}_{\mathbf{v}^2}$

$$\mathbf{u}_1 = (\mathbf{d}_{\mathbf{u}_0\times\mathbf{v}} \leq d^{max}_{\mathbf{v}^2})\ .*\ \mathbf{u}_0$$

This step eliminates some very high degree vertices that are chance neighbors of $\mathbf{v}$ (see gray vertical bar at far right of Figure 8.7).

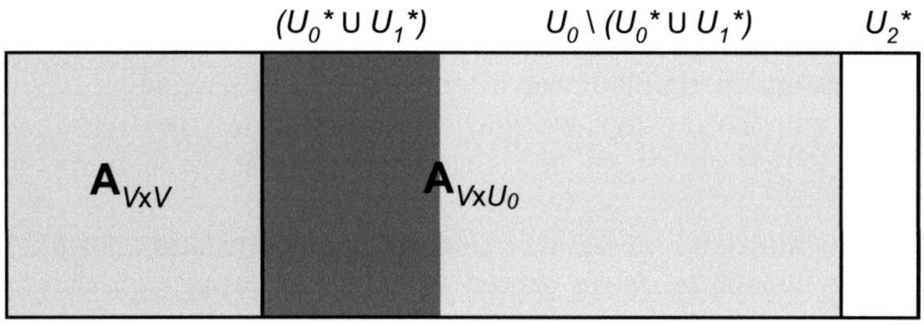

Figure 8.6. Tree filter step 0.
U_0 is the set of vertices that are neighbors of V. U_0 includes $U_0^\star \cup U_1^\star$ (first neighbors) but not $U_2^\star$ (second neighbors).

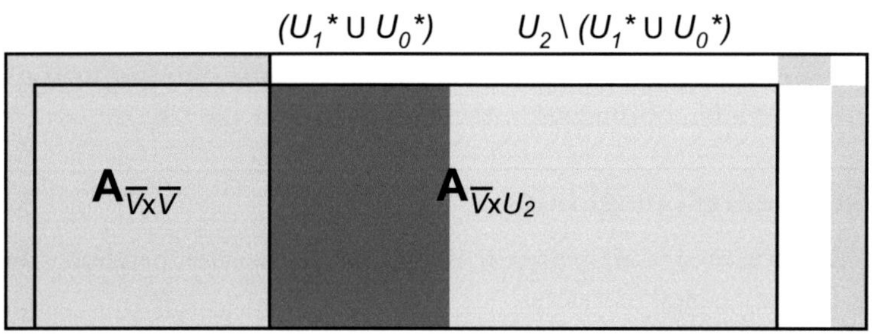

Figure 8.7. Tree filter steps 1a and 1b.
$\bar{V}$ are vertices with available edges (i.e., $\mathbf{d}_{V^2} < \mathbf{d}_{V^2}^{max}$). U_1 are vertices with available edges (i.e., $\mathbf{d}_{U_0 \times V} \leq d_{V^2}^{max}$). Gray vertices are eliminated by applying these constraints. U_2 are those vertices that remain.

8.6.3 Eliminate occupied nodes

Step 2 eliminates from $\mathbf{v}$ those vertices that are completely occupied

$$\bar{\mathbf{v}} = \mathbf{v} - (\mathbf{d}_{\mathbf{v}^2} \geq d_{\mathbf{v}^2}^{max})$$

Adding a neighbor to any node $\bar{\mathbf{v}}$ would cause the degree of the node to exceed $d_{\mathbf{v}^2}^{max}$. $\mathbf{u}_2$ is then computed by restricting the known tree vertices to $\bar{\mathbf{v}}$

$$\mathbf{u}_2 = (\mathbf{d}_{\mathbf{u}_1 \times \bar{\mathbf{v}}} > 0)$$

This step will eliminate the occasional interior tree vertex in $\mathbf{v}$, all of whose neighbors are already in $\mathbf{v}$.

8.6.4 Find high probability nodes

Step 3 computes the probability that a given neighbor is a part of the tree by computing how many neighbor slots each vertex in $\bar{V}$ has available $(d_{\mathbf{v}^2}^{max} - \mathbf{d}_{\mathbf{v}^2})$.

The number of available slots is then divided by the number of vertices in U_2 that could potentially fill those slots ($\mathbf{d}_{\bar{\mathbf{v}}\times\mathbf{u}_2}$)

$$\mathbf{w}_{\bar{\mathbf{v}}\times\mathbf{u}_2} = \max((d^{max}_{\mathbf{v}^2} - \mathbf{d}_{\mathbf{v}^2}) \; .* \; \bar{\mathbf{v}}, 0) \; .\&/ \; \mathbf{d}_{\bar{\mathbf{v}}\times\mathbf{u}_2}$$

Note: max and .&/ are used to deal with out-of-bounds numerators (i.e., negative) and denominators (i.e., equal to zero). Sorting the above weights can be used to make an ordered ranking of each vertex in U_2

$$\mathbf{w}^{sort}_{\bar{\mathbf{v}}\times\mathbf{u}_2} = \mathrm{sort}(\mathbf{w}_{\bar{\mathbf{v}}\times\mathbf{u}_2})$$

Determining how many vertices should be selected can be done by estimating the number of edges V should have that are in T ($N_V \; d^{max}_{\mathbf{v}^2}/2$) and subtracting the edges that are actually observed $2 \; nnz(\mathbf{A}_{\mathbf{v}^2})$. Further restricting the number of vertices to the highest 2/3 eliminates the lowest 1/3 of the probability distribution of weights (that are likely to be erroneous). The result is an estimate for the number of vertices in U_3

$$N_{U_3} = \lceil (2/3)[N_V \; d^{max}_{\mathbf{v}^2}/2 - 2 \; nnz(\mathbf{A}_{\mathbf{v}^2})] \rceil$$

Selecting the top N_{U_3} of the sorted weights gives a threshold weight that can be used to select U_3

$$\mathbf{u}_3 = (\mathbf{w}_{\bar{\mathbf{v}}\times\mathbf{u}_2} > \mathbf{w}^{sort}_{\bar{\mathbf{v}}\times\mathbf{u}_2}(N - N_{U_3}))$$

8.6.5 Find high degree nodes

U_3 should contain many vertices in $U^\star$, but it will also contain many vertices that are not. The probability of a random vertex having edges to two or more vertices in T is small. Therefore, selecting the vertices in U_3 that are connected to more than one vertex reduces the set to vertices that are nearly all in $U^\star$.

Let T_3 be the current estimate of vertices in the tree

$$\mathbf{t}_3 = \mathbf{v} + \mathbf{u}_3$$

Then, U_4 is the set of vertices where U_3 has more than one edge in T_3

$$\mathbf{u}_4 = (\mathbf{d}_{\mathbf{t}_3^2} . * \mathbf{u}_3 > 1)$$

Finally, it is worth noting that, given $U^\star$, the number of correctly selected vertices can be computed at any stage U_i in the above process via the equation

$$N_{U_i \in U^\star} = \mathbf{u}_i' \; \mathbf{u}^\star$$

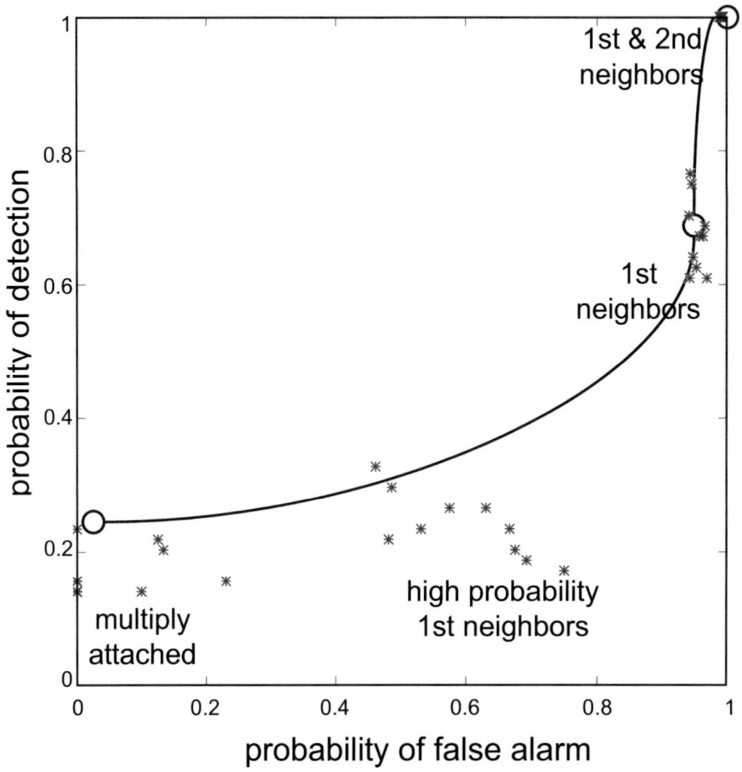

Figure 8.8. PD versus PFA.
Results from 10 Monte Carlo simulations of the tree detection algorithm are shown with crosses and the theoretical estimates are shown with circles.

8.7 Results and conclusions

The above tree-finding filter was applied to Monte Carlo simulations of a power law graph ($N = 2^{20}$, $M = 8N$) and a small tree ($|\mathbf{t}| = 2^7$), where half the vertices in the tree were given in V (i.e., $f = 1/2$). At each stage in the algorithm, the number of correct detections is calculated along with the estimated SNR. The PD and PFA are shown in Figure 8.8 along with the theoretical PD and PFA calculations. The results indicate that the algorithm is achieving close to what is theoretically possible. What is particularly interesting is that certain classes of vertices in the tree can be detected with virtually zero false alarms.

In conclusion, detecting subgraphs of interest in larger graphs is the goal of many graph analysis techniques. The basis of detection theory is computing the probability of a "foreground" with respect to a model of the "background" data. Combining these models allows estimates of the SNR, PD, and PFA for different classes of vertices in the foreground. These estimates can then be used to construct filters for computing the probability that a background graph contains a particular foreground graph. This approach is successfully applied to the problem of detecting a partially labeled tree graph in a power law background graph.

References

[Chakrabarti 2004] D. Chakrabarti, Y. Zhan, and C. Faloutsos. 2004. R-MAT: A recursive model for graph mining. In *Proceedings of the Fourth SIAM International Conference on Data Mining*, 442–446, SIAM, 2004. http://www.siam.org/proceedings/datamining/2004/dm04.php.

[Leskovec 2005] J. Leskovec, D. Chakrabarti, J. Kleinberg, C. Faloutsos. Realistic, mathematically tractable graph generation and evolution, using Kronecker multiplication. In *European Conference on Principles and Practice of Knowledge Discovery in Databases (ECML/PKDD 2005)*, Porto, Portugal, 2005.

[Furer & Raghavachari 1992] M. Furer and B. Raghavachari. Approximating the minimum degree spanning tree to within one from the optimal degree. In *Proceedings of the Third Annual ACM-SIAM Symposium on Discrete Algorithms* (SODA), 317–324, 1992.

[Van Loan 2000] C.F.V. Loan. The ubiquitous Kronecker product. *Journal of Computational and Applied Mathematics*, 123:85–100, 2000.

[Weinstein et al. 2009] C. Weinstein, W. Campbell, B. Delaney, and G. O'Leary. Modeling and detection techniques for counter-terror social network analysis and intent recognition. In *Proceedings of the 2009 IEEE Aerospace Conference*, 2009.

Part II

Data

Chapter 9

Kronecker Graphs

Jure Leskovec[*]

Abstract

How can we generate realistic networks? In addition, how can we do so with a mathematically tractable model that allows for rigorous analysis of network properties? Real networks exhibit a long list of surprising properties: heavy tails for the in- and out-degree distribution, heavy tails for the eigenvalues and eigenvectors, small diameters, and densification and shrinking diameters over time. We propose a generative model for networks that is mathematically tractable and produces the above-mentioned structural properties. Our model uses a standard matrix operation, the *Kronecker product*, to generate graphs, which we refer to as "Kronecker graphs," that are proved to naturally obey common network properties. Empirical evidence shows that Kronecker graphs can effectively model the structure of real networks. KRONFIT is a fast and scalable algorithm for fitting the Kronecker graph generation model to large real networks. KRONFIT takes *linear* time, by exploiting the structure of Kronecker matrix multiplication and by using statistical simulation techniques. Experiments on a range of networks show that KRONFIT finds parameters that mimic the properties of real networks. In fact, typically four parameters can accurately model several aspects of global network structure. Once fitted, the model parameters can be used to gain insights about the network structure, and the resulting synthetic graphs can be used for null-models, anonymization, extrapolations, and graph summarization.

*Computer Science Department, Stanford University, Stanford, CA 94305-9040 (`jure@cs.stanford.edu`).

Based on joint work with Deepayan Chakrabarti, Christos Faloutsos, Zoubin Ghahramani, and Jon Kleinberg.

9.1 Introduction

What do real graphs look like? How do they evolve over time? How can we generate synthetic, but realistic-looking, time-evolving graphs? Recently, network analysis has been attracting much interest, with an emphasis on finding patterns and abnormalities in social networks, computer networks, e-mail interactions, gene regulatory networks, and many more. Most of the work focuses on static snapshots of graphs, where fascinating "laws" have been discovered, including small diameters and heavy-tailed degree distributions.

In parallel with discoveries of such structural "laws," there has been work to find models of network formation that generate these structures. A good realistic network generation model is important for at least two reasons. The first is that such a model can generate graphs for extrapolations, hypothesis testing, "what-if" scenarios, and simulations, when real graphs are difficult or impossible to collect. For example, how well will a given protocol run on the Internet five years from now? Accurate network models can produce more realistic models for the future Internet, on which simulations can be run. The second reason is that these models require us to think about the network properties that generative models should obey to be realistic.

In this chapter, we introduce Kronecker graphs, a generative network model that obeys all the main static network patterns that have appeared in the literature; see, for instance, [Faloutsos et al. 1999, Albert et al. 1999, Chakrabarti et al. 2004, Farkas et al. 2001, Mihail & Papadimitriou 2002, Watts & Strogatz 1998]. Our model also obeys recently discovered temporal evolution patterns [Leskovec et al. 2005b, Leskovec et al. 2007a]. Contrary to other models that match this combination of network properties (as for example, [Bu & Towsley 2002, Klemm & Eguíluz 2002, Vázquez 2003, Leskovec et al. 2005b, Zheleva et al. 2009]), Kronecker graphs also lead to tractable analysis and rigorous proofs. Furthermore, the Kronecker graph generative process also has a nice natural interpretation and justification.

Our model is based on a matrix operation, the *Kronecker product*. There are several known theorems on Kronecker products. They correspond exactly to a significant portion of what we want to prove: heavy-tailed distributions for in-degree, out-degree, eigenvalues, and eigenvectors. We also demonstrate how Kronecker graphs can match the behavior of several real networks (social networks, citations, web, Internet, and others). While Kronecker products have been studied by the algebraic combinatorics community (see, e.g., [Chow 1997, Imrich 1998, Imrich & Klavžar 2000, Hammack 2009]), the present work is the first to employ this operation in the design of network models to match real data.

Then we also make a step further and tackle the following problem: given a large real network, we want to generate a synthetic graph, so that the resulting synthetic graph matches the properties of the real network as well as possible.

Ideally we would like (a) a graph generation model that *naturally* produces networks where many properties that are also found in real networks naturally emerge; (b) the model parameter estimation to be fast and scalable, so that we can handle networks with millions of nodes; and (c) the resulting set of parameters to generate realistic-looking networks that match the statistical properties of the target, real networks.

In general, the problem of modeling network structure presents several conceptual and engineering challenges. Which generative model should we choose, among the many in the literature? How do we measure the goodness of the fit? (Least squares do not work well for power laws.) If we use likelihood, how do we estimate it better than $O(N^2)$ complexity? How do we solve the node correspondence problem; i.e., which node of the real network corresponds to what node of the synthetic one?

To answer the above questions, we present KronFit, a fast and scalable algorithm for fitting Kronecker graphs by using the maximum-likelihood principle. When one calculates the likelihood, there are two challenges. First, one needs to solve the node correspondence problem by matching the nodes of the real and the synthetic network. Essentially, one has to consider all mappings of nodes of the network to the rows and columns of the graph adjacency matrix. This becomes intractable for graphs with more than tens of nodes. Even when given the "true" node correspondences, just evaluating the likelihood is still prohibitively expensive for large graphs. We present solutions to both of these problems. We develop a Metropolis sampling algorithm for sampling node correspondences, and we approximate the likelihood to obtain a *linear* time algorithm for Kronecker graph model parameter estimation that scales to large networks with millions of nodes and edges. KronFit gives orders of magnitude speedups against older methods.

Our extensive experiments on synthetic and real networks show that Kronecker graphs can efficiently model statistical properties of networks, such as degree distribution and diameter, while using only four parameters.

Once the model is fitted to the real network, there are several benefits and applications:

(a) *Network structure:* The parameters give us insight into the global structure of the network itself.

(b) *Null-model:* When working with network data, we would often like to assess the significance or the extent to which a certain network property is expressed. We can use a Kronecker graph as an accurate null-model.

(c) *Simulations:* Given an algorithm working on a graph, we would like to evaluate how its performance depends on various properties of the network. Using our model, one can generate graphs that exhibit various combinations of such properties and then evaluate the algorithm.

(d) *Extrapolations:* We can use the model to generate a larger graph to help us understand how the network will look in the future.

(e) *Sampling:* Conversely, we can also generate a smaller graph, which may be useful for running simulation experiments (e.g., simulating routing algorithms in computer networks, or virus/worm propagation algorithms) when these algorithms may be too slow to run on large graphs.

(f) *Graph similarity:* To compare the similarity of the structure of different networks (even of different sizes), one can use the differences in estimated parameters as a similarity measure.

(g) *Graph visualization and compression:* We can compress the graph, by storing just the model parameters, and the deviations between the real and the synthetic graph. Similarly, for visualization purposes, one can use the structure of the parameter matrix to visualize the backbone of the network, and then display the edges that deviate from the backbone structure.

(h) *Anonymization:* Suppose that the real graph cannot be publicized, e.g., corporate e-mail network or customer-product sales in a recommendation system. Yet, we would like to share our network. Our work gives ways to simulate such a realistic, "similar" network.

The present chapter builds on our previous work on Kronecker graphs (see [Leskovec et al. 2005a, Leskovec & Faloutsos 2007]) and is organized as follows. Section 9.2 briefly surveys the related literature. In Section 9.3, we introduce the Kronecker graph model and give formal statements about the properties of networks it generates. We investigate the model using simulations in Section 9.4 and continue by introducing KRONFIT, the Kronecker graphs parameter estimation algorithm, in Section 9.5. We present experimental results on a wide range of real and synthetic networks in Section 9.6. We close with discussion and conclusions in Sections 9.7 and 9.8.

9.2 Relation to previous work on network modeling

Networks across a wide range of domains present surprising regularities, such as power laws, small diameters, communities, and so on. We use these patterns as sanity checks; that is, our synthetic graphs should match those properties of the real target graph.

Most of the related work in this field has concentrated on two aspects: properties and patterns found in real-world networks, and then ways to find models to build understanding about the emergence of these properties. First, we will discuss the commonly found patterns in (static and temporally evolving) graphs, and finally, the state of the art in graph generation methods.

9.2.1 Graph patterns

Here we briefly introduce the network patterns (also referred to as properties or statistics) that we will later use to compare the similarity between the real networks and their synthetic counterparts produced by the Kronecker graphs model. While many patterns have been discovered, two of the principal ones are heavy-tailed degree distributions and small diameters.

Degree distribution: The degree distribution of a graph is a power law if the number of nodes N_d with degree d is given by $N_d \propto d^{-\gamma} (\gamma > 0)$, where

γ is called the power law exponent. Power laws have been found on the Internet [Faloutsos et al. 1999], the web [Kleinberg et al. 1999, Broder et al. 2000], citation graphs [Redner 1998], online social networks [Chakrabarti et al. 2004], and many others.

Small diameter: Most real-world graphs exhibit relatively small diameter (the "small- world" phenomenon, or "six degrees of separation" [Milgram 1967]). A graph has diameter D if every pair of nodes can be connected by a path of length at most D edges. The diameter D is susceptible to outliers. Thus, a more robust measure of the pairwise distances between nodes in a graph is the *integer effective diameter* [Tauro et al. 2001], which is the minimum number of links (steps/hops) in which some fraction (or quantile q, say $q = 0.9$) of all connected pairs of nodes can reach each other. Here we make use of *effective diameter*, which we define as follows [Leskovec et al. 2005b]. For each natural number h, let $g(h)$ denote the fraction of connected node pairs whose shortest connecting path has length at most h, i.e., at most h hops away. We then consider a function defined over all positive real numbers x by linearly interpolating between the points $(h, g(h))$ and $(h + 1, g(h+1))$ for each x, where $h = \lfloor x \rfloor$, and we define the *effective diameter* of the network to be the value x at which the function $g(x)$ achieves the value 0.9. The effective diameter has been found to be small for large real-world graphs, such as the Internet, web, and online social networks [Albert & Barabási 2002, Milgram 1967, Leskovec et al. 2005b].

Hop plot: It extends the notion of diameter by plotting the number of reachable pairs $g(h)$ within h hops, as a function of the number of hops h [Palmer et al. 2002]. It gives us a sense of how quickly nodes' neighborhoods expand with the number of hops.

Scree plot: This is a plot of the eigenvalues (or singular values) of the graph adjacency matrix, versus their rank, using the logarithmic scale. The scree plot is also often found to approximately obey a power law [Chakrabarti et al. 2004, Farkas et al. 2001]. Moreover, this pattern was also found analytically for random power law graphs [Chung et al. 2003, Mihail & Papadimitriou 2002].

Network values: The distribution of eigenvector components (indicators of "network value") associated with the largest eigenvalue of the graph adjacency matrix has also been found to be skewed [Chakrabarti et al. 2004].

Node triangle participation: Edges in real-world networks and especially in social networks tend to cluster [Watts & Strogatz 1998] and form triads of connected nodes. Node triangle participation is a measure of transitivity in networks. It counts the number of triangles a node participates in, i.e., the number of connections between the neighbors of a node. The plot of the number of triangles Δ versus the number of nodes that participate in Δ triangles has also been found to be skewed [Tsourakakis 2008].

Densification power law: The relation between the number of edges $M(t)$ and the number of nodes $N(t)$ in an evolving network at time t obeys the *densification power law* (DPL), which states that $M(t) \propto N(t)^a$. The *densification exponent* a is typically greater than 1, implying that the average degree of a node in the network is *increasing* over time (as the network gains more nodes and edges). Densification implies that real networks tend to sprout many more edges than nodes, and thus densify as they grow [Leskovec et al. 2005b, Leskovec et al. 2007a].

Shrinking diameter: The effective diameter of graphs tends to shrink or stabilize as the number of nodes in a network grows over time (see, for instance, [Leskovec et al. 2005b, Leskovec et al. 2007a]). Diameter shrinkage is somewhat counterintuitive since, from common experience, one would expect that as the volume of the object (a graph) grows, the size (i.e., the diameter) would also grow. But for real networks, this does not hold as the diameter shrinks and then seems to stabilize as the network grows.

9.2.2 Generative models of network structure

The earliest probabilistic generative model for graphs was the Erdős–Rényi random graph model [Erdős & A. Rényi 1960], where each pair of nodes has an identical, independent probability of being joined by an edge. The study of this model has led to a rich mathematical theory. However, as the model was not developed to model real-world networks, it produces graphs that fail to match real networks in a number of respects (for example, it does not produce heavy-tailed degree distributions).

The vast majority of recent network models involve some form of *preferential attachment* (see, for instance, [Barabási & Albert 1999, Albert & Barabási 2002, Winick & Jamin 2002, Kleinberg et al. 1999, Kumar et al. 1999, Flaxman et al. 2007]) that employs a simple rule: a new node joins the graph at each time step, and then creates a connection to an existing node u with a probability proportional to the degree of the node u. This rule creates a "rich get richer" phenomenon and power law tails in degree distribution. Typically, the diameter in this model grows slowly with the number of nodes N, violating the "shrinking diameter" property mentioned above.

There are many variations of preferential attachment model, all somehow employing the "rich get richer" type mechanism, e.g., the "copying model" (see [Kumar et al. 2000]), the "winner does not take all" model [Pennock et al. 2002], the "forest fire" model [Leskovec et al. 2005b], the "random surfer model" (see [Blum et al. 2006]), etc.

A different family of network methods strives for small diameter and local clustering in networks. Examples of such models include the Waxman generator [Waxman 1988] and the *small-world* model [Watts & Strogatz 1998]. Another family of models shows that heavy tails emerge if nodes try to optimize their connectivity under resource constraints; see [Carlson & Doyle 1999, Fabrikant et al. 2002].

In summary, most current models focus on modeling only one (static) network property. In addition, it is usually hard to analytically deduce properties of the network model.

9.2.3 Parameter estimation of network models

Until recently, relatively little effort was made to fit the above network models to real data. One of the difficulties is that most of the above models usually define a mechanism or a principle by which a network is constructed, and thus parameter estimation is either trivial or almost impossible.

Most work in estimating network models comes from the areas of social sciences, statistics, and social network analysis in which the *exponential random graphs*, also known as $p*$ model, were introduced [Wasserman & Pattison 1996]. The model essentially defines a log linear model over all possible graphs G, $p(G|\theta) \propto \exp(\theta^T s(G))$, where G is a graph, and s is a set of functions, that can be viewed as summary statistics for the structural features of the network. The $p*$ model usually focuses on "local" structural features of networks (e.g., characteristics of nodes that determine a presence of an edge, link reciprocity, etc.). As exponential random graphs have been very useful for modeling small networks, and individual nodes and edges, our goal here is different in the sense that we aim to accurately model the structure of the network as a whole. Moreover, we aim to model and estimate parameters of networks with millions of nodes while, even for graphs of small size (> 100 nodes), the number of model parameters in exponential random graphs usually becomes too large, and estimation is prohibitively expensive, both in terms of computational time and memory.

Regardless of a particular choice of a network model, a common theme when estimating the likelihood $P(G)$ of a graph G under some model is the challenge of finding the correspondence between the nodes of the true network and its synthetic counterpart. The node correspondence problem results in the factorially many possible matchings of nodes. One can think of the correspondence problem as a test of graph isomorphism. Two isomorphic graphs G and G' with differently assigned node IDs should have the same likelihood $P(G) = P(G')$, so we aim to find an accurate mapping between the nodes of the two graphs.

An ordering or a permutation defines the mapping of nodes in one network to nodes in the other network. For example, Butts [Butts 2005] used permutation sampling to determine similarity between two graph adjacency matrices, while Bezáková Kalai, and Santhanam [Bezáková et al. 2006] used permutations for graph model selection. Recently, an approach for estimating parameters of the "copying" model was introduced (see [Wiuf et al. 2006]); however, authors also noted that the class of "copying" models may not be rich enough to accurately model real networks. As we show later, the Kronecker graph model seems to have the necessary expressive power to mimic real networks well.

9.3 Kronecker graph model

The Kronecker graph model we propose here is based on a recursive construction. Defining the recursion properly is somewhat subtle, as a number of standard, related graph construction methods fail to produce graphs that densify according to the patterns observed in real networks, and they also produce graphs whose diameters increase. To produce densifying graphs with constant/shrinking diameter, and thereby match the qualitative behavior of a real network, we develop a procedure that is best described in terms of the *Kronecker product* of matrices.

9.3.1 Main idea

The main intuition behind the model (see Table 9.1) is to create self-similar graphs, recursively. We begin with an *initiator* graph $\mathbf{K}_1$, with N nodes and M edges,

Table 9.1. Table of symbols.

Symbol	Description
$\mathbf{G}$	Real network
N	Number of nodes in $\mathbf{G}$
M	Number of edges in $\mathbf{G}$
$\mathbf{K}$	Kronecker graph (synthetic estimate of G)
$\mathbf{K}_1$	Initiator of a Kronecker graphs
N	Number of nodes in initiator $\mathbf{K}_1$
M	Number of edges in $\mathbf{K}_1$ (the expected number of edges in $\mathcal{P}_1$, $M = \sum \theta_{ij}$)
$G \otimes H$	Kronecker product of adjacency matrices of graphs G and H
$\mathbf{K}_1^{\otimes k} = \mathbf{K}_k = K$	kth Kronecker power of $\mathbf{K}_1$
$\mathbf{K}_1[i,j]$	Entry at row i and column j of $\mathbf{K}_1$
$\Theta = \mathcal{P}_1$	Stochastic Kronecker initiator
$\mathcal{P}_1^{\otimes k} = \mathcal{P}_k = \mathcal{P}$	kth Kronecker power of $\mathcal{P}_1$
$\theta_{ij} = \mathcal{P}_1[i,j]$	Entry at row i and column j of $\mathcal{P}_1$
$p_{ij} = \mathcal{P}_k[i,j]$	Probability of an edge (i,j) in $\mathcal{P}_k$, i.e., entry at row i and column j of $\mathcal{P}_k$
$K = R(\mathcal{P})$	Realization of a stochastic Kronecker graph $\mathcal{P}$
$l(\Theta)$	Log-likelihood. Log-prob. that Θ generated real graph G, $\log P(\mathbf{G}\|\Theta)$
$\hat{\Theta}$	Parameters at maximum likelihood, $\hat{\Theta} = \operatorname{argmax}_{\Theta} P(G\|\Theta)$
σ	Permutation that maps node IDs of G to those of $\mathcal{P}$
a	Densification power law exponent, $M(t) \propto N(t)^a$
D	Diameter of a graph
N_c	Number of nodes in the largest weakly connected component of a graph
ω	Proportion of times `SwapNodes` permutation proposal distribution is used

and by recursion we produce successively larger graphs $\mathbf{K}_2, \mathbf{K}_3, \ldots$ such that the kth graph $\mathbf{K}_k$ is on $N_k = N^k$ nodes. If we want these graphs to exhibit a version of the densification power law [Leskovec et al. 2005b], then $\mathbf{K}_k$ should have $M_k = M^k$ edges. This is a property that requires some care in order to get right, as standard recursive constructions (for example, the traditional Cartesian product or the construction of Barabási [Barabási et al. 2001]) do not yield graphs satisfying the densification power law.

It turns out that the *Kronecker product* of two matrices is the right tool for this goal. The Kronecker product is defined as follows.

Definition 9.1 (Kronecker product of matrices). *Given two matrices* $\mathbf{A} = [a_{i,j}]$ *and* $\mathbf{B}$ *of sizes* $n \times m$ *and* $n' \times m'$*, respectively, the Kronecker product matrix* $\mathbf{C}$ *of dimensions* $(n \cdot n') \times (m \cdot m')$ *is given by*

$$\mathbf{C} = \mathbf{A} \otimes \mathbf{B} \doteq \begin{pmatrix} a_{1,1}\mathbf{B} & a_{1,2}\mathbf{B} & \ldots & a_{1,m}\mathbf{B} \\ a_{2,1}\mathbf{B} & a_{2,2}\mathbf{B} & \ldots & a_{2,m}\mathbf{B} \\ \vdots & \vdots & \ddots & \vdots \\ a_{n,1}\mathbf{B} & a_{n,2}\mathbf{B} & \ldots & a_{n,m}\mathbf{B} \end{pmatrix} \tag{9.1}$$

We then define the Kronecker product of two graphs simply as the Kronecker product of their corresponding adjacency matrices.

Definition 9.2 (Kronecker product of graphs [Weichsel 1962]). *If G and H are graphs with adjacency matrices $\boldsymbol{A}(G)$ and $\boldsymbol{A}(H)$, respectively, then the Kronecker product $G \otimes H$ is defined as the graph with adjacency matrix $\boldsymbol{A}(G) \otimes \boldsymbol{A}(H)$.*

Observation 1 (Edges in Kronecker-multiplied graphs).

$$\textit{Edge } (X_{ij}, X_{kl}) \in G \otimes H \textit{ iff } (X_i, X_k) \in G \textit{ and } (X_j, X_l) \in H$$

where X_{ij} and X_{kl} are nodes in $G \otimes H$, and X_i, X_j, X_k, and X_l are the corresponding nodes in G and H, as in Figure 9.1.

The last observation is crucial and deserves elaboration. Basically, each node in $G \otimes H$ can be represented as an ordered pair X_{ij}, with i a node of G and j a node of H, and with an edge joining X_{ij} and X_{kl} precisely when (X_i, X_k) is an edge of G and (X_j, X_l) is an edge of H. This is a direct consequence of the hierarchical nature of the Kronecker product. Figure 9.1(a)–(c) further illustrates this by showing the recursive construction of $G \otimes H$, when $G = H$ is a 3-node chain. Consider node $X_{1,2}$ in Figure 9.1(c): it belongs to the H graph that replaced node X_1 (see Figure 9.1(b)) and, in fact, is the X_2 node (i.e., the center) within this small H graph.

We propose to produce a growing sequence of matrices by iterating the Kronecker product.

Definition 9.3 (Kronecker power). *The kth power of $\mathbf{K}_1$ is defined as the matrix $\mathbf{K}_1^{\otimes k}$ (abbreviated to $\mathbf{K}_k$), such that*

$$\mathbf{K}_1^{\otimes k} = \mathbf{K}_k = \underbrace{\mathbf{K}_1 \otimes \mathbf{K}_1 \otimes \cdots \otimes \mathbf{K}_1}_{k \textit{ times}} = \mathbf{K}_{k-1} \otimes \mathbf{K}_1$$

Definition 9.4 (Kronecker graph). *Kronecker graph of order k is defined by the adjacency matrix $\mathbf{K}_1^{\otimes k}$, where $\mathbf{K}_1$ is the Kronecker initiator adjacency matrix.*

The self-similar nature of the Kronecker graph product is clear: to produce $\mathbf{K}_k$ from $\mathbf{K}_{k-1}$, we "expand" (replace) each node of $\mathbf{K}_{k-1}$ by converting it into a copy of $\mathbf{K}_1$, and we join these copies together according to the adjacencies in $\mathbf{K}_{k-1}$ (see Figures 9.1, 9.2, and 9.3). This process is very natural: one can imagine it as positing that communities within the graph grow recursively, with nodes in the community recursively getting expanded into miniature copies of the community. Nodes in the subcommunity then link among themselves and also to nodes from other communities.

Note that there are many different names to refer to the Kronecker product of graphs. Other names for the Kronecker product are tensor product, categorical product, direct product, cardinal product, relational product, conjunction, weak direct product or just product, and even Cartesian product [Imrich & Klavžar 2000].

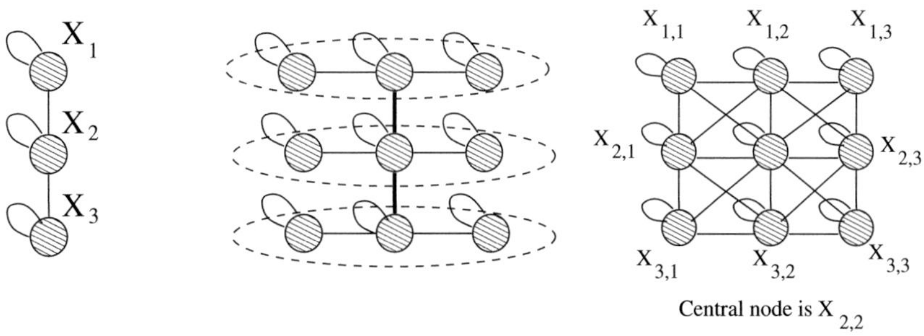

(a) Graph $\mathbf{K}_1$ (b) Intermediate stage (c) Graph $\mathbf{K}_2 = \mathbf{K}_1 \otimes \mathbf{K}_1$

1	1	0
1	1	1
0	1	1

(d) Adjacency matrix of $\mathbf{K}_1$

K_1	K_1	0
K_1	K_1	K_1
0	K_1	K_1

(e) Adjacency matrix of $\mathbf{K}_2 = \mathbf{K}_1 \otimes \mathbf{K}_1$

Figure 9.1. Example of Kronecker multiplication.
Top: a "3-chain" initiator graph and its Kronecker product with itself. Each of the X_i nodes gets expanded into 3 nodes, which are then linked using Observation 1. Bottom row: the corresponding adjacency matrices. See Figure 9.2 for adjacency matrices of $\mathbf{K}_3$ and $\mathbf{K}_4$.

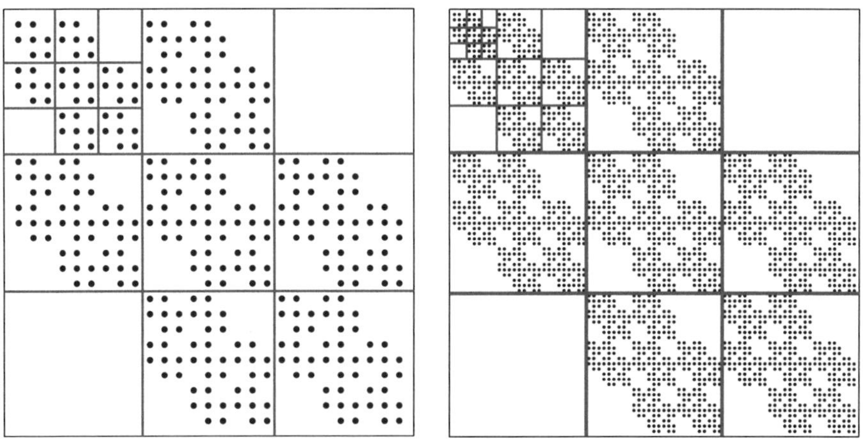

(a) $\mathbf{K}_3$ adjacency matrix (27×27) (b) $\mathbf{K}_4$ adjacency matrix (81×81)

Figure 9.2. Adjacency matrices of $\mathbf{K}_3$ and $\mathbf{K}_4$.
The third and fourth Kronecker power of $\mathbf{K}_1$ matrix as defined in Figure 9.1. Dots represent nonzero matrix entries, and white space represents zeros. Notice the recursive self-similar structure of the adjacency matrix.

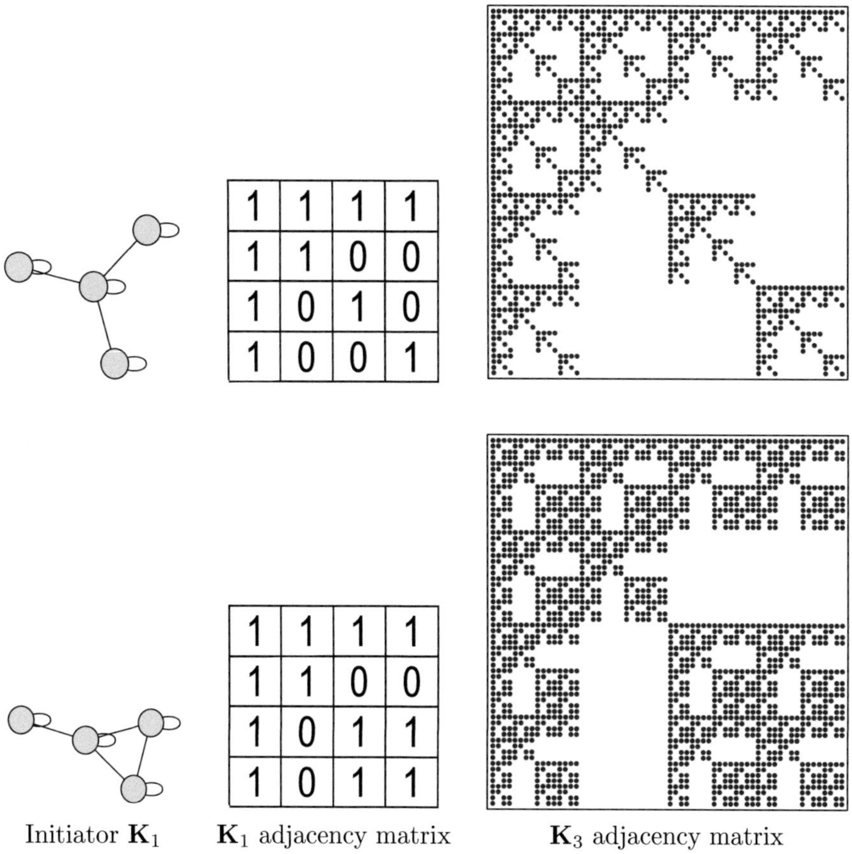

Figure 9.3. Self-similar adjacency matrices.
Two examples of Kronecker initiators on 4 nodes and the self-similar adjacency matrices they produce.

9.3.2 Analysis of Kronecker graphs

We shall now discuss the properties of Kronecker graphs, specifically, their degree distributions, diameters, eigenvalues, eigenvectors, and time evolution. The ability to prove analytical results about all of these properties is a major advantage of Kronecker graphs.

Degree distribution

The next few theorems prove that several distributions of interest are *multinomial* for our Kronecker graph model. This is important, because a careful choice of the initial graph $\mathbf{K}_1$ makes the resulting multinomial distribution behave like a power law or discrete Gaussian exponential (DGX) distribution [Bi et al. 2001, Clauset et al. 2007].

Theorem 9.5 (Multinomial degree distribution). *Kronecker graphs have multinomial degree distributions, for both in- and out-degrees.*

Proof. Let the initiator $\mathbf{K}_1$ have the degree sequence $d_1, d_2, \ldots, d_{N_1}$. Kronecker multiplication of a node with degree d expands it into N_1 nodes, with the corresponding degrees being $d \times d_1, d \times d_2, \ldots, d \times d_{N_1}$. After Kronecker powering, the degree of each node in graph $\mathbf{K}_k$ is of the form $d_{i_1} \times d_{i_2} \times \cdots \times d_{i_k}$, with $i_1, i_2, \ldots, i_k \in (1, \ldots, N_1)$, and there is one node for each ordered combination. This gives us the multinomial distribution on the degrees of $\mathbf{K}_k$. So, graph $\mathbf{K}_k$ will have multinomial degree distribution where the "events" (degrees) of the distribution will be combinations of degree products: $d_1^{i_1} d_2^{i_2} \cdots d_{N_1}^{i_{N_1}}$ (where $\sum_{j=1}^{N_1} i_j = k$) and event (degree) probabilities will be proportional to $\binom{k}{i_1 i_2 \ldots i_{N_1}}$. Note also that this is equivalent to noticing that the degrees of nodes in $\mathbf{K}_k$ can be expressed as the kth Kronecker power of the vector $(d_1, d_2, \ldots, d_{N_1})$. □

Spectral properties

Next we analyze the spectral properties of an adjacency matrix of a Kronecker graph. We show that both the distribution of eigenvalues and the distribution of component values of eigenvectors of the graph adjacency matrix follow multinomial distributions.

Theorem 9.6 (Multinomial eigenvalue distribution). *The Kronecker graph $\mathbf{K}_k$ has a multinomial distribution for its eigenvalues.*

Proof. Let $\mathbf{K}_1$ have the eigenvalues $\lambda_1, \lambda_2, \ldots, \lambda_{N_1}$. By properties of the Kronecker multiplication [Van Loan 2000, Langville & Stewart 2004], the eigenvalues of $\mathbf{K}_k$ are the kth Kronecker power of the vector of eigenvalues of the initiator matrix, $(\lambda_1, \lambda_2, \ldots, \lambda_{N_1})^{\otimes k}$. As in Theorem 9.5, the eigenvalue distribution is a multinomial. □

A similar argument using properties of Kronecker matrix multiplication shows the following.

Theorem 9.7 (Multinomial eigenvector distribution). *The components of each eigenvector of the Kronecker graph $\mathbf{K}_k$ follow a multinomial distribution.*

Proof. Let $\mathbf{K}_1$ have the eigenvectors $\vec{v}_1, \vec{v}_2, \ldots, \vec{v}_{N_1}$. By properties of the Kronecker multiplication [Van Loan 2000, Langville & Stewart 2004], the eigenvectors of $\mathbf{K}_k$ are given by the kth Kronecker power of the vector: $(\vec{v}_1, \vec{v}_2, \ldots, \vec{v}_N)$, which gives a multinomial distribution for the components of each eigenvector in $\mathbf{K}_k$. □

We have just covered several of the static graph patterns. Notice that the proofs were a direct consequences of the Kronecker multiplication properties.

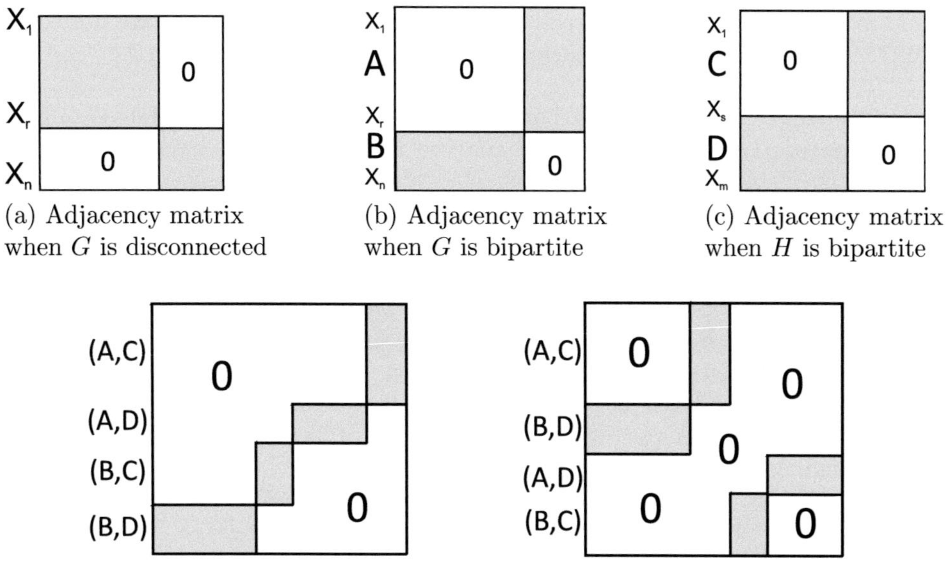

(a) Adjacency matrix when G is disconnected

(b) Adjacency matrix when G is bipartite

(c) Adjacency matrix when H is bipartite

(d) Kronecker product of two bipartite graphs G and H

(e) Rearranged adjacency matrix from panel (d)

Figure 9.4. Graph adjacency matrices.
Dark parts represent connected (filled with ones) and white parts represent empty (filled with zeros) parts of the adjacency matrix. (a) When G is disconnected, Kronecker multiplication with any matrix H will result in $G \otimes H$ being disconnected. (b) Adjacency matrix of a connected bipartite graph G with node partitions A and B. (c) Adjacency matrix of a connected bipartite graph G with node partitions C and D. (e) Kronecker product of two bipartite graphs G and H. (d) After rearranging the adjacency matrix $G \otimes H$, we clearly see that the resulting graph is disconnected.

Connectivity of Kronecker graphs

We now present a series of results on the connectivity of Kronecker graphs. We show, maybe a bit surprisingly, that even if a Kronecker initiator graph is connected, its Kronecker power can, in fact, be disconnected.

Lemma 9.8. *If at least one of G and H is a disconnected graph, then $G \otimes H$ is also disconnected.*

Proof. Without loss of generality, we can assume that G has two connected components, while H is connected. Figure 9.4(a) illustrates the corresponding adjacency matrix for G. Using the notation from Observation 1, let graph G have nodes $X_1, \ldots, X_n$, where nodes $\{X_1, \ldots X_r\}$ and $\{X_{r+1}, \ldots, X_n\}$ form the two connected components. Now, note that $(X_{ij}, X_{kl}) \notin G \otimes H$ for $i \in \{1, \ldots, r\}$, $k \in \{r+1, \ldots, n\}$, and for all j, l. This follows directly from Observation 1 as (X_i, X_k) are not edges in G. Thus, $G \otimes H$ must have at least two connected components. □

Actually it turns out that both G and H can be connected while $G \otimes H$ is disconnected. The following theorem analyzes this case.

Theorem 9.9. *If both G and H are connected but bipartite, then $G \otimes H$ is disconnected, and each of the two connected components is again bipartite.*

Proof. Without loss of generality, let G be bipartite with two partitions $A = \{X_1, \ldots, X_r\}$ and $B = \{X_{r+1}, \ldots, X_n\}$, where edges exist only between the partitions and no edges exist inside the partition: $(X_i, X_k) \notin G$ for $i, k \in A$ or $i, k \in B$. Similarly, let H also be bipartite with two partitions $C = \{X_1, \ldots, X_s\}$ and $D = \{X_{s+1}, \ldots, X_m\}$. Figures 9.4(b) and 9.4(c) illustrate the structure of the corresponding adjacency matrices.

Now, there will be two connected components in $G \otimes H$: the first component will be composed of nodes $\{X_{ij}\} \in G \otimes H$, where $(i \in A, j \in D)$ or $(i \in B, j \in C)$. And similarly, the second component will be composed of nodes $\{X_{ij}\}$, where $(i \in A, j \in C)$ or $(i \in B, j \in D)$. Basically, there exist edges between node sets (A, D) and (B, C), and similarly between (A, C) and (B, D) but not across the sets. To see this, we have to analyze the cases using Observation 1. For example, in $G \otimes H$ there exist edges between nodes (A, C) and (B, D) as there exist edges $(i, k) \in G$ for $i \in A, k \in B$, and $(j, l) \in H$ for $j \in C$ and $l \in D$. Similarly, it is true for nodes (A, C) and (B, D). However, no edges cross the two sets, e.g., nodes from (A, D) do not link to (A, C), as there are no edges between nodes in A (since G is bipartite). See Figures 9.4(d) and 9.4(e) for a visual proof. □

Note that bipartite graphs are triangle free and have no self-loops. Stars, chains, trees, and cycles of even length are all examples of bipartite graphs. In order to ensure that $\mathbf{K}_k$ is connected, for the remainder of the chapter we focus on initiator graphs $\mathbf{K}_1$ with self-loops on all of the vertices.

Temporal properties of Kronecker graphs

We continue with the analysis of temporal patterns of evolution of Kronecker graphs: the densification power law and shrinking/stabilizing diameter (see, for instance, [Leskovec et al. 2005b, Leskovec et al. 2007a]).

Theorem 9.10 (Densification power law). *Kronecker graphs follow the densification power law (DPL) with densification exponent* $a = \log(M)/\log(N)$.

Proof. Since the kth Kronecker power $\mathbf{K}_k$ has $N_k = N^k$ nodes and $M_k = M^k$ edges, it satisfies $M_k = N_k^a$, where $a = \log(M)/\log(N)$. The crucial point is that this exponent a is independent of k, and hence the sequence of Kronecker powers follows an exact version of the densification power law. □

We now show how the Kronecker product also preserves the property of constant diameter, a crucial ingredient for matching the diameter properties of many real-world network data sets. In order to establish this, we will assume that the initiator graph $\mathbf{K}_1$ has a self-loop on every node. Otherwise, its Kronecker powers may be disconnected.

Lemma 9.11. *If G and H each have diameter at most D and each has a self-loop on every node, then the Kronecker graph $G \otimes H$ also has diameter at most D.*

Proof. Each node in $G \otimes H$ can be represented as an ordered pair (v, w), with v a node of G and w a node of H, and with an edge joining (v, w) and (x, y) precisely when (v, x) is an edge of G and (w, y) is an edge of H. (Note this is exactly Observation 1.) Now, for an arbitrary pair of nodes (v, w) and (v', w'), we must show that there is a path of length at most D connecting them. Since G has diameter at most D, there is a path $v = v_1, v_2, \ldots, v_r = v'$, where $r \leq D$. If $r < D$, we can convert this into a path $v = v_1, v_2, \ldots, v_D = v'$ of length exactly D by simply repeating v' at the end for $D - r$ times. By an analogous argument, we have a path $w = w_1, w_2, \ldots, w_D = w'$. Now by the definition of the Kronecker product, there is an edge joining (v_i, w_i) and (v_{i+1}, w_{i+1}) for all $1 \leq i \leq D - 1$, and so $(v, w) = (v_1, w_1), (v_2, w_2), \ldots, (v_D, w_D) = (v', w')$ is a path of length D connecting (v, w) to (v', w'), as required. □

Theorem 9.12. *If $\mathbf{K}_1$ has diameter D and a self-loop on every node, then for every k, the graph $\mathbf{K}_k$ also has diameter D.*

Proof. This follows directly from the previous lemma, combined with induction on k. □

Define the q-effective diameter as the minimum D^* such that, for a q fraction of the reachable node pairs, the path length is at most D^*. The q-effective diameter is a more robust quantity than the diameter, the latter being prone to the effects of degenerate structures in the graph, e.g., very long chains). However, the q-effective diameter and diameter tend to exhibit qualitatively similar behavior. For reporting results in subsequent sections, we will generally consider the q-effective diameter with $q = 0.9$ and refer to this simply as the *effective diameter.*

Theorem 9.13 (Effective diameter). *If $\mathbf{K}_1$ has diameter D and a self-loop on every node, then for every q, the q-effective diameter of $\mathbf{K}_k$ converges to D (from below) as k increases.*

Proof. To prove this, it is sufficient to show that for two randomly selected nodes of $\mathbf{K}_k$, the probability that their distance is D converges to 1 as k goes to infinity.

We establish this as follows. Each node in $\mathbf{K}_k$ can be represented as an ordered sequence of k nodes from $\mathbf{K}_1$, and we can view the random selection of a node in $\mathbf{K}_k$ as a sequence of k independent random node selections from $\mathbf{K}_1$. Suppose that $v = (v_1, \ldots, v_k)$ and $w = (w_1, \ldots, w_k)$ are two such randomly selected nodes from $\mathbf{K}_k$. Now, if x and y are two nodes in $\mathbf{K}_1$ at distance D (such a pair (x, y) exists since $\mathbf{K}_1$ has diameter D), then with probability $1 - (1 - \frac{1}{N^2})^k$, there is some index j for which $\{v_j, w_j\} = \{x, y\}$. If there is such an index, then the distance between v and w is D. As the expression $1 - (1 - \frac{1}{N^2})^k$ converges to 1 as k increases, it follows that the q-effective diameter is converging to D. □

9.3.3 Stochastic Kronecker graphs

While the Kronecker power construction discussed so far yields graphs with a range of desired properties, its discrete nature produces "staircase effects" in the degrees and spectral quantities, simply because individual values have large multiplicities. For example, degree distribution and distribution of eigenvalues of graph adjacency matrix and the distribution of the principal eigenvector components (i.e., the "network" value) are all impacted by this. These quantities are multinomially distributed, leading to individual values with large multiplicities. Figure 9.5 illustrates the staircase effect.

Here we propose a stochastic version of Kronecker graphs that eliminates this effect. There are many possible ways one could introduce stochasticity into the Kronecker graph model. Before introducing the proposed model, we introduce two simple ways of introducing randomness to Kronecker graphs and describe why they do not work.

Probably the simplest (but wrong) idea is to generate a large deterministic Kronecker graph $\mathbf{K}_k$, and then uniformly at random flip some edges, i.e., uniformly at random select entries of the graph adjacency matrix and flip them ($1 \to 0, 0 \to 1$). However, this will not work, as it will essentially superimpose an Erdős–Rényi random graph, which would, for example, corrupt the degree distribution—real networks usually have heavy-tailed degree distributions while random graphs have binomial degree distributions. A second idea could be to allow a weighted initiator matrix, i.e., values of entries of $\mathbf{K}_1$ are not restricted to values $\{0, 1\}$ but rather can be any nonnegative real number. Using such $\mathbf{K}_1$, one would generate $\mathbf{K}_k$ and then threshold the $\mathbf{K}_k$ matrix to obtain a binary adjacency matrix K, i.e., for a chosen value of ϵ set $K[i, j] = 1$ if $\mathbf{K}_k[i, j] > \epsilon$ else $K[i, j] = 0$. This mechanism would selectively remove edges, and low degree nodes would get isolated first.

Now we define the *stochastic Kronecker graph model*, which overcomes the above issues. A more natural way to introduce stochasticity to Kronecker graphs is to relax the assumption that entries of the initiator matrix take only binary values. Instead, we allow entries of the initiator to take values on the interval $[0, 1]$. Now each entry of the initiator matrix encodes the probability of that particular edge appearing. We then Kronecker-power such an initiator matrix to obtain a large stochastic adjacency matrix, where again each entry of the large matrix gives the probability of that particular edge appearing in a big graph. Such a stochastic adjacency matrix defines a probability distribution over all graphs. To obtain a graph, we simply sample an instance from this distribution by sampling individual edges, where each edge appears independently with probability given by the entry of the large stochastic adjacency matrix. More formally, we define the Stochastic Kronecker graph as follows.

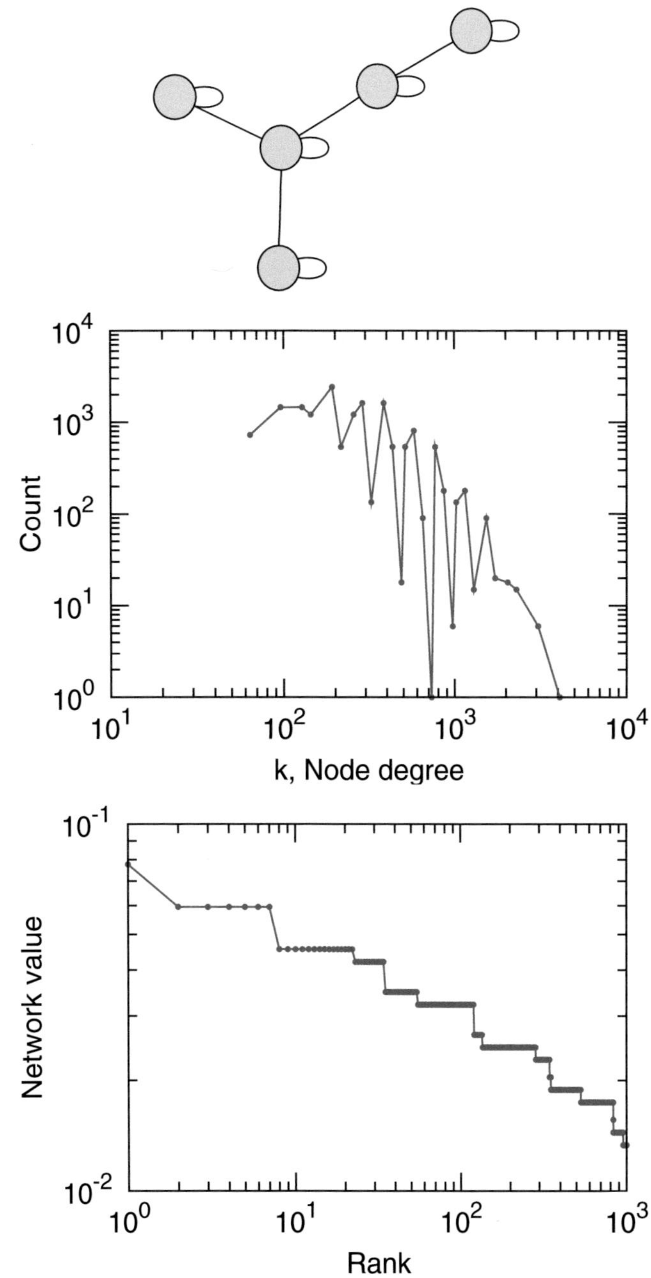

Figure 9.5. The "staircase" effect.
Top: Kronecker initiator $\mathbf{K}_1$. Middle: degree distribution of $\mathbf{K}_6$ (6th Kronecker power of $\mathbf{K}_1$). Bottom: network value of $\mathbf{K}_6$ (6th Kronecker power of $\mathbf{K}_1$). Notice the nonsmoothness of the curves.

Definition 9.14 (Stochastic Kronecker graph). *Let* $\mathcal{P}_1$ *be an* $N \times N$ probability matrix*: the value* $\theta_{ij} \in \mathcal{P}_1$ *denotes the probability that edge* (i,j) *is present,* $\theta_{ij} \in [0,1]$.

Then the kth Kronecker power $\mathcal{P}_1^{\otimes k} = \mathcal{P}_k$*, where each entry* $p_{uv} \in \mathcal{P}_k$ *encodes the probability of an edge* (u,v).

To obtain a graph, an instance *(or* realization*)* $K = R(\mathcal{P}_k)$*, we include edge* (u,v) *in* K *with probability* p_{uv}*,* $p_{uv} \in \mathcal{P}_k$.

First, note that the sum of the entries of $\mathcal{P}_1$, $\sum_{ij} \theta_{ij}$, can be greater than 1. Second, notice that in principle it takes $O(N^{2k})$ time to generate an instance K of a stochastic Kronecker graph from the probability matrix $\mathcal{P}_k$. This means the time to get a realization K is quadratic in the size of $\mathcal{P}_k$ as one has to flip a coin for each possible edge in the graph. Later we show how to generate stochastic Kronecker graphs much faster, in the time *linear* in the expected number of edges in $\mathcal{P}_k$.

Probability of an edge

For the size of graphs we aim to model and generate here, taking $\mathcal{P}_1$ (or $\mathbf{K}_1$) and then explicitly performing the Kronecker product of the initiator matrix is infeasible. The reason is that $\mathcal{P}_1$ is usually dense, so $\mathcal{P}_k$ is also dense and one cannot explicitly store it in memory to directly iterate the Kronecker product. However, due to the structure of Kronecker multiplication, one can easily compute the probability of an edge in $\mathcal{P}_k$.

The probability p_{uv} of an edge (u,v) occurring in kth Kronecker power $\mathcal{P} = \mathcal{P}_k$ can be calculated in $O(k)$ time as follows

$$p_{uv} = \prod_{i=0}^{k-1} \mathcal{P}_1 \left[\left\lfloor \frac{u-1}{N^i} \right\rfloor (\mathrm{mod} N) + 1, \left\lfloor \frac{v-1}{N^i} \right\rfloor (\mathrm{mod} N) + 1 \right] \tag{9.2}$$

The equation imitates recursive descent into the matrix $\mathcal{P}$, where at every level i the appropriate entry of $\mathcal{P}_1$ is chosen. Since $\mathcal{P}$ has N^k rows and columns, it takes $O(k \log N)$ to evaluate the equation. Refer to Figure 9.6 for the illustration of the recursive structure of $\mathcal{P}$.

9.3.4 Additional properties of Kronecker graphs

Stochastic Kronecker graphs with initiator matrix of size $N = 2$ were studied by Mahdian and Xu [Mahdian & Xu 2007]. The authors showed a phase transition for the emergence of the giant component and another phase transition for connectivity, and they proved that such graphs have constant diameters beyond the connectivity threshold, but are not searchable using a decentralized algorithm [Kleinberg 1999].

A general overview of the Kronecker product is given in [Imrich & Klavžar 2000], and properties of Kronecker graphs related to graph minors, planarity, cut vertex,

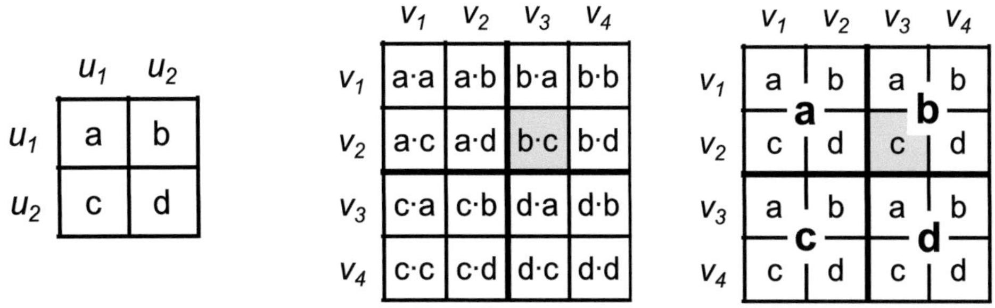

(a) 2×2 stochastic Kronecker initiator $\mathcal{P}_1$

(b) Probability matrix $\mathcal{P}_2 = \mathcal{P}_1 \otimes \mathcal{P}_1$

(c) Alternative view of $\mathcal{P}_2 = \mathcal{P}_1 \otimes \mathcal{P}_1$

Figure 9.6. Stochastic Kronecker initiator. The initiator matrix $\mathcal{P}_1$ and the corresponding 2nd Kronecker power $\mathcal{P}_2$. Notice the recursive nature of the Kronecker product, with edge probabilities in $\mathcal{P}_2$ simply being products of entries of $\mathcal{P}_1$.

and cut edge have been explored in [Bottreau & Metivier 1998]. Moreover, recently Tsourakakis [Tsourakakis 2008] gave a closed form expression for the number of triangles in a Kronecker graph that depends on the eigenvalues of the initiator graph $\mathbf{K}_1$.

9.3.5 Two interpretations of Kronecker graphs

Next, we present two natural interpretations of the generative process behind the Kronecker graphs that go beyond the purely mathematical construction of Kronecker graphs as introduced so far.

We already mentioned the first interpretation when we first defined Kronecker graphs. One intuition is that networks are hierarchically organized into communities (clusters). Communities then grow recursively, creating miniature copies of themselves. Figure 9.1 depicts the process of the recursive community expansion. In fact, several researchers have argued that real networks are hierarchically organized (see, for instance, [Ravasz et al. 2002, Ravasz & Barabási 2003]), and algorithms to extract the network hierarchical structure have also been developed [Sales-Pardo et al. 2007, Clauset et al. 2009]. Moreover, especially web graphs [Dill et al. 2002, Dorogovtsev et al. 2002, Crovella & Bestavros 1997] and biological networks [Ravasz & Barabási 2003] were found to be self-similar and "fractal."

The second intuition comes from viewing every node of $\mathcal{P}_k$ as being described with an ordered sequence of k nodes from $\mathcal{P}_1$. (This is similar to Observation 1 and the proof of Theorem 9.13.)

Let's label nodes of the initiator matrix $\mathcal{P}_1$ as $u_1, \ldots, u_N$, and nodes of $\mathcal{P}_k$ as $v_1, \ldots, v_{N^k}$. Then every node v_i of $\mathcal{P}_k$ is described with a sequence $(v_i(1), \ldots, v_i(k))$ of node labels of $\mathcal{P}_1$, where $v_i(l) \in \{u_1, \ldots, u_k\}$. Similarly, consider also a second node v_j with the label sequence $(v_j(1), \ldots, v_j(k))$. Then the probability p_e of an edge (v_i, v_j) in $\mathcal{P}_k$ is exactly

$$p_e(v_i, v_j) = \mathcal{P}_k[v_i, v_j] = \prod_{l=1}^{k} \mathcal{P}_1[v_i(l), v_j(l)]$$

(Note this is exactly equation (9.2).)

Now one can look at the description sequence of node v_i as a k-dimensional vector of attribute values $(v_i(1), \ldots, v_i(k))$. Then $p_e(v_i, v_j)$ is exactly the coordinate-wise product of appropriate entries of $\mathcal{P}_1$, where the node description sequence selects which entries of $\mathcal{P}_1$ to multiply. Thus, the $\mathcal{P}_1$ matrix can be thought of as the attribute similarity matrix; i.e., it encodes the probability of linking given that two nodes agree/disagree on the attribute value. Then the probability of an edge is simply a product of individual attribute similarities over the k N-valued attributes that describe each of the two nodes.

This intuition gives us a very natural interpretation of stochastic Kronecker graphs. Each node is described by a sequence of categorical attribute values or features, and then the probability of two nodes linking depends on the product of individual attribute similarities. This way, Kronecker graphs can effectively model homophily (nodes with similar attribute values are more likely to link) by $\mathcal{P}_1$ having high-value entries on the diagonal, or heterophily (nodes that differ are more likely to link) by $\mathcal{P}_1$ having high entries off the diagonal.

Figure 9.6 shows an example. Let's label nodes of $\mathcal{P}_1$ as u_1, u_2 as in Figure 9.6(a). Then every node of $\mathcal{P}_k$ is described with an ordered sequence of k binary attributes. For example, Figure 9.6(b) shows an instance for $k = 2$ where node v_2 of $\mathcal{P}_2$ is described by (u_1, u_2), and similarly v_3 by (u_2, u_1). Then as shown in Figure 9.6(b), the probability of edge $p_e(v_2, v_3) = b \cdot c$, which is exactly $\mathcal{P}_1[u_2, u_1] \cdot \mathcal{P}_1[u_1, u_2] = b \cdot c$—the product of entries of $\mathcal{P}_1$, where the corresponding elements of the description of nodes v_2 and v_3 act as selectors of which entries of $\mathcal{P}_1$ to multiply.

Figure 9.6(c) further illustrates the recursive nature of Kronecker graphs. One can see the Kronecker product as recursive descent into the big adjacency matrix where at each stage one of the entries or blocks is chosen. For example, to get to entry (v_2, v_3), one first needs to dive into quadrant b followed by quadrant c. This intuition will help us in Section 9.3.6 to devise a fast algorithm for generating Kronecker graphs.

However, there are also two notes to make here. First, using a single initiator $\mathcal{P}_1$, we are implicitly assuming that there is one single and universal attribute similarity matrix that holds across all k N-ary attributes. One can easily relax this assumption by taking a different initiator matrix for each attribute (initiator matrices can even be of different sizes as attributes are of different arity), and then Kronecker-multiplying them to obtain a large network. Here each initiator matrix plays the role of attribute similarity matrix for that particular attribute.

For simplicity and convenience, we will work with a single initiator matrix, but all our methods can be trivially extended to handle multiple initiator matrices. Moreover, as we will see later in Section 9.6, even a single 2×2 initiator matrix seems to be enough to capture large-scale statistical properties of real-world networks.

The second assumption is harder to relax. When describing every node v_i with a sequence of attribute values, we are implicitly assuming that the values of all attributes are uniformly distributed (have same proportions) and that every node has a unique combination of attribute values. So, all possible combinations of attribute values are taken. For example, node v_1 in a large matrix $\mathcal{P}_k$ has attribute sequence $(u_1, u_1, \ldots, u_1)$, and v_N has $(u_1, u_1, \ldots, u_1, u_N)$, while the "last" node v_{N^k} has attribute values $(u_N, u_N, \ldots, u_N)$. One can think of this as counting in the N-ary number system, where node attribute descriptions range from 0 (i.e., "leftmost" node with attribute description $(u_1, u_1, \ldots, u_1)$) to N^k (i.e., "rightmost" node attribute description $(u_N, u_N, \ldots, u_N)$).

A simple way to relax the above assumption is to take a larger initiator matrix with a smaller number of parameters than the number of entries. This means that multiple entries of $\mathcal{P}_1$ will share the same value (parameter). For example, if attribute u_1 takes one value 66% of the time, and the other value 33% of the time, then one can model this by taking a 3×3 initiator matrix with only four parameters. Adopting the naming convention of Figure 9.6, we see that parameter a now occupies a 2×2 block, which then also makes b and c occupy 2×1 and 1×2 blocks, and d a single cell. This way one gets a four-parameter model with uneven feature value distribution.

We note that the view of Kronecker graphs in which every node is described with a set of features and the initiator matrix encodes the probability of linking given the attribute values of two nodes somewhat resembles the random dot product graph model [Young & Scheinerman 2007, Nickel 2008]. The important difference here is that we multiply individual linking probabilities, while in random dot product graphs one takes the sum of individual probabilities, which seems somewhat less natural.

9.3.6 Fast generation of stochastic Kronecker graphs

The intuition for fast generation of stochastic Kronecker graphs comes from the recursive nature of the Kronecker product and is closely related to the R-MAT graph generator [Chakrabarti et al. 2004]. Generating a stochastic Kronecker graph K on N nodes naively takes $O(N^2)$ time. Here we present a linear time $O(E)$ algorithm, where E is the (expected) number of edges in K.

Figure 9.6(c) shows the recursive nature of the Kronecker product. To "arrive" to a particular edge (v_i, v_j) of $\mathcal{P}_k$, one has to make a sequence of k (in our case $k = 2$) decisions among the entries of $\mathcal{P}_1$, multiply the chosen entries of $\mathcal{P}_1$, and then place the edge (v_i, v_j) with the obtained probability.

Instead of flipping $O(N^2) = O(N^{2k})$ biased coins to determine the edges, we can place E edges by directly simulating the recursion of the Kronecker product. Basically, we recursively choose subregions of matrix K with probability proportional to θ_{ij}, $\theta_{ij} \in \mathcal{P}_1$, until in k steps we descend to a single cell of the big adjacency matrix K and place an edge. For example, for (v_2, v_3) in Figure 9.6(c), we first have to choose b followed by c.

The probability of each individual edge of $\mathcal{P}_k$ follows a Bernoulli distribution, as the edge occurrences are independent. By the Central Limit Theorem [Petrov 1995], the number of edges in $\mathcal{P}_k$ tends to a normal distribution with mean $(\sum_{i,j=1}^{N_1} \theta_{ij})^k = M^k$, where $\theta_{ij} \in \mathcal{P}_1$. So, given a stochastic initiator matrix $\mathcal{P}_1$, we first sample the expected number of edges E in $\mathcal{P}_k$. Then we place E edges in a graph $\mathbf{K}$ by applying the recursive descent for k steps where at each step we choose entry (i, j) with probability θ_{ij}/M where $M = \sum_{ij} \theta_{ij}$. Since we add $E = M^k$ edges, the probability that edge (v_i, v_j) appears in $\mathbf{K}$ is exactly $\mathcal{P}_k[v_i, v_j]$. In stochastic Kronecker graphs, the initiator matrix encodes both the total number of edges in a graph and their structure. $\sum \theta_{ij}$ encodes the number of edges in the graph, while the proportions (ratios) of values θ_{ij} define how many edges each part of a graph adjacency matrix will contain.

In practice, it can happen that more than one edge lands in the same (v_i, v_j) entry of big adjacency matrix K. If an edge lands in an already occupied cell, we insert it again. Even though values of $\mathcal{P}_1$ are usually skewed, adjacency matrices of real networks are so sparse that this is not really a problem in practice. Empirically we note that around 1% of edges collide.

9.3.7 Observations and connections

Next, we describe several observations about the properties of Kronecker graphs and make connections to other network models.

- *Bipartite graphs:* Kronecker graphs can naturally model bipartite graphs. Instead of starting with a square $N \times N$ initiator matrix, one can choose an arbitrary $N \times M_1$ initiator matrix, where rows define the "left" and columns the "right" side of the bipartite graph. Kronecker multiplication will then generate bipartite graphs with partition sizes N^k and M_1^k.

- *Graph distributions:* $\mathcal{P}_k$ defines a distribution over all graphs as it encodes the probability of all possible N^{2k} edges appearing in a graph by using an exponentially smaller number of parameters (just N^2). As we will later see, even a very small number of parameters, e.g., 4 (2×2 initiator matrix) or 9 (3×3 initiator), is enough to accurately model the structure of large networks.

- *Extension of Erdős–Rényi random graph model:* Stochastic Kronecker graphs represent an extension of Erdős–Rényi random graphs [Erdős & A. Rényi 1960]. If one takes $\mathcal{P}_1 = [\theta_{ij}]$, where every $\theta_{ij} = p$, then we obtain exactly the Erdős–Rényi model of random graphs $G_{n,p}$, where every edge appears independently with probability p.

- *Relation to the R-MAT model:* The recursive nature of stochastic Kronecker graphs makes them related to the R-MAT generator [Chakrabarti et al. 2004]. The difference between the two models is that in R-MAT one needs to separately specify the number of edges, while in stochastic Kronecker graphs initiator matrix $\mathcal{P}_1$ also encodes the number of edges in the graph. Section 9.3.6 built on this similarity to devise a fast algorithm for generating stochastic Kronecker graphs.

- *Densification:* Similarly as with deterministic Kronecker graphs, the number of nodes in a stochastic Kronecker graph grows as N^k, and the expected number of edges grows as $(\sum_{ij} \theta_{ij})^k$. This means one would want to choose values θ_{ij} of the initiator matrix $\mathcal{P}_1$ so that $\sum_{ij} \theta_{ij} > N$ in order for the resulting network to densify.

9.4 Simulations of Kronecker graphs

Next we perform a set of simulation experiments to demonstrate the ability of Kronecker graphs to match the patterns of real-world networks. In the next section, we will tackle the problem of estimating the Kronecker graph model from real data, i.e., finding the most likely initiator $\mathcal{P}_1$. Here we present simulation experiments using Kronecker graphs to explore the parameter space and to compare properties of Kronecker graphs to those found in large real networks.

9.4.1 Comparison to real graphs

We observe two kinds of graph patterns—"static" and "temporal." As mentioned earlier, common static patterns include degree distribution, scree plot (eigenvalues of graph adjacency matrix versus rank), and distribution of components of the principal eigenvector of a graph adjacency matrix. Temporal patterns include the diameter over time and the densification power law. For the diameter computation, we use the effective diameter as defined in Section 9.2.

For the purpose of this section, consider the following setting. Given a real graph $\mathbf{G}$, we want to find the Kronecker initiator that produces a qualitatively similar graph. In principle, one could try choosing each of the N^2 parameters for the matrix $\mathcal{P}_1$ separately. However, we reduce the number of parameters from N^2 to just two: α and β. Let $\mathbf{K}_1$ be the initiator matrix (binary, deterministic). Then we create the corresponding stochastic initiator matrix $\mathcal{P}_1$ by replacing each "1" and "0" of $\mathbf{K}_1$ with α and β, respectively ($\beta \leq \alpha$). The resulting probability matrices maintain—with some random noise—the self-similar structure of the Kronecker graphs in the previous section (which, for clarity, we call *deterministic Kronecker graphs*). We defer the discussion of how to automatically estimate $\mathcal{P}_1$ from data G to the next section.

The data sets we use here are the following:

- CIT-HEP-TH: This is a citation graph for high-energy physics theory research papers from preprint archive ArXiv, with a total of $N = 29{,}555$ papers and $E = 352{,}807$ citations [Gehrke et al. 2003]. We follow the citation graph's evolution from January 1993 to April 2003, with one data point per month.

- AS-ROUTEVIEWS: We also analyze a static data set consisting of a single snapshot of connectivity among Internet autonomous systems [RouteViews 1997] from January 2000, with $N = 6474$ and $E = 26{,}467$.

Results are shown in Figure 9.7 for the CIT-HEP-TH graph which evolves over time. We show the plots of one static and one temporal pattern. We see that

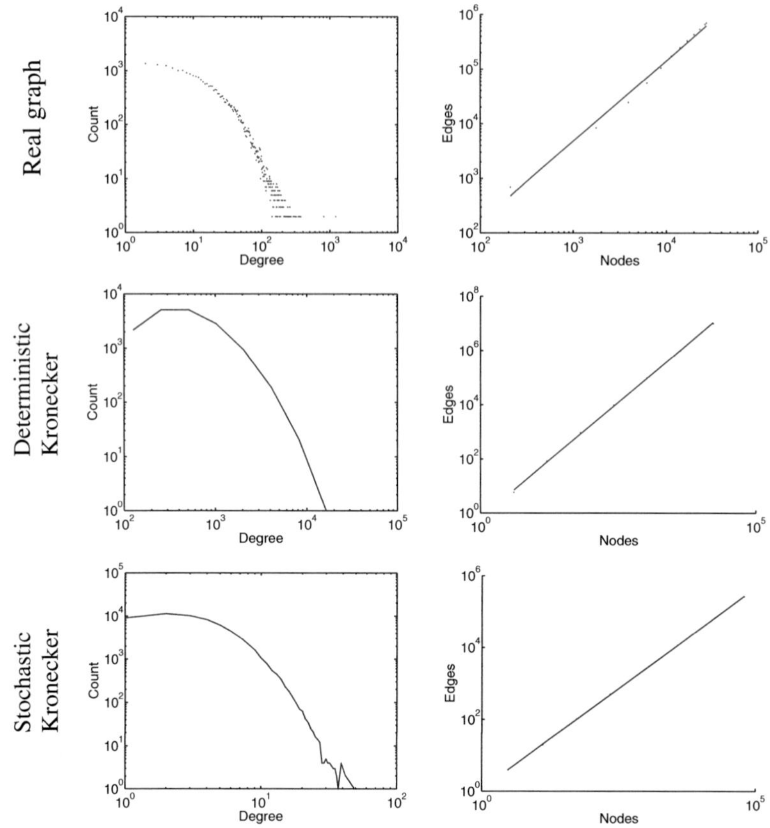

(a) Degree (b) DPL

Figure 9.7. Citation network (CIT-HEP-TH).
Patterns from the real graph (top row), the deterministic Kronecker graph with $\mathbf{K}_1$ being a star graph on four nodes (center + three satellites) (middle row), and the stochastic Kronecker graph ($\alpha = 0.41$, $\beta = 0.11$, bottom row). (a) is the PDF of degrees in the graph (log-log scale), and (b) is the number of edges versus number of nodes over time (log-log scale). Notice that the stochastic Kronecker graph qualitatively matches all the patterns very well.

the deterministic Kronecker model already to some degree captures the qualitative structure of the degree distribution, as well as the temporal pattern represented by the densification power law. However, the deterministic nature of this model results in discrete behavior, as shown in the degree distribution plot for the deterministic Kronecker graph of Figure 9.7. We see that the stochastic Kronecker graphs smooth out these distributions, further matching the qualitative structure of the real data.

Similarly, Figure 9.8 shows plots for the static patterns in the *autonomous systems* (AS-ROUTEVIEWS) graph. Recall that we analyze a single, static network

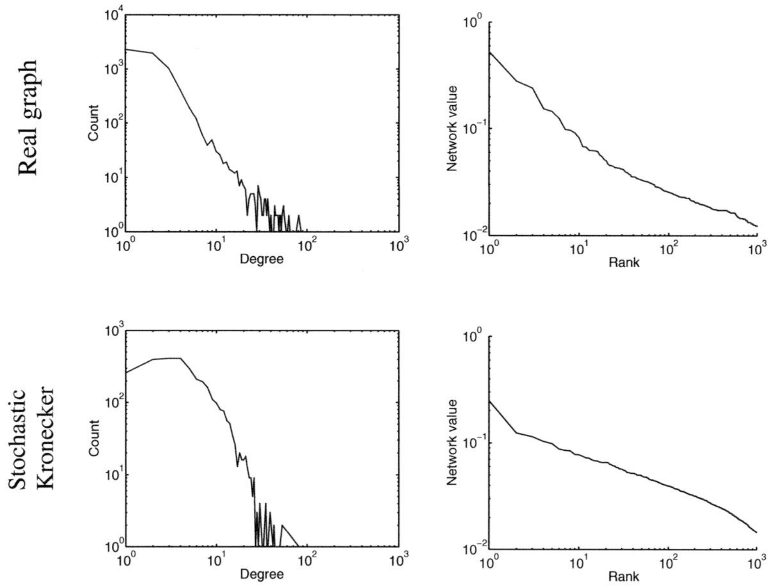

(a) Degree (b) "Network value"

Figure 9.8. Autonomous systems (AS-ROUTEVIEWS). Real (top) versus Kronecker (bottom). Column (a) shows the degree distribution. Column (b) shows a more static pattern. Notice that, again, the stochastic Kronecker graph matches well the properties of the real graph.

snapshot in this case. In addition to the degree distribution, we show a typical plot [Chakrabarti et al. 2004] of the distribution of *network values* (principal eigenvector components, sorted, versus rank). Notice that, again, the stochastic Kronecker graph matches well the properties of the real graph.

9.4.2 Parameter space of Kronecker graphs

Last, we present simulation experiments that investigate the parameter space of stochastic Kronecker graphs.

First, in Figure 9.9, we show the ability of Kronecker graphs to generate networks with increasing, constant, and decreasing/stabilizing effective diameter. We start with a four-node chain initiator graph (shown in the top row of Figure 9.3), setting each "1" of $\mathbf{K}_1$ to α and each "0" to $\beta = 0$ to obtain $\mathcal{P}_1$ that we then use to generate a growing sequence of graphs. We plot the effective diameter of each $R(\mathcal{P}_k)$ as we generate a sequence of growing graphs $R(\mathcal{P}_2), R(\mathcal{P}_3), \ldots, R(\mathcal{P}_{10})$. $R(\mathcal{P}_{10})$ has exactly 1,048,576 nodes. Notice that stochastic Kronecker graphs are very flexible models. When the generated graph is very sparse (low value of α), we obtain graphs with slowly increasing effective diameter (Figure 9.9 (top)). For intermediate values

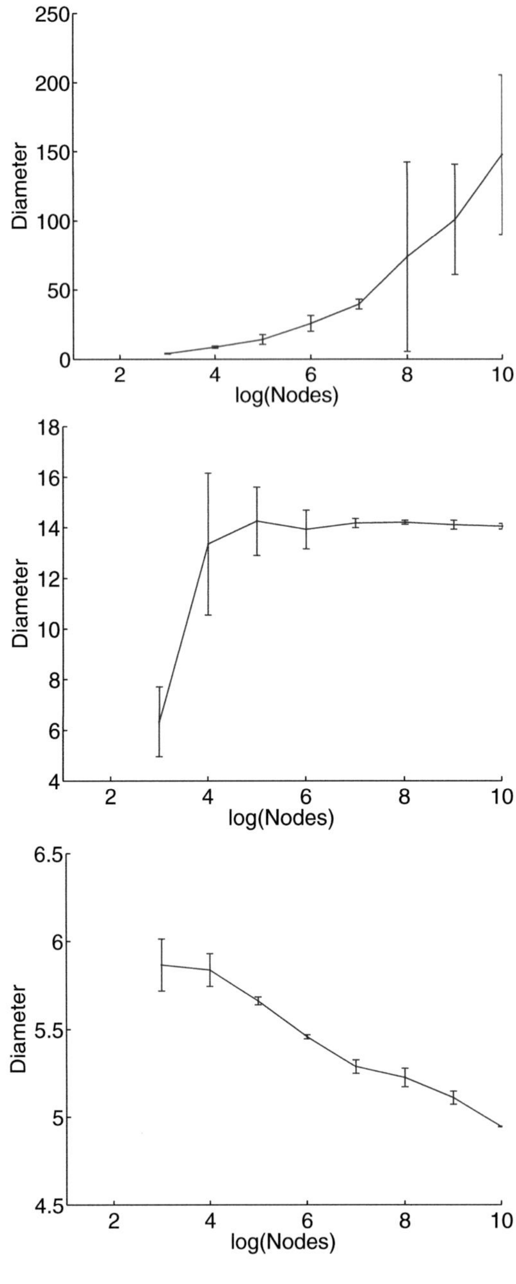

Figure 9.9. Effective diameter over time.
Time evolution of a 4-node chain initiator graph. Top: increasing diameter ($\alpha = 0.38, \beta = 0$). Middle: constant diameter ($\alpha = 0.43, \beta = 0$). Bottom: decreasing diameter ($\alpha = 0.54, \beta = 0$). After each consecutive Kronecker power, we measure the effective diameter.

of α, we get graphs with constant diameter (Figure 9.9 (middle)) that, in our case, also slowly densify with densification exponent $a = 1.05$. Lastly, we see an example of a graph with shrinking/stabilizing effective diameter. Here we set $\alpha = 0.54$, which results in a densification exponent of $a = 1.2$. Note that these observations are not contradicting Theorem 9.11. Actually, these simulations agree well with the analysis of [Mahdian & Xu 2007].

Next, we examine the parameter space of a stochastic Kronecker graph for which we choose a star on four nodes as an initiator graph and parameterize with α and β as before. The initiator graph and the structure of the corresponding (deterministic) Kronecker graph adjacency matrix is shown in the top row of Figure 9.3.

Figure 9.10 (top) shows the sharp transition in the fraction of the number of nodes that belong to the largest weakly connected component as we fix $\beta = 0.15$ and slowly increase α. Such phase transitions on the size of the largest connected component also occur in Erdős–Rényi random graphs. Figure 9.10 (middle) further explores this by plotting the fraction of nodes in the largest connected component (N_c/N) over the full parameter space. Notice the sharp transition between disconnected (white area) and connected graphs (dark).

Last, Figure 9.10 (bottom) shows the effective diameter over the parameter space (α, β) for the four-node star initiator graph. Notice that when parameter values are small, the effective diameter is small since the graph is disconnected and not many pairs of nodes can be reached. The shape of the transition between low-high diameter closely follows the shape of the emergence of the connected component. Similarly, when parameter values are large, the graph is very dense and the diameter is small. There is a narrow band in parameter space where we get graphs with interesting diameters.

9.5 Kronecker graph model estimation

In previous sections, we investigated various properties of networks generated by the (stochastic) Kronecker graphs model. Many of these properties were also observed in real networks. Moreover, we also gave closed form expressions (parametric forms) for values of these statistical network properties, allowing us to calculate a property (e.g., diameter, eigenvalue spectrum) of a network directly from just the initiator matrix. So in principle, one could invert these equations and directly get from a property (e.g., shape of degree distribution) to the values of initiator matrix.

However, in previous sections, we did not say anything about how various network properties of a Kronecker graph correlate and interdepend. For example, it could be the case that two network properties are mutually exclusive. For instance, perhaps one could only match the network diameter but not the degree distribution or vice versa. However, as we show later, this is not the case.

Now we turn our attention to automatically estimating the Kronecker initiator graph. The setting is that we are given a real network G and would like to find a stochastic Kronecker initiator $\mathcal{P}_1$ that produces a synthetic Kronecker graph K that is "similar" to G. One way to measure similarity is to compare statistical network properties, such as diameter and degree distribution, of graphs G and K.

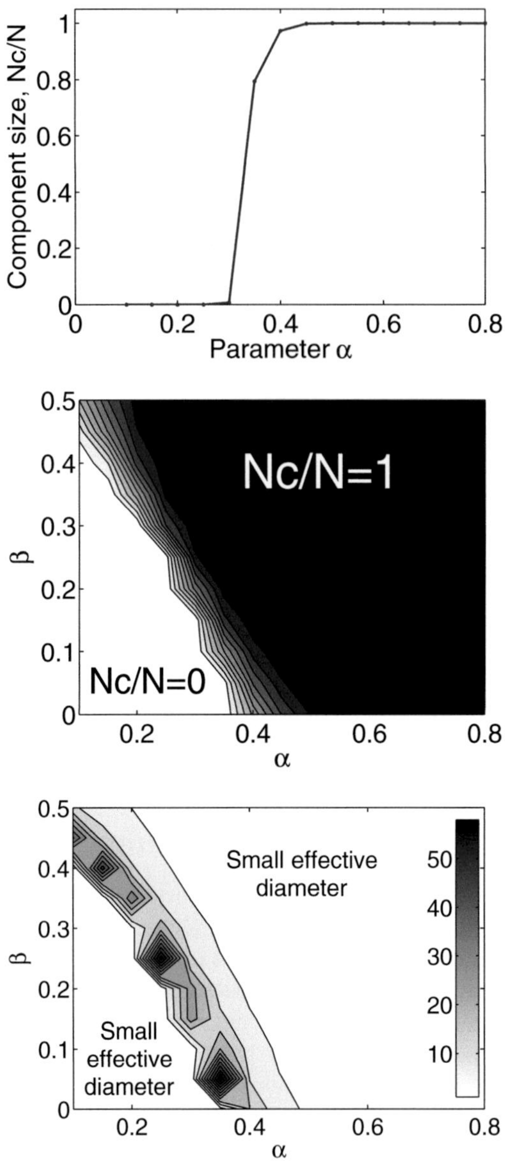

Figure 9.10. Largest weakly connected component. Fraction of nodes in the largest weakly connected component (N_c/N) and the effective diameter for four-star initiator graph. Top: largest component size; we fix $\beta = 0.15$ and vary α. Middle: largest component size; we vary both α and β. Bottom: effective diameter of the network. If the network is disconnected or has very dense path lengths that are short, then the diameter is large when the network is barely connected.

Comparing statistical properties already suggests a very direct approach to this problem: one could first identify the set of network properties (statistics) to match, then define a quality of fit metric and somehow optimize over it. For example, one could use the KL divergence [Kullback & Leibler 1951] or the sum of squared differences between the degree distribution of the real network G and its synthetic counterpart K. Moreover, as we are interested in matching several such statistics between the networks, one would have to meaningfully combine these individual error metrics into a global error metric. So, one would have to specify what kind of properties he or she cares about and then combine them accordingly. This would be a hard task as the patterns of interest have very different magnitudes and scales. Moreover, as new network patterns are discovered, the error functions would have to be changed and models re-estimated. And even then it is not clear how to define the optimization procedure to maximize the quality of fit and how to perform optimization over the parameter space.

Our approach here is different. Instead of committing to a set of network properties ahead of time, we try to directly match the adjacency matrices of the real network G and its synthetic counterpart K. The idea is that if the adjacency matrices are similar, then the global statistical properties (statistics computed over K and G) will also match. Moreover, by directly working with the graph itself (and not summary statistics), we do not commit to any particular set of network statistics (network properties/patterns), and as new statistical properties of networks are discovered, our models and estimated parameters will still hold.

9.5.1 Preliminaries

Stochastic graph models induce probability distributions over graphs. A generative model assigns a probability $P(\mathbf{G})$ to every graph $\mathbf{G}$. $P(\mathbf{G})$ is the *likelihood* that a given model (with a given set of parameters) generates the graph $\mathbf{G}$. We concentrate on the stochastic Kronecker graph model and consider fitting it to a real graph $\mathbf{G}$, our data. We use the maximum-likelihood approach, i.e., we aim to find parameter values, the initiator $\mathcal{P}_1$, that maximize $P(\mathbf{G})$ under the stochastic Kronecker graph model.

This approach presents several challenges:

- **Model selection:** A graph is a single structure and not a set of items drawn independently and identically distributed (i.i.d.) from some distribution. So, one cannot split it into independent training and test sets. The fitted parameters will thus be best to generate a *particular* instance of a graph. Also, overfitting could be an issue since a more complex model generally fits better.

- **Node correspondence:** The second challenge is the node correspondence or node labeling problem. The graph $\mathbf{G}$ has a set of N nodes, and each node has a unique label (index, ID). Labels do not carry any particular meaning, they just uniquely denote or identify the nodes. One can think of this as the graph is first generated and then the labels (node IDs) are randomly assigned. This means that two isomorphic graphs that have different node labels should have the same likelihood. A permutation σ is sufficient to describe the node correspondences as it maps labels (IDs) to nodes of the graph. To compute the likelihood $P(\mathbf{G})$, one has to consider all node correspondences $P(\mathbf{G}) = \sum_\sigma P(\mathbf{G}|\sigma)P(\sigma)$, where the sum is over all $N!$ permutations σ of N nodes. Calculating this *superexponential* sum explicitly is infeasible for any graph with more than a handful of nodes. Intuitively, one can think of this summation as some kind of graph isomorphism test where we are searching for the best correspondence (mapping) between nodes of G and $\mathcal{P}$.

- **Likelihood estimation:** Even if we assume one can efficiently solve the node correspondence problem, calculating $P(\mathbf{G}|\sigma)$ naively takes $O(N^2)$ as one has to evaluate the probability of each of the N^2 possible edges in the graph adjacency matrix. Again, for graphs of the size we want to model here, approaches with quadratic complexity are infeasible.

To develop our solution, we use sampling to avoid the superexponential sum over the node correspondences. By exploiting the structure of the Kronecker matrix multiplication, we develop an algorithm to evaluate $P(\mathbf{G}|\sigma)$ in *linear* time $O(M)$. Since real graphs are *sparse*, i.e., the number of edges is roughly of the same order as the number of nodes, this makes fitting of Kronecker graphs to large networks feasible.

9.5.2 Problem formulation

Suppose we are given a graph $\mathbf{G}$ on $N = N^k$ nodes (for some positive integer k) and an $N \times N$ stochastic Kronecker graphs initiator matrix $\mathcal{P}_1$. Here $\mathcal{P}_1$ is a parameter matrix, a set of parameters that we aim to estimate. For now, also assume N, the size of the initiator matrix, is given. Later, we will show how to automatically select it. Next, using $\mathcal{P}_1$, we create a stochastic Kronecker graph probability matrix $\mathcal{P}_k$, where every entry p_{uv} of $\mathcal{P}_k$ contains a probability that node u links to node v. We then evaluate the probability that $\mathbf{G}$ is a realization of $\mathcal{P}_k$. The task is to find such $\mathcal{P}_1$ that has the highest probability of realizing (generating) $\mathbf{G}$.

Formally, we are solving

$$\arg\max_{\mathcal{P}_1} P(\mathbf{G}|\mathcal{P}_1) \tag{9.3}$$

To keep the notation simpler, we use standard symbol Θ to denote the parameter matrix $\mathcal{P}_1$ that we are trying to estimate. We denote entries of $\Theta = \mathcal{P}_1 = [\theta_{ij}]$, and similarly we denote $\mathcal{P} = \mathcal{P}_k = [p_{ij}]$. Note that here we slightly simplified the notation: we use Θ to refer to $\mathcal{P}_1$, and θ_{ij} are elements of Θ. Similarly, p_{ij} are elements of $\mathcal{P}$ ($\equiv \mathcal{P}_k$). Moreover, we denote $\mathbf{K} = R(\mathcal{P})$; i.e., K is a realization of the stochastic Kronecker graph sampled from probabilistic adjacency matrix $\mathcal{P}$.

As noted before, because the node IDs are assigned arbitrarily and they carry no significant information, we have to consider all the mappings of nodes from $\mathbf{G}$ to rows and columns of stochastic adjacency matrix $\mathcal{P}$. A priori all labelings are equally likely. A permutation σ of the set $\{1, \ldots, N\}$ defines this mapping of nodes from $\mathbf{G}$ to stochastic adjacency matrix $\mathcal{P}$. To evaluate the likelihood of $\mathbf{G}$, one needs to consider all possible mappings of N nodes of $\mathbf{G}$ to rows (columns) of $\mathcal{P}$.

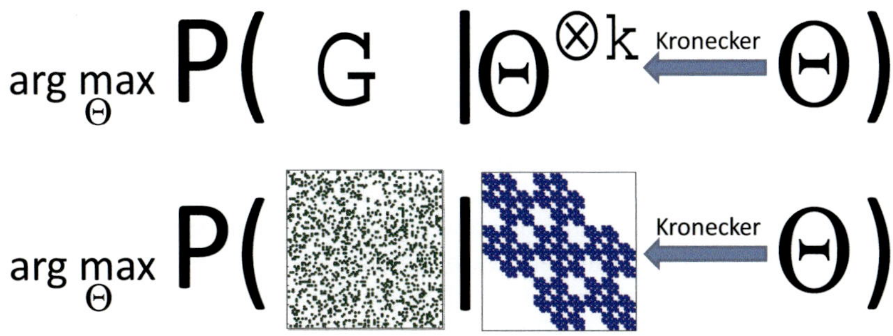

Figure 9.11. Kronecker parameter estimation as an optimization problem. We search over the initiator matrices Θ ($\equiv \mathcal{P}_1$). Using Kronecker multiplication, we create probabilistic adjacency matrix $\Theta^{\otimes k}$ that is of the same size as real network G. Now, we evaluate the likelihood by simultaneously traversing and multiplying entries of $\mathbf{G}$ and $\Theta^{\otimes k}$ (see equation (9.5)). As shown by the figure, permutation σ plays an important role, as permuting rows and columns of $\mathbf{G}$ could make it look more similar to $\Theta^{\otimes k}$ and thus increase the likelihood.

For convenience, we work with *log-likelihood* $l(\Theta)$, and solve $\hat{\Theta} = \arg\max_\Theta l(\Theta)$, where $l(\Theta)$ is defined as

$$\begin{aligned} l(\Theta) = \log P(\mathbf{G}|\Theta) &= \log \sum_\sigma P(\mathbf{G}|\Theta,\sigma)P(\sigma|\Theta) \\ &= \log \sum_\sigma P(\mathbf{G}|\Theta,\sigma)P(\sigma) \end{aligned} \tag{9.4}$$

The likelihood that a given initiator matrix Θ and permutation σ gave rise to the real graph $\mathbf{G}$, $P(\mathbf{G}|\Theta,\sigma)$ is calculated naturally as follows. First, by using Θ, we create the stochastic Kronecker graph adjacency matrix $\mathcal{P} = \mathcal{P}_k = \Theta^{\otimes k}$. Permutation σ defines the mapping of nodes of $\mathbf{G}$ to the rows and columns of stochastic adjacency matrix $\mathcal{P}$. (See Figure 9.11 for the illustration.)

We then model edges as independent Bernoulli random variables parameterized by the parameter matrix Θ. So, each entry p_{uv} of $\mathcal{P}$ gives exactly the probability of edge (u,v) appearing.

We then define the likelihood

$$P(\mathbf{G}|\mathcal{P},\sigma) = \prod_{(u,v)\in\mathbf{G}} \mathcal{P}[\sigma_u,\sigma_v] \prod_{(u,v)\notin\mathbf{G}} (1-\mathcal{P}[\sigma_u,\sigma_v]) \tag{9.5}$$

where we denote σ_i as the ith element of the permutation σ, and $\mathcal{P}[i,j]$ is the element at row i and column j of matrix $\mathcal{P} = \Theta^{\otimes k}$.

The likelihood is defined very naturally. We traverse the entries of adjacency matrix G and then, on the basis of whether a particular edge appeared in G or not, we take the probability of the edge occurring (or not) as given by $\mathcal{P}$ and multiply these probabilities. As one has to touch all the entries of the stochastic adjacency matrix $\mathcal{P}$, evaluating equation (9.5) takes $O(N^2)$ time.

We further illustrate the process of estimating stochastic Kronecker initiator matrix Θ in Figure 9.11. We search over initiator matrices Θ to find the one that maximizes the likelihood $P(\mathbf{G}|\Theta)$. To estimate $P(\mathbf{G}|\Theta)$, we are given a concrete Θ, and now we use Kronecker multiplication to create probabilistic adjacency matrix $\Theta^{\otimes k}$ that is of same size as real network G. Now, we evaluate the likelihood by traversing the corresponding entries of $\mathbf{G}$ and $\Theta^{\otimes k}$. Equation (9.5) basically traverses the adjacency matrix of $\mathbf{G}$ and maps every entry (u, v) of G to a corresponding entry (σ_u, σ_v) of $\mathcal{P}$. Then, in case that edge (u, v) exists in G (i.e., $\mathbf{G}[u, v] = 1$), the likelihood of that particular edge existing is $\mathcal{P}[\sigma_u, \sigma_v]$, and similarly, in case the edge (u, v) does not exist, the likelihood is simply $1 - \mathcal{P}[\sigma_u, \sigma_v]$. This also demonstrates the importance of permutation σ, as permuting rows and columns of $\mathbf{G}$ could make the adjacency matrix look more "similar" to $\Theta^{\otimes k}$ and would increase the likelihood.

So far, we showed how to assess the quality (likelihood) of a particular Θ. So, naively one could perform some kind of grid search to find best Θ. However, this is very inefficient. A better way is to compute the gradient of the log-likelihood $\frac{\partial}{\partial \Theta} l(\Theta)$ and then use the gradient to update the current estimate of Θ and move towards a solution of higher likelihood. Algorithm 9.1 gives an outline of the optimization procedure.

Algorithm 9.1. Kronecker fitting.
Input consists of size of parameter matrix N, graph $\mathbf{G}$ on $N = N^k$ nodes, and learning rate λ. Output consists of Maximum Likelihood Estimation (MLE) parameters $\hat{\Theta}$ ($N \times N$ probability matrix).

$\hat{\Theta} = \textsc{KronFit}(N, \mathbf{G}, N, \lambda)$

initialize $\hat{\Theta}_1$
while not converged
  do evaluate gradient: $\frac{\partial}{\partial \hat{\Theta}_t} l(\hat{\Theta}_t)$
    update parameter estimates: $\hat{\Theta}_{t+1} = \hat{\Theta}_t + \lambda \frac{\partial}{\partial \hat{\Theta}_t} l(\hat{\Theta}_t)$
return $\hat{\Theta} = \hat{\Theta}_t$

However, there are several difficulties with this algorithm. First, we are assuming gradient-descent-type optimization will find a good solution, i.e., the problem does not have (too many) local minima. Second, we are summing over exponentially many permutations in equation (9.4). Third, the evaluation of equation (9.5) as it is written now takes $O(N^2)$ time and needs to be evaluated $N!$ times. So, given a concrete Θ, just naively calculating the likelihood takes $O(N!N^2)$ time, and then one also has to optimize over Θ.

Observation 2. *The complexity of naively calculating the likelihood $P(\mathbf{G}|\Theta)$ of the graph $\mathbf{G}$ is $O(N!N^2)$, where N is the number of nodes in $\mathbf{G}$.*

Next, we show that all this can be done in *linear time.*

9.5.3 Summing over the node labelings

To maximize equation (9.3) using Algorithm 9.1, we need to obtain the gradient of the log-likelihood $\frac{\partial}{\partial\Theta}l(\Theta)$. We can write

$$
\begin{aligned}
\frac{\partial}{\partial\Theta}l(\Theta) &= \frac{\sum_{\sigma}\frac{\partial}{\partial\Theta}P(G|\sigma,\Theta)P(\sigma)}{\sum_{\sigma'}P(G|\sigma',\Theta)P(\sigma')} \\
&= \frac{\sum_{\sigma}\frac{\partial \log P(G|\sigma,\Theta)}{\partial\Theta}P(\mathbf{G}|\sigma,\Theta)P(\sigma)}{P(G|\Theta)} \\
&= \sum_{\sigma}\frac{\partial \log P(\mathbf{G}|\sigma,\Theta)}{\partial\Theta}P(\sigma|\mathbf{G},\Theta)
\end{aligned}
\tag{9.6}
$$

Note that we are still summing over all $N!$ permutations σ, so calculating equation (9.6) is computationally intractable for graphs with more than a handful of nodes. However, the equation has a nice form that allows for use of simulation techniques to avoid the summation over superexponentially many node correspondences. Thus, we simulate draws from the permutation distribution $P(\sigma|\mathbf{G},\Theta)$, and then evaluate the quantities at the sampled permutations to obtain the expected values of log-likelihood and gradient. Algorithm 9.2 gives the details.

Algorithm 9.2. Calculating log-likelihood and gradient.
Input consists of parameter matrix Θ and graph $\mathbf{G}$. Output consists of log-likelihood $l(\Theta)$ and gradient $\frac{\partial}{\partial\Theta}l(\Theta)$.

$(l(\Theta), \frac{\partial}{\partial\Theta}l(\Theta)) = \textsc{KronCalcGrad}(\Theta, \mathbf{G})$
for t = 1 **to** T
  do $\sigma_t = \textsc{SamplePermutation}(\mathbf{G}, \Theta)$
    $l_t = \log P(\mathbf{G}|\sigma^{(t)}, \Theta)$
    $grad_t = \frac{\partial}{\partial\Theta}\log P(\mathbf{G}|\sigma^{(t)}, \Theta)$
return $l(\Theta) = \frac{1}{T}\sum_t l_t$, and $\frac{\partial}{\partial\Theta}l(\Theta) = \frac{1}{T}\sum_t grad_t$

Note that we can also permute the rows and columns of the parameter matrix Θ to obtain equivalent estimates. Therefore, Θ is not strictly identifiable because of these permutations. Since the space of permutations on N nodes is very large (grows as $N!$), the permutation sampling algorithm will explore only a small fraction of the space of all permutations and may converge to one of the global maxima (but may not explore all $N!$ of them) of the parameter space. As we empirically show later, our results are not sensitive to this and multiple restarts result in equivalent (but often permuted) parameter estimates.

Sampling permutations

Next, we describe the Metropolis algorithm to simulate draws from the permutation distribution $P(\sigma|\mathbf{G},\Theta)$, which is given by

$$
P(\sigma|\mathbf{G},\Theta) = \frac{P(\sigma,\mathbf{G},\Theta)}{\sum_{\tau}P(\tau,\mathbf{G},\Theta)} = \frac{P(\sigma,\mathbf{G},\Theta)}{Z}
$$

where Z is the normalizing constant that is hard to compute since it involves the sum over $N!$ elements. However, if we compute the likelihood ratio between permutations σ and σ' (equation (9.7)), the normalizing constants nicely cancel out

$$\frac{P(\sigma'|\mathbf{G},\Theta)}{P(\sigma|\mathbf{G},\Theta)} = \prod_{(u,v)\in\mathbf{G}} \frac{\mathcal{P}[\sigma'_u,\sigma'_v]}{\mathcal{P}[\sigma_u,\sigma_v]} \prod_{(u,v)\notin\mathbf{G}} \frac{(1-\mathcal{P}[\sigma'_u,\sigma'_v])}{(1-\mathcal{P}[\sigma_u,\sigma_v])} \tag{9.7}$$

$$= \prod_{\substack{(u,v)\in\mathbf{G}\\(\sigma_u,\sigma_v)\neq(\sigma'_u,\sigma'_v)}} \frac{\mathcal{P}[\sigma'_u,\sigma'_v]}{\mathcal{P}[\sigma_u,\sigma_v]} \prod_{\substack{(u,v)\notin\mathbf{G}\\(\sigma_u,\sigma_v)\neq(\sigma'_u,\sigma'_v)}} \frac{(1-\mathcal{P}[\sigma'_u,\sigma'_v])}{(1-\mathcal{P}[\sigma_u,\sigma_v])} \tag{9.8}$$

The above formula suggests the use of a Metropolis sampling algorithm (see [Gamerman 1997]) to simulate draws from the permutation distribution since Metropolis is solely based on such ratios (where normalizing constants cancel out). In particular, suppose that in the Metropolis algorithm (Algorithm 9.3) we consider a move from permutation σ to a new permutation σ'. The probability of accepting the move to σ' is given by equation (9.7) (if $\frac{P(\sigma'|\mathbf{G},\Theta)}{P(\sigma|\mathbf{G},\Theta)} \leq 1$) or 1 otherwise.

Algorithm 9.3. Sample permutation.
Metropolis sampling of the node permutation. Input consists of the Kronecker initiator matrix Θ and a graph $\mathbf{G}$ on N nodes. Output consists of permutation $\sigma^{(i)} \sim P(\sigma|\mathbf{G},\Theta)$. $U(0,1)$ is a uniform distribution on $[0,1]$, and $\sigma' :=$ `SwapNodes`(σ, j, k) is the permutation σ' obtained from σ by swapping elements at positions j and k.

$\sigma^{(i)} = \text{SamplePermutation}(\mathbf{G},\Theta)$
$\sigma^{(0)} = (1,\ldots,N)$
$i = 1$
repeat
  Draw j and k uniformly from $(1,\ldots,N)$
  $\sigma^{(i)} =$ `SwapNodes`$(\sigma^{(i-1)}, j, k)$
  Draw u from $U(0,1)$
  if $u > \frac{P(\sigma^{(i)}|\mathbf{G},\Theta)}{P(\sigma^{(i-1)}|\mathbf{G},\Theta)}$
    then $\sigma^{(i)} = \sigma^{(i-1)}$
  i = i + 1
until $\sigma^{(i)} \sim P(\sigma|\mathbf{G},\Theta)$
return $\sigma^{(i)}$

Now we have to devise a way to sample permutations σ from the proposal distribution. One way to do this would be to simply generate a random permutation σ' and then check the acceptance condition. This approach would be very inefficient as we expect the distribution $P(\sigma|\mathbf{G},\Theta)$ to be heavily skewed; i.e., there will be a relatively small number of good permutations (node mappings). Even more so as the degree distributions in real networks are skewed, there will be many bad permutations with low likelihood and few good ones that do a good job in matching nodes of high degree.

To make the sampling process "smoother," i.e., sample permutations that are not that different (and thus are not randomly jumping across the permutation space), we design a Markov chain. The idea is to stay in the high-likelihood part of the permutation space longer. We do this by making samples dependent; i.e., given σ', we want to generate next candidate permutation σ'' to then evaluate the likelihood ratio. When designing the Markov chain step, one has to be careful so that the proposal distribution satisfies the detailed balance condition: $\pi(\sigma')P(\sigma'|\sigma'') = \pi(\sigma'')P(\sigma''|\sigma')$, where $P(\sigma'|\sigma'')$ is the transition probability of obtaining permutation σ' from σ'', and $\pi(\sigma')$ is the stationary distribution.

In Algorithm 9.3, we use a simple proposal where, given permutation σ', we generate σ'' by swapping elements at two uniformly at random chosen positions of σ'. We refer to this proposal as `SwapNodes`. While this is simple and clearly satisfies the detailed balance condition, it is also inefficient in a way that most of the times low degree nodes will get swapped (a direct consequence of heavy-tailed degree distributions). This has two consequences: (a) we will slowly converge to good permutations (accurate mappings of high degree nodes), and (b) once we reach a good permutation, very few permutations will get accepted as most proposed permutations σ' will swap low degree nodes (as they form the majority of nodes).

A possibly more efficient way would be to swap elements of σ based on corresponding node degree, so that high degree nodes would get swapped more often. However, doing this directly does not satisfy the detailed balance condition. A way of sampling labels biased by node degrees that at the same time satisfies the detailed balance condition is the following: we pick an edge in $\mathbf{G}$ uniformly at random and swap the labels of the nodes at the edge endpoints. Notice this is biased towards swapping labels of nodes with high degrees simply as they have more edges. The detailed balance condition holds as edges are sampled uniformly at random. We refer to this proposal as `SwapEdgeEndpoints`.

However, the issue with this proposal is that if the graph $\mathbf{G}$ is disconnected, we will only be swapping labels of nodes that belong to the same connected component. This means that some parts of the permutation space will never get visited. To overcome this problem, we execute `SwapNodes` with some probability ω and `SwapEdgeEndpoints` with probability $1 - \omega$.

To summarize, we consider the following two permutation proposal distributions:

- $\sigma'' =$ `SwapNodes`(σ'): we obtain σ'' by taking σ', uniformly at random selecting a pair of elements and swapping their positions.

- $\sigma'' =$ `SwapEdgeEndpoints`(σ'): we obtain σ'' from σ' by first sampling an edge (j, k) from $\mathbf{G}$ uniformly at random, then we take σ' and swap the labels at positions j and k.

Speeding up the likelihood ratio calculation

We further speed up the algorithm by using the following observation. As written, equation (9.7) takes $O(N^2)$ to evaluate since we have to consider N^2 possible edges. However, notice that permutations σ and σ' differ only at two positions, i.e., elements at position j and k are swapped, i.e., σ and σ' map all nodes except the two to the same locations. This means those elements of equation (9.7) cancel out. Thus to update the likelihood, we only need to traverse two rows and columns of matrix $\mathcal{P}$, namely rows and columns j and k, since everywhere else the mapping of the nodes to the adjacency matrix is the same for both permutations. This results in equation (9.8), where the products now range only over the two rows/columns of $\mathcal{P}$ where σ and σ' differ.

Graphs we are working with here are too large to allow us to explicitly create and store the stochastic adjacency matrix $\mathcal{P}$ by Kronecker-powering the initiator matrix Θ. Every time probability $\mathcal{P}[i,j]$ of edge (i,j) is needed, equation (9.2) is evaluated, which takes $O(k)$. So a single iteration of Algorithm 9.3 takes $O(kN)$.

Observation 3. *Sampling a permutation σ from $P(\sigma|\mathbf{G},\Theta)$ takes $O(kN)$.*

So far, we have shown how to obtain a permutation, but we still need to evaluate the likelihood and find the gradients that will guide us in finding a good initiator matrix. Naively evaluating the network likelihood (gradient) as written in equation (9.6) takes $O(N^2)$ time.

9.5.4 Efficiently approximating likelihood and gradient

We just showed how to efficiently sample node permutations. Now, given a permutation, we show how to efficiently evaluate the likelihood and its gradient. Similar to evaluating the likelihood ratio, naively calculating the log-likelihood $l(\Theta)$ or its gradient $\frac{\partial}{\partial\Theta}l(\Theta)$ takes time quadratic in the number of nodes. Next, we show how to compute this in linear time $O(M)$.

We begin with the observation that real graphs are sparse, that is, the number of edges is not quadratic but rather almost linear in the number of nodes, $M \ll N^2$. This means that the majority of entries of the graph adjacency matrix are zero, i.e., most of the edges are not present. We exploit this fact. The idea is to first calculate the likelihood (gradient) of an empty graph, i.e., a graph with zero edges, and then correct for the edges that actually appear in G.

To naively calculate the likelihood for an empty graph, one needs to evaluate every cell of the graph adjacency matrix. We consider Taylor approximation to the likelihood, and exploit the structure of matrix $\mathcal{P}$ to devise a constant-time algorithm.

First, consider the second order Taylor approximation to the log-likelihood of an edge that succeeds with probability x but does not appear in the graph

$$\log(1-x) \approx -x - \frac{1}{2}x^2$$

Calculating $l_e(\Theta)$, the log-likelihood of an empty graph, becomes

$$l_e(\Theta) = \sum_{i=1}^{N}\sum_{j=1}^{N}\log(1-p_{ij}) \approx -\left(\sum_{i=1}^{N}\sum_{j=1}^{N}\theta_{ij}\right)^k - \frac{1}{2}\left(\sum_{i=1}^{N}\sum_{j=1}^{N}{\theta_{ij}}^2\right)^k \tag{9.9}$$

Notice that while the first pair of sums ranges over N elements, the last pair only ranges over N elements ($N = \log_k N$). Equation (9.9) holds due to the recursive structure of matrix $\mathcal{P}$ generated by the Kronecker product. We substitute the $\log(1 - p_{ij})$ with its Taylor approximation, which gives a sum over elements of $\mathcal{P}$ and their squares. Next, we notice that the sum of elements of $\mathcal{P}$ forms a multinomial series, and thus $\sum_{i,j} p_{ij} = (\sum_{i,j} \theta_{ij})^k$, where θ_{ij} denotes an element of Θ, and p_{ij} an element of $\Theta^{\otimes k}$.

Calculating the log-likelihood of $\mathbf{G}$ now takes $O(M)$. First, we approximate the likelihood of an empty graph in constant time and then account for the edges that are actually present in $\mathbf{G}$; i.e., we subtract the "no-edge" likelihood and add the "edge" likelihoods

$$l(\Theta) = l_e(\Theta) + \sum_{(u,v)\in\mathbf{G}} - \log(1 - \mathcal{P}[\sigma_u, \sigma_v]) + \log(\mathcal{P}[\sigma_u, \sigma_v])$$

We note that by using the second order Taylor approximation to the log-likelihood of an empty graph, the error term of the approximation is $\frac{1}{3}(\sum_i {\theta_{ij}}^3)^k$, which can diverge for large k. For typical values of initiator matrix $\mathcal{P}_1$ (that we present in Section 9.6.5), we note that one needs about a fourth- or fifth-order Taylor approximation for the error of the approximation to actually go to zero as k approaches infinity, i.e., $\sum_{ij} {\theta_{ij}}^{n+1} < 1$, where n is the order of Taylor approximation employed.

9.5.5 Calculating the gradient

Calculation of the gradient of the log-likelihood follows exactly the same pattern as described above. First, by using the Taylor approximation, we calculate the gradient as if graph $\mathbf{G}$ would have no edges. Then, we correct the gradient for the edges that are present in $\mathbf{G}$. As in the previous section, we speed up the calculations of the gradient by exploiting the fact that two consecutive permutations σ and σ' differ only at two positions, and thus given the gradient from the previous step, one only needs to account for the swap of the two rows and columns of the gradient matrix $\partial\mathcal{P}/\partial\Theta$ to update to the gradients of individual parameters.

9.5.6 Determining the size of an initiator matrix

The question we answer next is how to determine the right number of parameters, i.e., what is the right size of matrix Θ? This is a classical question of model selection in which there is a tradeoff between the complexity of the model and the quality of the fit. A bigger model with more parameters usually fits better; however, it is also more likely to overfit the data.

For model selection to find the appropriate value of N, the size of matrix Θ, and to choose the right tradeoff between the complexity of the model and the quality of the fit, we propose to use the Bayes Information Criterion (BIC) [Schwarz 1978]. Stochastic Kronecker graph models the presence of edges with independent Bernoulli random variables, where the canonical number of parameters is N^{2k}, which is a function of a lower-dimensional parameter Θ. This is then a *curved exponential family* [Efron 1975], and BIC naturally applies

$$\mathrm{BIC}(N) = -l(\hat{\Theta}_N) + \frac{1}{2}N^2 \log(N^2)$$

where $\hat{\Theta}_N$ are the maximum-likelihood parameters of the model with $N \times N$ parameter matrix, and N is the number of nodes in G. Note that one could also add an additional term to the above formula to account for multiple global maxima of the likelihood space, but as N is small, the additional term would make no real difference.

As an alternative to BIC, one could also consider the Minimum Description Length (MDL) principle [Rissanen 1978] in which the model is scored by the quality of the fit plus the size of the description that encodes the model and the parameters.

9.6 Experiments on real and synthetic data

Next, we describe our experiments on a range of real and synthetic networks. We divide the experiments into several subsections. First, we examine the convergence and mixing of the Markov chain of our permutation sampling scheme. Then, we consider estimating the parameters of synthetic Kronecker graphs to see whether KronFit is able to recover the parameters used to generate the network. Last, we consider fitting stochastic Kronecker graphs to large real-world networks.

9.6.1 Permutation sampling

In our experiments, we considered both synthetic and real graphs. Unless mentioned otherwise, all synthetic Kronecker graphs were generated using $\mathcal{P}_1^* = [0.8, 0.6; 0.5, 0.3]$, and $k = 14$, which gives us a graph $\mathbf{G}$ on N =16,384 nodes and M =115,741 edges. We chose this particular $\mathcal{P}_1^*$ as it resembles the typical initiator for real networks analyzed later in this section.

Convergence of the log-likelihood and the gradient

First, we examine the convergence of Metropolis permutation sampling, where permutations are sampled sequentially. A new permutation is obtained by modifying the previous one, which creates a Markov chain. We want to assess the convergence and mixing of the chain. We aim to determine how many permutations one needs to draw to reliably estimate the likelihood and the gradient, and also how long it takes until the samples converge to the stationary distribution. For the experiment, we generated a synthetic stochastic Kronecker graph using $\mathcal{P}_1^*$ as defined above. Then, starting with a random permutation, we ran Algorithm 9.3 and measured how the likelihood and the gradients converge to their true values.

In this particular case, we first generated a stochastic Kronecker graph $\mathbf{G}$ as described above, but then calculated the likelihood and the parameter gradients for $\Theta' = [0.8, 0.75; 0.45, 0.3]$. We averaged the likelihoods and gradients over buckets of 1000 consecutive samples and plotted how the log-likelihood, calculated over the

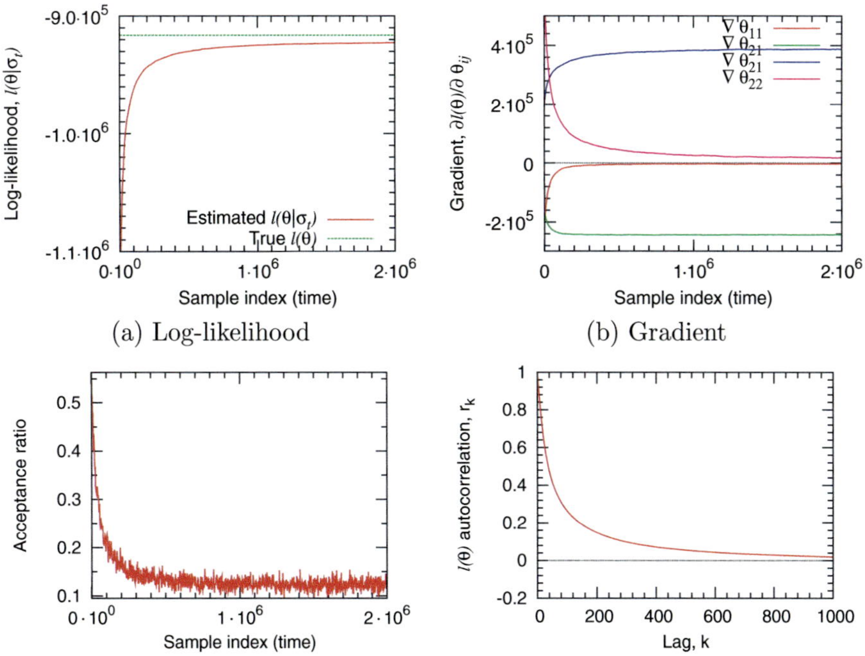

(a) Log-likelihood (b) Gradient

(c) Acceptance ratio (d) Autocorrelation

Figure 9.12. Convergence of the log-likelihood.
Components of the gradient toward their true values for Metropolis permutation sampling (Algorithm 9.3) with the number of samples.

sampled permutations, approached the true log-likelihood (that we can compute since **G** is a stochastic Kronecker graph).

First, we present experiments that aim to answer how many samples (i.e., permutations) does one need to draw to obtain a reliable estimate of the gradient (see equation (9.6)). Figure 9.12(a) shows how the estimated log-likelihood approaches the true likelihood. Notice that estimated values quickly converge to their true values, i.e., Metropolis sampling quickly moves towards "good" permutations. Similarly, Figure 9.12(b) plots the convergence of the gradients. Notice that θ_{11} and θ_{22} of Θ' and $\mathcal{P}_1^*$ match, so gradients of these two parameters should converge to zero and indeed they do. On the other hand, θ_{12} and θ_{21} differ between Θ' and $\mathcal{P}_1^*$. Notice that the gradient for one is positive as the parameter θ_{12} of Θ' should be decreased, and similarly for θ_{21}, the gradient is negative as the parameter value should be increased to match the Θ'. In summary, this shows that log-likelihood and gradients rather quickly converge to their true values.

In Figures 9.12(c) and (d), we also investigate the properties of the Markov chain Monte Carlo sampling procedure and assess convergence and mixing criteria. First, we plot the fraction of accepted proposals. It stabilizes at around 15%, which is quite close to the rule of thumb of 25%. Second, Figure 9.12(d) plots the autocorrelation of the log-likelihood as a function of the lag. Autocorrelation r_k of a signal X is a function of the lag k where r_k is defined as the correlation of signal X at time t with X at $t+k$, i.e., correlation of the signal with itself at lag k. High autocorrelations within chains indicate slow mixing and, usually, slow convergence. On the other hand, fast decay of autocorrelation implies better mixing, and thus one needs fewer samples to accurately estimate the gradient or the likelihood. Notice the rather fast autocorrelation decay.

All in all, these experiments show that one needs to sample on the order of tens of thousands of permutations for the estimates to converge. We also verified that the variance of the estimates is sufficiently small. In our experiments, we start with a random permutation and use long burn-in time. Then, when performing optimization, we use the permutation from the previous step to initialize the permutation at the current step of the gradient descent. Intuitively, small changes in parameter space Θ also mean small changes in $P(\sigma|G,\Theta)$.

Different proposal distributions

In Section 9.5.3, we defined two permutation sampling strategies: `SwapNodes`, where we pick two nodes uniformly at random and swap their labels (node IDs), and `SwapEdgeEndpoints`, where we pick a random edge in a graph and then swap the labels of the edge endpoints. We also discussed that one can interpolate between the two strategies by executing `SwapNodes` with probability ω and `SwapEdgeEndpoints` with probability $1-\omega$.

So, given a stochastic Kronecker graph $\mathbf{G}$ on N =16,384 and M =115,741 generated from $\mathcal{P}_1^* = [0.8, 0.7; 0.5, 0.3]$, we evaluate the likelihood of $\Theta' = [0.8, 0.75; 0.45, 0.3]$. As we sample permutations, we observe how the estimated likelihood converges to the true likelihood. Moreover, we also vary parameter ω, which interpolates between the two permutation proposal distributions. The quicker the convergence toward the true log-likelihood, the better the proposal distribution.

Figure 9.13 plots the convergence of the log-likelihood with the number of sampled permutations. We plot the average over nonoverlapping buckets of 1000 consecutive permutations. Faster convergence implies better permutation proposal distribution. When we use only `SwapNodes` ($\omega = 1$) or `SwapEdgeEndpoints` ($\omega = 0$), convergence is rather slow. We obtain the best convergence for ω around 0.6.

Similarly, Figure 9.14(a) plots the autocorrelation as a function of the lag k for different choices of ω. Faster autocorrelation decay means better mixing of the Markov chain. Again, notice that we get the best mixing for $\omega \approx 0.6$. (Notice logarithmic y-axis.)

Last, we diagnose how long the sampling procedure must be run before the generated samples can be considered to be drawn (approximately) from the stationary distribution. We call this the *burn-in time* of the chain. There are various procedures for assessing convergence. Here we adopt the approach of Gelman et al. [Gelman et al. 2003], which is based on running multiple Markov chains each from a different starting point and then comparing the variance within the chain and between the chains. The sooner the within- and between-chain variances become equal, the shorter the burn-in time, i.e., the sooner the samples are drawn from the stationary distribution.

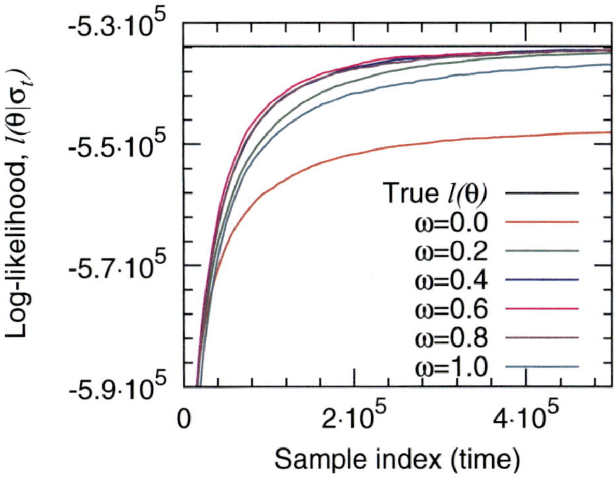

Figure 9.13. Convergence as a function of ω.
Convergence of the log-likelihood and gradients for Metropolis permutation sampling (Algorithm 9.3) for different choices of ω that span `SwapNodes` ($\omega = 1$) and `SwapEdgeEndpoints` ($\omega = 0$). Notice fastest convergence of log-likelihood for $\omega = 0.6$.

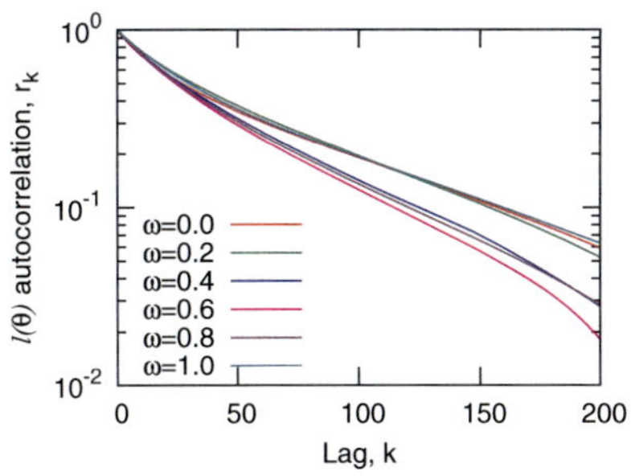

(a) Autocorrelation

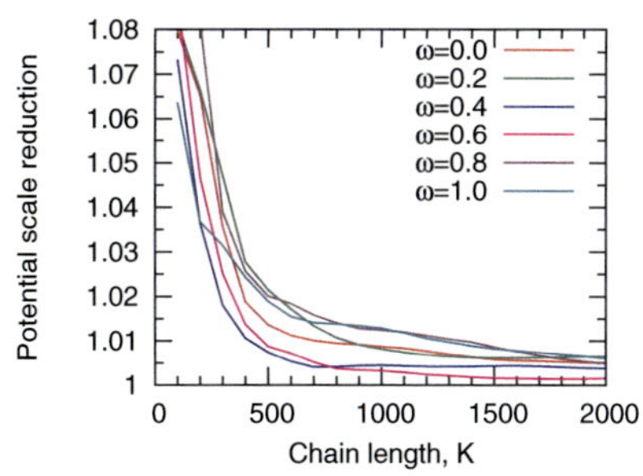

(b) Potential scale reduction

Figure 9.14. Autocorrelation as a function of ω.
(a) Autocorrelation plot of the log-likelihood for the different choices of parameter ω. Notice we get best mixing with $\omega \approx 0.6$. (b) The potential scale reduction that compares the variance inside and across independent Markov chains for different values of parameter ω.

Let l be the parameter that is being simulated with J different chains, and then let $l_j^{(k)}$ denote the kth sample of the jth chain, where $j = 1, \ldots, J$ and $k = 1, \ldots, K$. More specifically, in our case we run separate permutation sampling chains. So, we first sample permutation $\sigma_j^{(k)}$ and then calculate the corresponding log-likelihood $l_j^{(k)}$.

First, we compute between- and within-chain variances $\hat{\sigma}_B^2$ and $\hat{\sigma}_W^2$, where between-chain variance is obtained by

$$\hat{\sigma}_B^2 = \frac{K}{J-1}\sum_{j=1}^{J}(\bar{l}_{\cdot j} - \bar{l}_{\cdot\cdot})^2$$

where $\bar{l}_{\cdot j} = \frac{1}{K}\sum_{k=1}^{K} l_j^{(k)}$ and $\bar{l}_{\cdot\cdot} = \frac{1}{J}\sum_{j=1}^{J}\bar{l}_{\cdot j}$.

Similarly, the within-chain variance is defined by

$$\hat{\sigma}_W^2 = \frac{1}{J(K-1)}\sum_{j=1}^{J}\sum_{k=1}^{K}(l_j^{(k)} - \bar{l}_{\cdot j})^2$$

Then, the marginal posterior variance of $\hat{l}$ is calculated using

$$\hat{\sigma}^2 = \frac{K-1}{K}\hat{\sigma}_W^2 + \frac{1}{K}\hat{\sigma}_B^2$$

Finally, we estimate the *potential scale reduction* [Gelman et al. 2003] of l by

$$\sqrt{\hat{R}} = \sqrt{\frac{\hat{\sigma}^2}{\hat{\sigma}_W^2}}$$

Note that as the length of the chain $K \to \infty$, $\sqrt{\hat{R}}$ converges to 1 from above. The recommendation for convergence assessment from [Gelman et al. 2003] is that the potential scale reduction is below 1.2.

Figure 9.14(b) gives the Gelman–Rubin–Brooks plot, where we plot the potential scale reduction $\sqrt{\hat{R}}$ over the increasing chain length K for different choices of parameter ω. Notice that the potential scale reduction quickly decays towards 1. Similarly, as in Figure 9.14, the extreme values of ω give slow decay, while we obtain the fastest potential scale reduction when $\omega \approx 0.6$.

Properties of the permutation space

Next, we explore the properties of the permutation space. We would like to quantify what fraction of permutations are "good" (have high likelihood) and how quickly are they discovered. For the experiment, we took a real network G (AS-ROUTEVIEWS network) and the MLE parameters $\hat{\Theta}$ for it that we estimated beforehand ($l(\hat{\Theta}) \approx -150{,}000$). The network G has 6474 nodes, which means the space of all permutations has $\approx 10^{22{,}000}$ elements.

First, we sampled 1 billion (10^9) permutations σ_i uniformly at random, i.e., $P(\sigma_i) = 1/(6474!)$ and for each we evaluated its log-likelihood $l(\sigma_i|\Theta) = \log P(\Theta|G, \sigma_i)$. We ordered the permutations in deceasing log-likelihood and plotted $l(\sigma_i|\Theta)$ versus rank. Figure 9.15(a) gives the plot. Notice that very few random permutations are very bad (i.e., they give low likelihood); similarly, few permutations are very good, while most of them are somewhere in between. Notice that best "random"

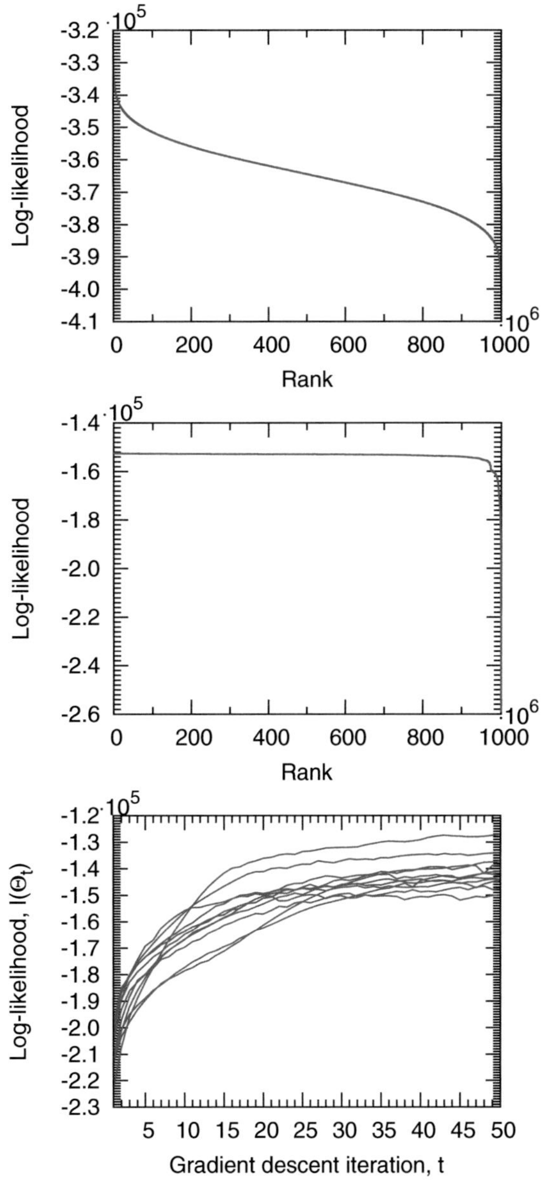

Figure 9.15. Distribution of log-likelihood.
Distribution of log-likelihood of permutations sampled uniformly at random ($l(\Theta|\sigma_i)$ where $\sigma_i \sim P(\sigma)$) (top), and when sampled from $P(\sigma|\Theta, G)$ (middle). Notice the space of good permutations is rather small, but our sampling quickly finds permutations of high likelihood. Convergence of log-likelihood for 10 runs of gradient descent, each from a different random starting point (bottom).

permutation has log-likelihood of $\approx -320{,}000$, which is far below true likelihood $l(\hat{\Theta}) \approx -150{,}000$. This suggests that only a very small fraction of all permutations gives good node labelings.

On the other hand, we also repeated the same experiment but now using permutations sampled from the permutation distribution $\sigma_i \sim P(\sigma|\Theta, G)$ via our Metropolis sampling scheme. Figure 9.15(b) gives the plot. Notice the radical difference. Now the $l(\sigma|\Theta_i)$ very quickly converges to the true likelihood of $\approx -150,000$. This suggests that while the number of "good" permutations (accurate node mappings) is rather small, our sampling procedure quickly converges to the "good" part of the permutation space, where node mappings are accurate, and spends the most time there.

9.6.2 Properties of the optimization space

In maximizing the likelihood, we use a stochastic approximation to the gradient. This adds variance to the gradient and makes efficient optimization techniques, like conjugate gradient, highly unstable. Thus, we use gradient descent, which is slower but easier to control. First, we make the following observation.

Observation 4. *Given a real graph* $\mathbf{G}$ *then finding the maximum-likelihood stochastic Kronecker initiator matrix* $\hat{\Theta}$

$$\hat{\Theta} = \arg\max_{\Theta} P(\mathbf{G}|\Theta)$$

is a nonconvex optimization problem.

Proof. By definition, permutations of the Kronecker graphs initiator matrix Θ all have the same log-likelihood. This means that we have several global minima that correspond to permutations of parameter matrix Θ, and then between them the log-likelihood drops. This means that the optimization problem is nonconvex. □

The above observation does not seem promising for estimating $\hat{\Theta}$ using gradient descent as it is prone to finding local minima. To test for this behavior, we ran the following experiment. We generated 100 synthetic Kronecker graphs on 16,384 (2^{14}) nodes and 1.4 million edges on the average, each with a randomly chosen 2×2 parameter matrix Θ^*. For each of the 100 graphs, we ran a single trial of gradient descent starting from a random parameter matrix Θ' and try to recover Θ^*. In 98% of the cases, the gradient descent converged to the true parameters. Many times the algorithm converged to a different global minima, i.e., $\hat{\Theta}$ is a permuted version of original parameter matrix Θ^*. Moreover, the median number of gradient descent iterations was only 52.

These results suggest a surprisingly nice structure of our optimization space: it seems to behave like a convex optimization problem with many equivalent global minima. Moreover, this experiment is also a good sanity check as it shows that, given a Kronecker graph, we can recover and identify the parameters that were used to generate it.

Moreover, Figure 9.15(c) plots the log-likelihood $l(\Theta_t)$ of the current parameter estimate Θ_t over the iterations t of the stochastic gradient descent. We plot the log-likelihood for 10 different runs of gradient descent, each time starting from a different random set of parameters Θ_0. Notice that in all runs, gradient descent always converges toward the optimum, and none of the runs gets stuck in some local maxima.

9.6.3 Convergence of the graph properties

We approached the problem of estimating stochastic Kronecker initiator matrix Θ by defining the likelihood over the individual entries of the graph adjacency matrix. However, what we would really like is to be given a real graph $\mathbf{G}$ and then generate a synthetic graph $\mathbf{K}$ that has similar properties as the real $\mathbf{G}$. By properties we mean network statistics that can be computed from the graph, e.g., diameter, degree distribution, clustering coefficient, etc. A priori it is not clear that our approach, which tries to match individual entries of the graph adjacency matrix, will also be able to reproduce these global network statistics. However, as is shown next, this is not the case.

To get some understanding of the convergence of the gradient descent in terms of the network properties, we performed the following experiment. After every step t of stochastic gradient descent, we compared the true graph $\mathbf{G}$ with the synthetic Kronecker graph $\mathbf{K}_t$ generated using the current parameter estimates $\hat{\Theta}_t$. Figure 9.16(a) gives the convergence of log-likelihood, and Figure 9.16(b) gives absolute error in parameter values ($\sum |\hat{\theta}_{ij} - \theta^*_{ij}|$, where $\hat{\theta}_{ij} \in \hat{\Theta}_t$ and $\theta^*_{ij} \in \Theta^*$). Similarly, Figure 9.16(c) plots the effective diameter, and Figure 9.16(d) gives the largest singular value of graph adjacency matrix K as it converges to the largest singular value of G.

The properties of $\mathbf{K}_t$ quickly converge to those of $\mathbf{G}$ even though we are not directly optimizing to the network properties of $\mathbf{G}$. The log-likelihood increases, the absolute error of parameters decreases, and the diameter and the largest singular value of $\mathbf{K}_t$ all converge to $\mathbf{G}$. This is a nice result as it shows that, through maximizing the likelihood, the resulting graphs become more and more similar also in their structural properties (even though we are not directly optimizing over them).

9.6.4 Fitting to real-world networks

Next, we present experiments of fitting a Kronecker graph model to real-world networks. Given a real network G, we aim to discover the most likely parameters $\hat{\Theta}$ that ideally would generate a synthetic graph $\mathbf{K}$ having similar properties to real G. This assumes that Kronecker graphs are a good model of the network structure, and that KRONFIT is able to find good parameters. In the previous section, we showed that KRONFIT can efficiently recover the parameters. Now we examine how well a Kronecker graph can model the structure of real networks.

We consider several different networks, such as a graph of connectivity among Internet autonomous systems (AS-ROUTEVIEWS) with $N = 6474$ and $M = 26{,}467$ a who-trusts-whom type social network from Epinions [Richardson et al. 2003]

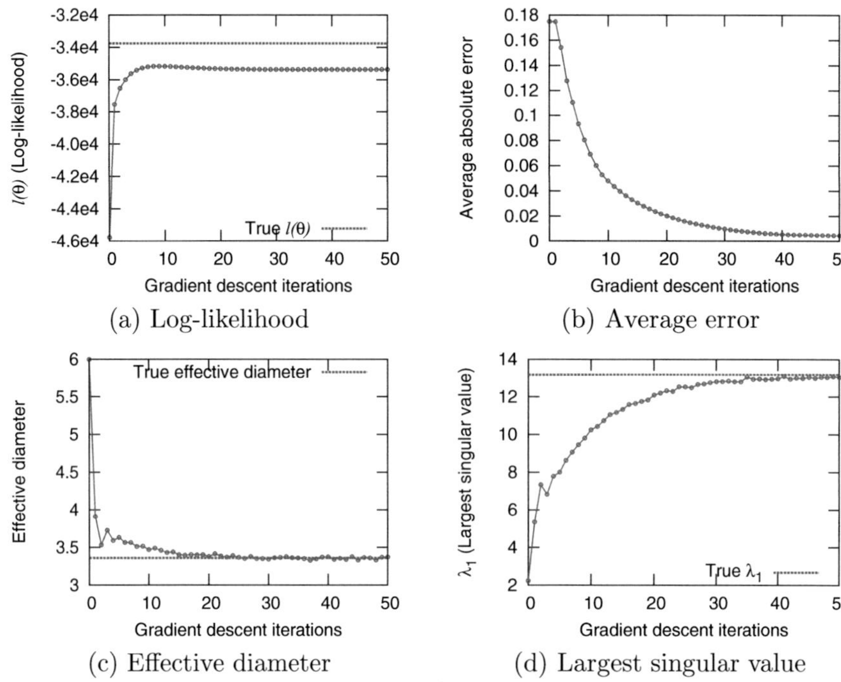

Figure 9.16. Convergence of graph properties.
Convergence with the number of iterations of gradient descent using the synthetic data set. We start with a random choice of parameters, and with steps of gradient descent, the Kronecker graph better and better matches network properties of the target graph.

(Epinions) with $N = 75{,}879$ and $M = 508{,}960$ and many others. The largest network we consider for fitting is Flickr, a photo-sharing online social network with 584,207 nodes and 3,555,115 edges.

For the purpose of this section, we take a real network G, find parameters $\hat{\Theta}$ using KronFit, generate a synthetic graph $\mathbf{K}$ using $\hat{\Theta}$, and then compare G and $\mathbf{K}$ by comparing their properties that we introduced in Section 9.2. In all experiments, we start from a random point (random initiator matrix) and run gradient descent for 100 steps. At each step, we estimate the likelihood and the gradient on the basis of 510,000 sampled permutations from which we discard the first 10,000 samples to allow the chain to burn in.

Fitting to autonomous systems network

First, we focus on the autonomous systems (AS) network obtained from the University of Oregon Route Views project [RouteViews 1997]. Given the AS network G, we run KronFit to obtain parameter estimates $\hat{\Theta}$. Using the $\hat{\Theta}$, we then generate a synthetic Kronecker graph K and compare the properties of G and K.

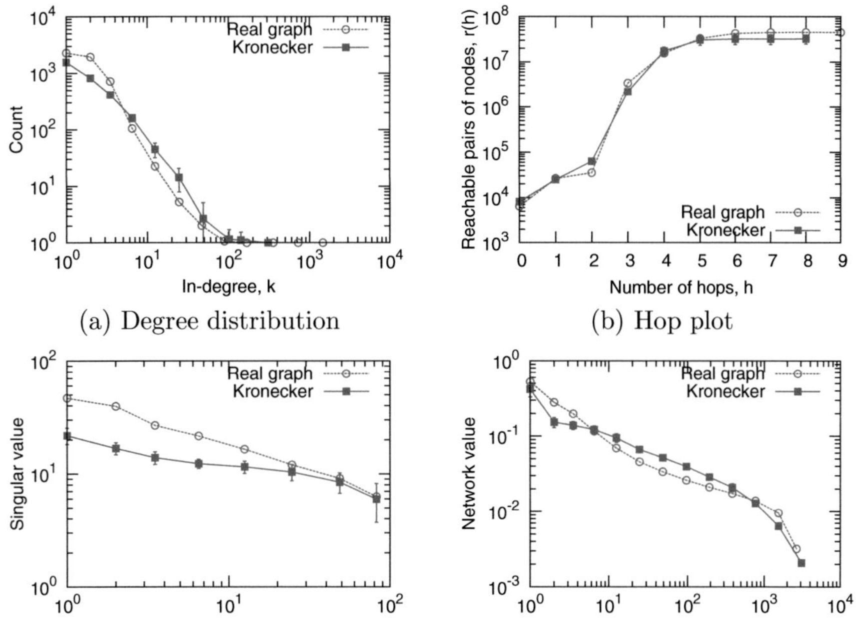

(a) Degree distribution (b) Hop plot

(c) Scree plot (d) "Network" value

Figure 9.17. Autonomous systems (AS-ROUTEVIEWS)**.** Overlaid patterns of the real graph and the fitted Kronecker graph. Notice that the fitted Kronecker graph matches patterns of the real graph while using only four parameters (2×2 initiator matrix).

Figure 9.17 shows the properties of AS-ROUTEVIEWS and compares them with the properties of a synthetic Kronecker graph generated using the fitted parameters $\hat{\Theta}$ of size 2×2. Notice that the properties of both graphs match really well. The estimated parameters are $\hat{\Theta} = [0.987, 0.571; 0.571, 0.049]$.

Figure 9.17(a) compares the degree distributions of the AS-ROUTEVIEWS network and its synthetic Kronecker estimate. In this and all other plots, we use the exponential binning, which is a standard procedure to de-noise the data when plotting on log-log scales. Notice a very close match in degree distribution between the real graph and its synthetic counterpart.

Figure 9.17(b) plots the cumulative number of pairs of nodes $g(h)$ that can be reached in $\leq h$ hops. The hop plot gives a sense about the distribution of the shortest path lengths in the network and about the network diameter. Last, Figures 9.17(c) and (d) plot the spectral properties of the graph adjacency matrix. Figure 9.17(c) plots largest singular values versus rank, and Figure 9.17(d) plots the components of the left singular vector (the network value) versus the rank. Again notice the good agreement with the real graph while using only four parameters.

Moreover, on all plots, the error bars of two standard deviations show the variance of the graph properties for different realizations $R(\hat{\Theta}^{\otimes k})$. To obtain the error bars, we took the same $\hat{\Theta}$ and generated 50 realizations of a Kronecker graph. For most of the plots, the error bars are so small as to be practically invisible; this shows that the variance of network properties when generating a stochastic Kronecker graph is indeed very small.

Also notice that the AS-ROUTEVIEWS is an undirected graph and that the fitted parameter matrix $\hat{\Theta}$ is in fact symmetric. This means that without a priori biasing the fitting toward undirected graphs, the recovered parameters obey this aspect of the network. Fitting the AS-ROUTEVIEWS graph from a random set of parameters, performing gradient descent for 100 iterations, and at each iteration sampling half a million permutations took less than 10 minutes on a standard desktop PC. This is a significant speedup over [Bezáková et al. 2006], where using a similar permutation sampling approach for calculating the likelihood of a preferential attachment model on a similar AS-ROUTEVIEWS graph took about two days on a cluster of 50 machines.

Choice of the initiator matrix size N

As mentioned earlier, for finding the optimal number of parameters, i.e., selecting the size of the initiator matrix, BIC naturally applies to the case of Kronecker graphs. Figure 9.23(b) shows BIC scores for the following experiment. We generated a Kronecker graph with $N = 2187$ and $M = 8736$ using $N = 3$ (9 parameters) and $k = 7$. For $1 \leq N \leq 9$, we find the MLE parameters using gradient descent and calculate the BIC scores. The model with the lowest score is chosen. As Figure 9.23(b) shows, we recovered the true model, i.e., BIC score is the lowest for the model with the true number of parameters, $N = 3$.

Intuitively we expect a more complex model with more parameters to fit the data better. Thus, we expect larger N to generally give better likelihood. On the other hand, the fit will also depend on the size of the graph $\mathbf{G}$. Kronecker graphs can only generate graphs on N^k nodes, while real graphs do not necessarily have N^k nodes (for some, preferably small, integers N and k). To solve this problem, we choose k so that $N^{k-1} < N(G) \leq N^k$, and then augment G by adding $N^k - N$ isolated nodes. Or equivalently, we pad the adjacency matrix of G with zeros until it is of the appropriate size, $N^k \times N^k$. While this solves the problem of requiring the integer power of the number of nodes, it also makes the fitting problem harder; for example, when $N \ll N^k$, we are basically fitting G plus a large number of isolated nodes.

Table 9.2 shows the results of fitting Kronecker graphs to AS-ROUTEVIEWS while varying the size of the initiator matrix N. First, notice that, in general, larger N results in higher log-likelihood $l(\hat{\Theta})$ at MLE. Similarly, notice (column N^k) that while AS-ROUTEVIEWS has 6474 nodes, Kronecker estimates have up to 16,384 nodes ($16,384 = 4^7$, which is the first integer power of 4 greater than 6474). However, we also show the number of nonzero-degree (nonisolated) nodes in the Kronecker graph (column $|\{\deg(u) > 0\}|$). Notice that the number of nonisolated nodes well corresponds to the number of nodes in the AS-ROUTEVIEWS network. This shows that KRONFIT is actually fitting the graph well, and it successfully fits

Table 9.2. Log-likelihood at MLE.
MLE for different choices of the size of the initiator matrix N for the AS-ROUTEVIEWS graph. Notice that the log-likelihood $l(\hat{\theta})$ generally increases with the model complexity N. Also notice the effect of zero-padding, i.e., for $N = 4$ and $N = 5$, the constraint of the number of nodes being an integer power of N decreases the log-likelihood. However, the column $|\{\deg(u) > 0\}|$ gives the number of nonisolated nodes in the network, which is much less than N^k and is, in fact, very close to the true number of nodes in the AS-ROUTEVIEWS. Using the BIC scores, we see that $N = 3$ or $N = 6$ is the best choice for the size of the initiator matrix.

N	$l(\hat{\Theta})$	N^k	M^k	$\|\{\deg(u) > 0\}\|$	BIC score
2	−152,499	8192	25,023	5675	152,506
3	−127,066	6561	28,790	5683	127,083
4	−153,260	16,384	24,925	8222	153,290
5	−149,949	15,625	29,111	9822	149,996
6	−128,241	7776	26,557	6623	128,309
AS-ROUTEVIEWS			26,467	6474	

the structure of the graph plus a number of isolated nodes. Last, column M^k gives the number of edges in the corresponding Kronecker graph, which is close to the true number of edges of the AS-ROUTEVIEWS graph.

Last, comparing the log-likelihood at the MLE and the BIC score in Table 9.2, we notice that the log-likelihood heavily dominates the BIC score. This means that the size of the initiator matrix (number of parameters) is so small that overfitting is not a concern. Thus, we can just choose the initiator matrix that maximizes the likelihood. A simple calculation shows that one would need to take initiator matrices with thousands of entries before the model complexity part of the BIC score would start to play a significant role.

We further examine the sensitivity of the choice of the initiator size by the following experiment. We generate a stochastic Kronecker graph K on nine parameters ($N = 3$), and then fit a Kronecker graph K' with a smaller number of parameters (four instead of nine, $N' = 2$), and also a Kronecker graph K'' of the same complexity as K ($N'' = 3$).

Figure 9.18 plots the properties of all three graphs. Not surprisingly, K'' (blue) fits the properties of K (red) perfectly as the initiator is of the same size. On the other hand, K' (green) is a simpler model with only four parameters (instead of nine as in K and K'') and still generally fits well: hop plot and degree distribution match well, while spectral properties of graph adjacency matrix, especially scree plot, are not matched that well. This shows that nothing drastic happens and that even a bit too simple model still fits the data well. In general, we observe empirically that by increasing the size of the initiator matrix, one does not gain radically better fits for degree distribution and hop plot. On the other hand, there is usually an

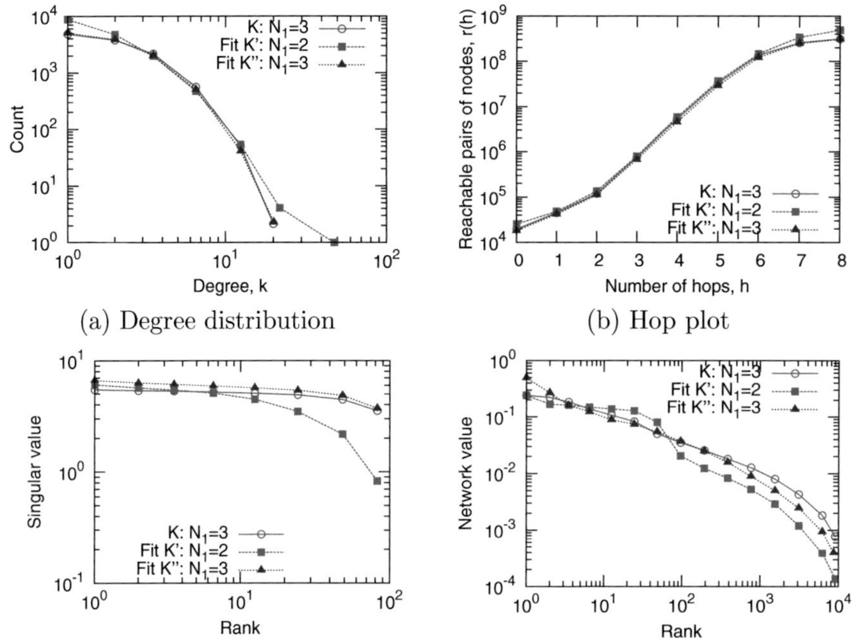

(a) Degree distribution (b) Hop plot

(c) Scree plot (d) "Network" value

Figure 9.18. 3 × 3 stochastic Kronecker graphs. Given a stochastic Kronecker graph G generated from $N = 3$ (circles), we fit a Kronecker graph K' with $N' = 2$ (squares) and K'' with $N'' = 3$ (triangles). Not surprisingly, K'' fits the properties of K perfectly as the model is of the same complexity. On the other hand, K' has only four parameters (instead of nine as in K and K'') and still fits well.

improvement in the scree plot and the plot of network values when one increases the initiator size.

Network parameters over time

Next, we briefly examine the evolution of the Kronecker initiator for a temporally evolving graph. The idea is that, given parameter estimates of a real graph G_t at time t, we can forecast the future structure of the graph G_{t+x} at time $t + x$, i.e., using parameters obtained from G_t, we can generate a larger synthetic graph K that will be similar to G_{t+x}.

As we have the information about the evolution of the AS-ROUTEVIEWS network, we estimated parameters for three snapshots of the network when it had about 2^k nodes. Table 9.3 gives the results of the fitting for the three temporal snapshots of the AS-ROUTEVIEWS network. Notice that the parameter estimates

Table 9.3. Parameter estimates of temporal snapshots. Parameter estimates of the three temporal snapshots of the AS-ROUTEVIEWS network. Notice that estimates stay remarkably stable over time.

Snapshot at time	N	M	$l(\hat{\Theta})$	Estimates at MLE, $\hat{\Theta}$
T_1	2048	8794	−40,535	[0.981, 0.633; 0.633, 0.048]
T_2	4088	15,711	−82,675	[0.934, 0.623; 0.622, 0.044]
T_3	6474	26,467	−152,499	[0.987, 0.571; 0.571, 0.049]

$\hat{\Theta}$ remain remarkably stable over time. This stability means that Kronecker graphs can be used to estimate the structure of the networks in the future, i.e., parameters estimated from the historic data can extrapolate the graph structure in the future.

Figure 9.19 further explores this. It overlays the graph properties of the real AS-ROUTEVIEWS network at time T_3 and the synthetic graphs for which we used the parameters obtained on historic snapshots of AS-ROUTEVIEWS at times T_1 and T_2. The agreements are good, demonstrating that Kronecker graphs can forecast the structure of the network in the future.

Moreover, this experiment also shows that parameter estimates do not suffer much from the zero-padding of a graph adjacency matrix (i.e., adding isolated nodes to make G have N^k nodes). Snapshots of AS-ROUTEVIEWS at T_1 and T_2 have close to 2^k nodes, while we had to add 26% (1718) isolated nodes to the network at T_3 to make the number of nodes be 2^k. Regardless of this, we see that the parameter estimates $\hat{\Theta}$ remain basically constant over time, which seems to be independent of the number of isolated nodes added. This means that the estimated parameters are not biased too much from zero-padding the adjacency matrix of G.

9.6.5 Fitting to other large real-world networks

Last, we present results of fitting stochastic Kronecker graphs to 20 large real-world networks: large online social networks, (EPINIONS, FLICKR, and DELICIOUS), web and blog graphs (WEB-NOTREDAME, BLOG-NAT05-6M, BLOG-NAT06ALL), Internet and peer-to-peer networks (AS-NEWMAN, GNUTELLA-25, GNUTELLA-30), collaboration networks of coauthorships from DBLP (CA-DBLP) and various areas of physics (CA-HEP-TH, CA-HEP-PH, CA-GR-QC), physics citation networks (CIT-HEP-PH, CIT-HEP-TH), an email network (EMAIL-INSIDE), a protein-interaction network (BIO-PROTEINS), and a bipartite affiliation network (authors-to-papers, ATP-GR-QC). Refer to Table 9.5 in the appendix for the description and basic properties of these networks. They are available for download at http://snap.stanford.edu.

For each data set, we started gradient descent from a random point (random initiator matrix) and ran it for 100 steps. At each step, we estimated the likelihood and the gradient on the basis of 510,000 sampled permutations where we discard the first 10,000 samples to allow the chain to burn in.

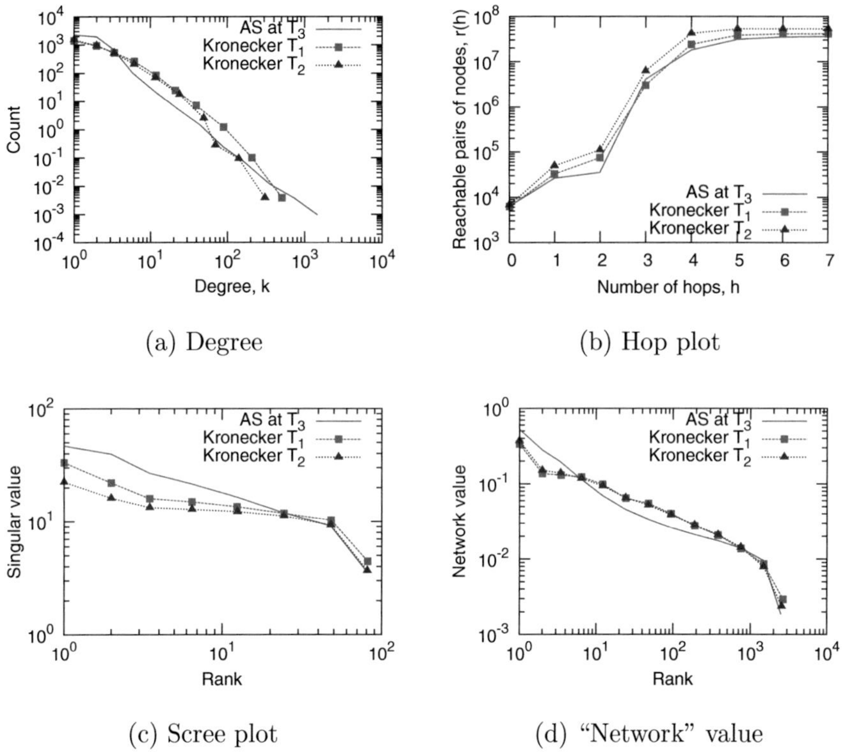

Figure 9.19. Autonomous Systems (AS) network over time (AS-ROUTEVIEWS)**.** Overlaid patterns of real AS-ROUTEVIEWS network at time T_3 and the Kronecker graphs with parameters estimated from AS-ROUTEVIEWS at time T_1 and T_2. Notice the good fits which mean that parameters estimated on historic snapshots can be used to estimate the graph in the future.

Table 9.4 gives the estimated parameters, the corresponding log-likelihoods, and the wall clock times. All experiments were carried out on a standard desktop computer. Notice that the estimated initiator matrices $\hat{\Theta}$ seem to have almost universal structure with a large value in the top left entry, a very small value at the bottom right corner, and intermediate values in the other two corners. We further discuss the implications of such a structure of a Kronecker initiator matrix on the global network structure in the next section.

Last, Figures 9.20 and 9.21 show overlays of various network properties of real and estimated synthetic networks. In addition to the network properties we plotted in Figure 9.18, we also separately plot in- and out-degree distributions (as both networks are directed) and plot the node triangle participation in panel (c), where we plot the number of triangles a node participates in versus the number of such

Table 9.4. Results of parameter estimation.

Parameters for 20 different networks. Table 9.5 gives the description and basic properties of the network data sets. Networks are available for download at http://snap.stanford.edu.

Network	N	M	Estimated MLE parameters $\hat{\Theta}$	$l(\hat{\Theta})$	Time
As-RouteViews	6474	26,467	$[0.987, 0.571; 0.571, 0.049]$	−152,499	8m15s
AtP-gr-qc	19,177	26,169	$[0.902, 0.253; 0.221, 0.582]$	−242,493	7m40s
Bio-Proteins	4626	29,602	$[0.847, 0.641; 0.641, 0.072]$	−185,130	43m41s
Email-Inside	986	32,128	$[0.999, 0.772; 0.772, 0.257]$	−107,283	1h07m
CA-gr-qc	5242	28,980	$[0.999, 0.245; 0.245, 0.691]$	−160,902	14m02s
As-Newman	22,963	96,872	$[0.954, 0.594; 0.594, 0.019]$	−593,747	28m48s
Blog-nat05-6m	31,600	271,377	$[0.999, 0.569; 0.502, 0.221]$	−1,994,943	47m20s
Blog-nat06all	32,443	318,815	$[0.999, 0.578; 0.517, 0.221]$	−2,289,009	52m31s
CA-hep-ph	12,008	237,010	$[0.999, 0.437; 0.437, 0.484]$	−1,272,629	1h22m
CA-hep-th	9877	51,971	$[0.999, 0.271; 0.271, 0.587]$	−343,614	21m17s
Cit-hep-ph	30,567	348,721	$[0.994, 0.439; 0.355, 0.526]$	−2,607,159	51m26s
Cit-hep-th	27,770	352,807	$[0.990, 0.440; 0.347, 0.538]$	−2,507,167	15m23s
Epinions	75,879	508,837	$[0.999, 0.532; 0.480, 0.129]$	−3,817,121	45m39s
Gnutella-25	22,687	54,705	$[0.746, 0.496; 0.654, 0.183]$	−530,199	16m22s
Gnutella-30	36,682	88,328	$[0.753, 0.489; 0.632, 0.178]$	−919,235	14m20s
Delicious	205,282	436,735	$[0.999, 0.327; 0.348, 0.391]$	−4,579,001	27m51s
Answers	598,314	1,834,200	$[0.994, 0.384; 0.414, 0.249]$	−20,508,982	2h35m
CA-DBLP	425,957	2,696,489	$[0.999, 0.307; 0.307, 0.574]$	−26,813,878	3h01m
Flickr	584,207	3,555,115	$[0.999, 0.474; 0.485, 0.144]$	−32,043,787	4h26m
Web-NotreDame	325,729	1,497,134	$[0.999, 0.414; 0.453, 0.229]$	−14,588,217	2h59m

nodes. (Again the error bars show the variance of network properties over different realizations $R(\hat{\Theta}^{\otimes k})$ of a stochastic Kronecker graph.)

Notice that, for both networks and in all cases, the properties of the real network and the synthetic Kronecker coincide very well. Using stochastic Kronecker graphs with just four parameters, we match the scree plot, degree distributions, triangle participation, hop plot, and network values.

Given the previous experiments from the autonomous systems graph, we only present the results for the simplest model with initiator size $N = 2$. Empirically, we also observe that $N = 2$ gives surprisingly good fits and the estimation procedure is the most robust and converges the fastest. Using larger initiator matrices $N > 2$ generally helps improve the likelihood but not dramatically. In terms of matching the network properties, we also get a slight improvement by making the model more complex. Figure 9.22 gives the percent improvement in log-likelihood as we make the model more complex. We use the log-likelihood of a 2×2 model as a baseline and estimate the log-likelihood at the MLE for larger initiator matrices. Again, models with more parameters tend to fit better. However, sometimes due to zero-padding of a graph adjacency matrix, they actually have lower log-likelihood (as seen in Table 9.2).

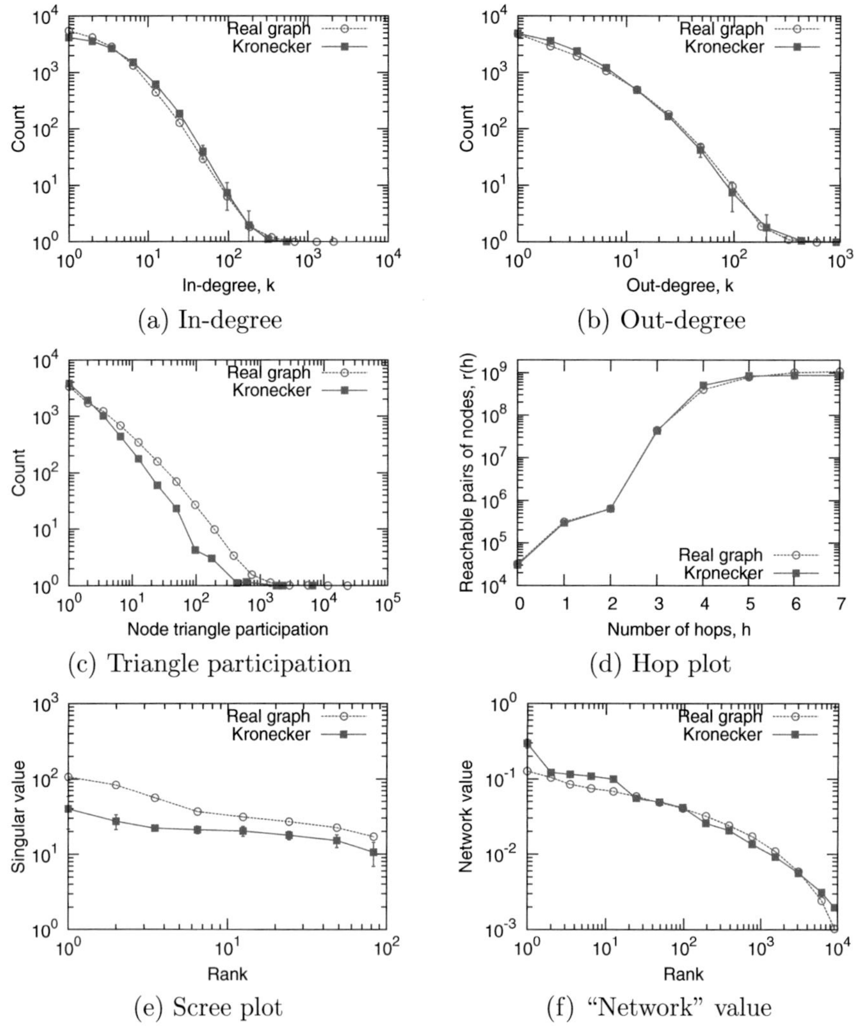

Figure 9.20. Blog network (BLOG-NAT06ALL).
Overlaid patterns of the real network and the estimated Kronecker graph using four parameters (2×2 initiator matrix). Notice that the Kronecker graph matches all properties of the real network.

9.6.6 Scalability

Last, we also empirically evaluate the scalability of the KRONFIT. The experiment confirms that KRONFIT runtime scales linearly with the number of edges M in a graph G. More precisely, we performed the following experiment.

We generated a sequence of increasingly larger synthetic graphs on N nodes and $8N$ edges, and measured the time of one iteration of gradient descent, i.e.,

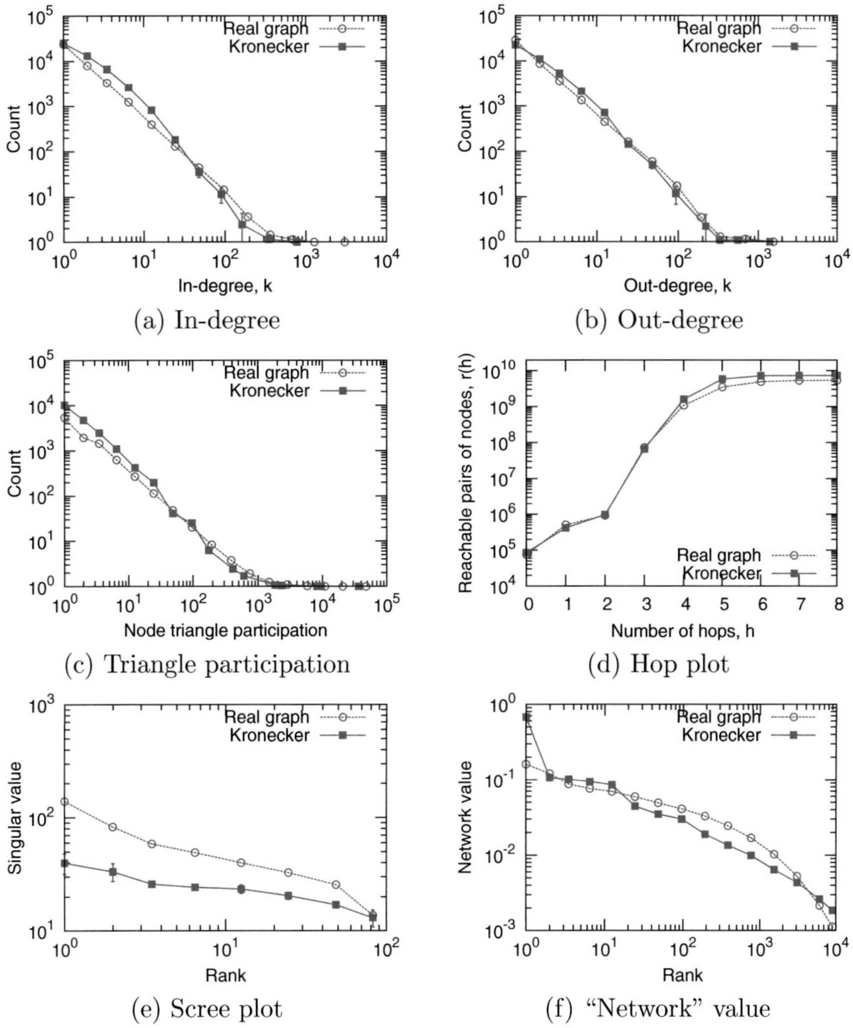

Figure 9.21. Who-trusts-whom social network (Epinions). Overlaid patterns of the real network and the fitted Kronecker graph using only four parameters (2×2 initiator matrix). Again, the synthetic Kronecker graph matches all the properties of the real network.

sample one million permutations and evaluate the gradients. We started with a graph on 1000 nodes and finished with a graph on 8 million nodes and 64 million edges. Figure 9.23(a) shows that KronFit scales *linearly* with the size of the network. We plot wall-clock time versus size of the graph. The dashed line gives a linear fit to the data points.

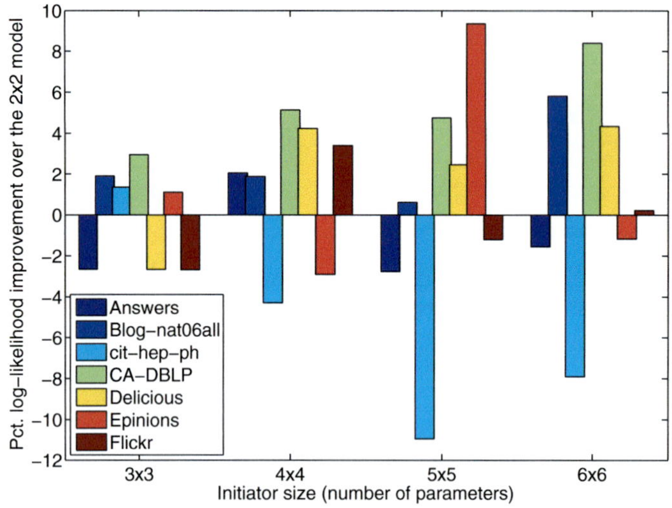

Figure 9.22. Improvement in log-likelihood.
Percent improvement in log-likelihood over the 2×2 model as we increase the model complexity (size of initiator matrix). In general, larger initiator matrices that have more degrees of freedom help improve the fit of the model.

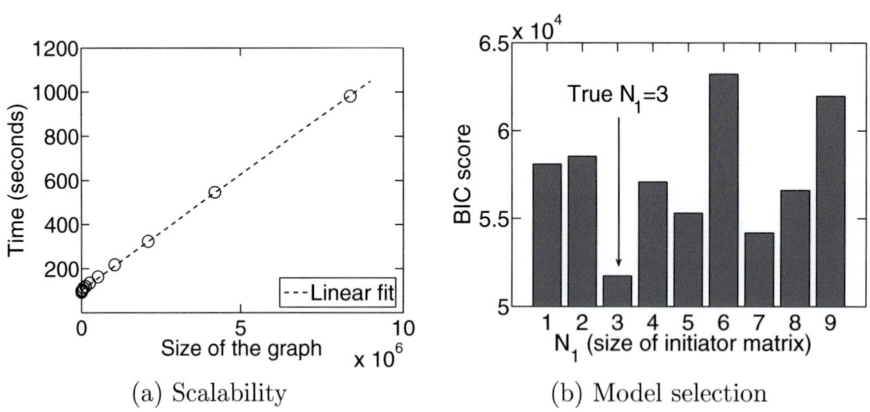

Figure 9.23. Performance.
(a) Processor time to sample one million gradients as the graph grows. Notice the algorithm scales linearly with the graph size. (b) BIC score for model selection.

9.7 Discussion

Here we discuss several of the desirable properties of the proposed Kronecker graphs.

Generality: Stochastic Kronecker graphs include several other generators as special cases. For $\theta_{ij} = c$, we obtain the classical Erdős–Rényi random graph model. For $\theta_{i,j} \in \{0, 1\}$, we obtain a deterministic Kronecker graph. Setting the $\mathbf{K}_1$ matrix to a 2×2 matrix, we obtain the R-MAT generator [Chakrabarti et al. 2004]. In contrast to Kronecker graphs, the R-MAT cannot extrapolate into the future since it needs to know the number of edges to insert. Thus, it is incapable of obeying the densification power law.

Phase transition phenomena: The Erdős–Rényi graphs exhibit phase transitions [Erdős & A. Rényi 1960]. Several researchers argue that real systems are "at the edge of chaos" or phase transition [Bak 1996, Sole & Goodwin 2000]. Stochastic Kronecker graphs also exhibit phase transitions [Mahdian & Xu 2007] for the emergence of the giant component and another phase transition for connectivity.

Implications to the structure of the large-real networks: Empirically, we found that 2×2 initiator ($N = 2$) fits well the properties of real-world networks. Moreover, given a 2×2 initiator matrix, one can look at it as a recursive expansion of two groups into subgroups. We introduced this recursive view of Kronecker graphs back in Section 9.3. So, one can then interpret the diagonal values of Θ as the proportion of edges inside each of the groups, and the off-diagonal values give the fraction of edges connecting the groups. Figure 9.24 illustrates the setting for two groups.

For example, as shown in Figure 9.24, large a, d and small b, c would imply that the network is composed of hierarchically nested communities, where there are many edges inside each community and few edges crossing them [Leskovec 2009]. One could think of this structure as some kind of organizational or university hierarchy, where one expects the most friendships between people within same lab, a bit less between people in the same department, less across different departments, and the least friendships to be formed across people from different schools of the university.

However, parameter estimates for a wide range of networks presented in Table 9.4 suggest a very different picture of the network structure. Notice that for most networks $a \gg b > c \gg d$. Moreover, $a \approx 1$, $b \approx c \approx 0.6$ and $d \approx 0.2$. We empirically observed that the same structure of initiator matrix $\hat{\Theta}$ also holds when fitting 3×3 or 4×4 models. Always the top left element is the largest, and then the values on the diagonal decay faster than the values off the diagonal [Leskovec 2009]. This suggests a network structure that is also known as *core-periphery* [Borgatti & Everett 2000, Holme 2005], the *jellyfish* [Tauro et al. 2001, Siganos et al. 2006], or the *octopus* [Chung & Lu 2006] structure of the network, as illustrated in Figure 9.24(c).

All of the above basically say that the network is composed of a densely linked network core and the periphery. In our case, this would imply the following structure of the initiator matrix. The core is modeled by parameter a and the periphery by d. Most edges are inside the core (large a), and very few between the nodes of periphery (small d). Then there are many more edges between the core and the periphery

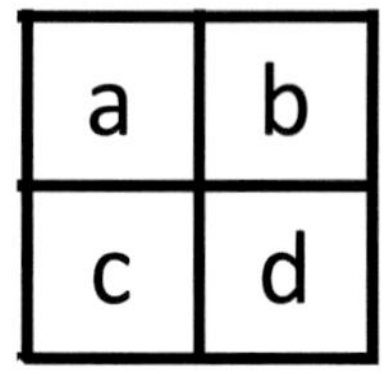

(a) 2×2 initiator matrix

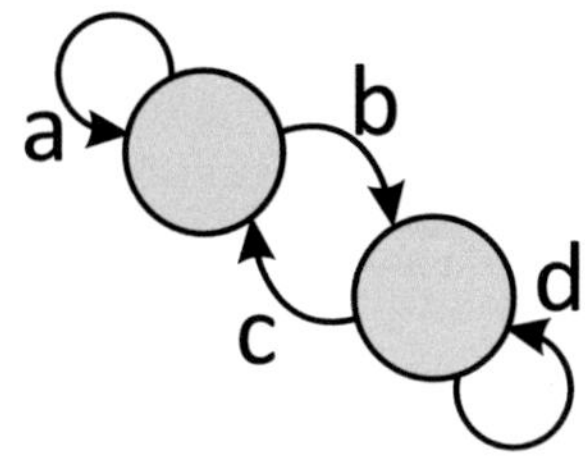

(b) Two recursive communities

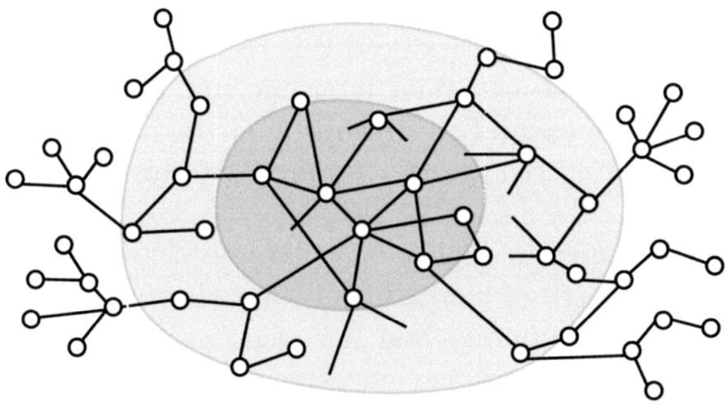

(c) Core-periphery

Figure 9.24. Kronecker communities.
2×2 Kronecker initiator matrix (a) can be thought of as two communities where there are a and d edges inside each of the communities and b and c edges crossing the two communities as illustrated in (b). Each community can then be recursively divided using the same pattern. (c) The onionlike core-periphery structure where the network gets denser and denser as we move towards the center of the network.

than inside the periphery ($b, c > d$) [Leskovec 2009]. This is exactly what we see as well. In the spirit of Kronecker graphs, the structure repeats recursively—the core again has the dense core and the periphery, and so on. Similarly, the periphery itself has the core and the periphery.

This structure suggests an onionlike *nested core-periphery* [Leskovec et al. 2008b, Leskovec et al. 2008a] network structure as illustrated in Figure 9.24(c), where the network is composed of denser and denser layers as one moves towards the center of the network. We also observe a similar structure of the Kronecker initiator when fitting a 3×3 or 4×4 initiator matrix. The diagonal elements have large but decreasing values with off-diagonal elements following the same decreasing pattern.

These Kronecker initiators imply that networks do not break nicely into hierarchically organized sets of communities that lend themselves to graph partitioning and community detection algorithms. On the contrary, it appears that large networks can be decomposed into a densely linked core with many small periphery pieces hanging off the core. Our recent results [Leskovec et al. 2008b, Leskovec et al. 2008a] make a similar observation (but based on a completely different methodology based on graph partitioning) about the clustering and community structure of large real-world networks.

9.8 Conclusion

In conclusion, the main contribution of this work is a family of models of network structure that uses a nontraditional matrix operation, the *Kronecker product*. The resulting graphs have (a) all the static properties (heavy-tailed degree distribution, small diameter, etc.) and (b) all the temporal properties (densification, shrinking diameter) that are found in real networks. In addition, we can formally prove all of these properties.

Several of the proofs are extremely simple, thanks to the rich theory of Kronecker multiplication. We also provide proofs about the diameter and effective diameter, and we show that stochastic Kronecker graphs can mimic real graphs well.

Moreover, we also presented KRONFIT, a fast, scalable algorithm to estimate the stochastic Kronecker initiator, which can then be used to create a synthetic graph that mimics the properties of a given real network.

In contrast to earlier work, our work has the following novelties: (a) it is among the few that estimates the parameters of the chosen generator in a principled way, (b) it is among the few that has a concrete measure of goodness of the fit (namely, the likelihood), (c) it avoids the quadratic complexity of computing the likelihood by exploiting the properties of the Kronecker graphs, and (d) it avoids the factorial explosion of the node correspondence problem by using the Metropolis sampling.

The resulting algorithm matches well all the known properties of real graphs. As we show with the Epinions graph and the AS graph, it scales linearly on the number of edges, and it is orders of magnitudes faster than earlier graph-fitting attempts: 20 minutes on a commodity PC versus two days on a cluster of 50 workstations [Bezáková et al. 2006].

The benefits of fitting a Kronecker graph model into a real graph are several:

- *Extrapolation*: Once we have the Kronecker generator Θ for a given real matrix $\mathbf{G}$ (such that $\mathbf{G}$ is mimicked by $\Theta^{\otimes k}$), a larger version of $\mathbf{G}$ can be generated by $\Theta^{\otimes k+1}$.

- *Null-model*: When analyzing a real network G, one often needs to asses the significance of the observation. $\Theta^{\otimes k}$ that mimics G can be used as an accurate model of G.

- *Network structure*: Estimated parameters give insight into the global network and community structure of the network.
- *Forecasting*: As we demonstrated, one can obtain Θ from a graph G_t at time t such that $\mathbf{G}$ is mimicked by $\Theta^{\otimes k}$. Then Θ can be used to model the structure of G_{t+x} in the future.
- *Sampling*: Similarly, if we want a realistic sample of the real graph, we could use a smaller exponent in the Kronecker exponentiation, like $\Theta^{\otimes k-1}$.
- *Anonymization*: Since $\Theta^{\otimes k}$ mimics $\mathbf{G}$, we can publish $\Theta^{\otimes k}$, without revealing information about the nodes of the real graph $\mathbf{G}$.

Future work could include extensions of Kronecker graphs to evolving networks. We envision formulating a dynamic Bayesian network with first order Markov dependencies, where parameter matrix at time t depends on the graph G_t at current time t and the parameter matrix at time $t-1$. Given a series of network snapshots, one would then aim to estimate initiator matrices at individual time steps and the parameters of the model governing the evolution of the initiator matrix. We expect that, on the basis of the evolution of the initiator matrix, one would gain greater insight into the evolution of large networks.

A second direction for future work is to explore connections between Kronecker graphs and Random Dot Product graphs [Young & Scheinerman 2007, Nickel 2008]. This also nicely connects with the "attribute view" of Kronecker graphs as described in Section 9.3.5. It would be interesting to design methods to estimate the individual node attribute values as well as the attribute-attribute similarity matrix (i.e., the initiator matrix). If some networks node attributes are already given, one could then try to infer "hidden" or missing node attribute values and this way gain insight into individual nodes as well as individual edge formations. Moreover, this would be interesting as one could further evaluate how realistic is the "attribute view" of Kronecker graphs.

Last, we also mention possible extensions of Kronecker graphs for modeling weighted and labeled networks. Currently stochastic Kronecker graphs use a Bernoulli edge generation model, i.e., an entry of big matrix $\mathcal{P}$ encodes the parameter of a Bernoulli coin. In a similar spirit, one could consider entries of $\mathcal{P}$ to encode parameters of different edge generative processes. For example, to generate networks with weights on edges, an entry of $\mathcal{P}$ could encode the parameter of an exponential distribution, or in the case of labeled networks, one could use several initiator matrices in parallel and this way encode parameters of a multinomial distribution over different node attribute values.

Appendix: Table of networks

Table 9.5 lists all the network data sets that were used in this chapter. We also computed some of the structural network properties. Most of the networks are available for download at http://snap.stanford.edu.

Table 9.5. Network data sets analyzed.

Statistics of networks considered: number of nodes N; number of edges E, number of nodes in largest connected component N_c, fraction of nodes in largest connected component N_c/N, average clustering coefficient $\bar{C}$; diameter D, and average path length $\bar{D}$. Networks are available for download at http://snap.stanford.edu.

Network	N	E	N_c	N_c/N	$\bar{C}$	D	$\bar{D}$	Description
Social networks								
ANSWERS	598,314	1,834,200	488,484	0.82	0.11	22	5.72	Yahoo! Answers social network [Leskovec et al. 2008b]
DELICIOUS	205,282	436,735	147,567	0.72	0.3	24	6.28	del.icio.us social network [Leskovec et al. 2008b]
EMAIL-INSIDE	986	32,128	986	1.00	0.45	7	2.6	European research organization email network [Leskovec et al. 2007a]
EPINIONS	75,879	508,837	75,877	1.00	0.26	15	4.27	Who-trusts-whom graph of epinions.com [Richardson et al. 2003]
FLICKR	584,207	3,555,115	404,733	0.69	0.4	18	5.42	Flickr photo-sharing social network [Kumar et al. 2006]
Information (citation) networks								
BLOG-NAT05-6M	31,600	271,377	29,150	0.92	0.24	10	3.4	Blog-to-blog citation network (6 months of data) [Leskovec et al. 2007b]
BLOG-NAT06ALL	32,443	318,815	32,384	1.00	0.2	18	3.94	Blog-to-blog citation network (1 year of data) [Leskovec et al. 2007b]
CIT-HEP-PH	30,567	348,721	34,401	1.13	0.3	14	4.33	Citation network of ArXiv hep-th papers [Gehrke et al. 2003]
CIT-HEP-TH	27,770	352,807	27,400	0.99	0.33	15	4.2	Citations network of ArXiv hep-ph papers [Gehrke et al. 2003]
Collaboration networks								
CA-DBLP	425,957	2,696,489	317,080	0.74	0.73	23	6.75	DBLP coauthorship network [Backstrom et al. 2006]
CA-GR-QC	5242	28,980	4158	0.79	0.66	17	6.1	Coauthorship network in gr-qc ArXiv [Leskovec et al. 2005b]
CA-HEP-PH	12,008	237,010	11,204	0.93	0.69	13	4.71	Coauthorship network in hep-ph ArXiv [Leskovec et al. 2005b]
CA-HEP-TH	9877	51,971	8638	0.87	0.58	18	5.96	Coauthorship network in hep-th ArXiv [Leskovec et al. 2005b]
Web graphs								
WEB-NOTREDAME	325,729	1,497,134	325,729	1.00	0.47	46	7.22	Web graph of University of Notre Dame [Albert et al. 1999]
Internet networks								
AS-NEWMAN	22,963	96,872	22,963	1.00	0.35	11	3.83	AS graph from Newman [newman07netdata]
AS-ROUTEVIEWS	6474	26,467	6474	1.00	0.4	9	3.72	AS from Oregon Route View [Leskovec et al. 2005b]
GNUTELLA-25	22,687	54,705	22,663	1.00	0.01	11	5.57	Gnutella P2P network on 3/25 2000 [Ripeanu et al. 2002]
GNUTELLA-30	36,682	88,328	36,646	1.00	0.01	11	5.75	Gnutella P2P network on 3/30 2000 [Ripeanu et al. 2002]
Bipartite networks								
ATP-GR-QC	19,177	26,169	14,832	0.77	0	35	11.08	Affiliation network of gr-qc category in ArXiv [Leskovec et al. 2007b]
Biological networks								
BIO-PROTEINS	4626	29,602	4626	1.00	0.12	12	4.24	Yeast protein interaction network [Colizza et al. 2005]

References

[newman07netdata] Network data. http://www-personal.umich.edu/~mejn/netdata, July 16, 2007.

[Albert & Barabási 2002] R. Albert and A.-L. Barabási. Statistical mechanics of complex networks. *Reviews of Modern Physics*, 74:47–97, 2002.

[Albert et al. 1999] R. Albert, H. Jeong, and A.-L. Barabási. Diameter of the world-wide web. *Nature*, 401:130–131, 1999.

[Backstrom et al. 2006] L. Backstrom, D. Huttenlocher, J. Kleinberg, and X. Lan. Group formation in large social networks: membership, growth, and evolution. In *KDD '06: Proceedings of the 12th ACM SIGKDD international conference on knowledge discovery and data mining*, 44–54, 2006.

[Bak 1996] P. Bak. *How Nature Works: The Science of Self-Organized Criticality.* New York: Springer, 1996.

[Barabási & Albert 1999] A.-L. Barabási and R. Albert. Emergence of scaling in random networks. *Science*, 286:509–512, 1999.

[Barabási et al. 2001] A.-L. Barabási, E. Ravasz, and T. Vicsek. Deterministic scale-free networks. *Physica A*, 299:559–564, 2001.

[Bezáková et al. 2006] I. Bezáková, A. Kalai, and R. Santhanam. Graph model selection using maximum likelihood. In *ICML '06: Proceedings of the 23rd International Conference on Machine Learning*, 105–112, 2006.

[Bi et al. 2001] Z. Bi, C. Faloutsos, and F. Korn. The DGX distribution for mining massive, skewed data. In *KDD '01: Proceedings of the 6th ACM SIGKDD International Conference on Knowledge Discovery and Data Mining*, 17–26, 2001.

[Blum et al. 2006] A. Blum, H. Chan, and M. Rwebangira. A random-surfer web-graph model. In *ANALCO '06: Proceedings of the 3rd Workshop on Analytic Algorithmics and Combinatorics*, 2006.

[Borgatti & Everett 2000] S.P. Borgatti and M.G. Everett. Models of core/periphery structures. *Social Networks*, 21:375–395, 2000.

[Bottreau & Metivier 1998] A. Bottreau and Y. Metivier. Some remarks on the Kronecker product of graphs. *Information Processing Letters*, 68:55–61, 1998.

[Broder et al. 2000] A. Broder, R. Kumar, F. Maghoul, P. Raghavan, S. Rajagopalan, R. Stata, A. Tomkins, and J. Wiener. Graph structure in the web: Experiments and models. In *WWW '00: Proceedings of the 9th International World Wide Web Conference*, 2000.

[Bu & Towsley 2002] T. Bu and D.F. Towsley. On distinguishing between internet power law topology generators. In *INFOCOM*, 2002.

[Butts 2005] C.T. Butts. Permutation models for relational data. (Technical Report MBS 05-02, Univ. of California, Irvine, 2005.

[Carlson & Doyle 1999] J.M. Carlson and J. Doyle. Highly optimized tolerance: A mechanism for power laws in designed systems. *Physical Review E*, 60:1412–1427, 1999.

[Chakrabarti et al. 2004] D. Chakrabarti, Y. Zhan, and C. Faloutsos. R-MAT: A recursive model for graph mining. In Proceedings of the 4th *SDM04: SIAM International Conference on Data Mining*, 2004. http://www.siam.org/proceedings/datemining/2004/dmot.php.

[Chow 1997] T. Chow. The Q-spectrum and spanning trees of tensor products of bipartite graphs. *Proceedings of the American Mathematical Society*, 125:3155–3161, 1997.

[Chung & Lu 2006] F.R.K. Chung and L. Lu. *Complex Graphs and Networks*, volume 107 of *CBMS Regional Conference Series in Mathematics.* American Mathematical Society, 2006.

[Chung et al. 2003] F.R.K. Chung, L. Lu, and V. Vu. Eigenvalues of random power law graphs. *Annals of Combinatorics*, 7:21–33, 2003.

[Clauset et al. 2009] A. Clauset, C. Moore, and M.E.J. Newman. Hierarchical structure and the prediction of missing links in networks. *Nature*, 453:98–101, 2008.

[Clauset et al. 2007] A. Clauset, C.R. Shalizi, and M.E.J. Newman. Power-law distributions in empirical data. *ArXiv*, http://arxiv.org/abs/0706.1062, 2007.

[Colizza et al. 2005] V. Colizza, A. Flammini, M.A. Serrano, and A. Vespignani. Characterization and modeling of protein protein interaction networks. *Physica. A. Statistical Mechanics and Its Applications*, 352:1–27, 2005.

[Crovella & Bestavros 1997] M.E. Crovella and A. Bestavros. Self-similarity in World Wide Web traffic: Evidence and possible causes. *IEEE /ACM Transactions on Networking*, 5:835–846, 1997.

[Dill et al. 2002] S. Dill, R. Kumar, K.S. Mccurley, S. Rajagopalan, D. Sivakumar, and A. Tomkins. Self-similarity in the web. *ACM Trans. Internet Technology*, 2:205–223, 2002.

[Dorogovtsev et al. 2002] S.N. Dorogovtsev, A.V. Goltsev, and J.F.F. Mendes. Pseudofractal scale-free web. *Physical Review E*, 65:066122, 2002.

[Efron 1975] B. Efron. Defining the curvature of a statistical problem (with applications to second order efficiency). *The Annals of Statistics*, 3:1189–1242, 1975.

[Erdős & A. Rényi 1960] P. Erdős and A. Rényi. On the evolution of random graphs. *Publication of the Mathematical Institute of the Hungarian Academy of Science*, 5:17–67, 1960.

[Fabrikant et al. 2002] A. Fabrikant, E. Koutsoupias, and C.H. Papadimitriou. Heuristically optimized trade-offs: A new paradigm for power laws in the internet. In *ICALP '02: Proceedings of the 29th International Colloquium on Automata, Languages, and Programming*, volume 2380, 110–122, Berlin: Springer, 2002.

[Faloutsos et al. 1999] M. Faloutsos, P. Faloutsos, and C. Faloutsos. On power-law relationships of the internet topology. In *SIGCOMM '99: Proceedings of the Conference on Applications, Technologies, Architectures, and Protocols for Computer Communication*, 251–262, 1999.

[Farkas et al. 2001] I. Farkas, I. Derényi, A.-L. Barabási, and T. Vicsek. Spectra of "real-world" graphs: Beyond the semicircle law. *Physical Review E*, 64:026704, 2001.

[Flaxman et al. 2007] A.D. Flaxman, A.M. Frieze, and J. Vera. A geometric preferential attachment model of networks II. In *WAW '07: Proceedings of the 5th Workshop on Algorithms and Models for the Web-Graph*, 41–55, 2007.

[Gamerman 1997] D. Gamerman. *Markov Chain Monte Carlo, Stochastic Simulation for Bayesian Inference.* London: Chapman & Hall, 1997.

[Gehrke et al. 2003] J. Gehrke, P. Ginsparg, and J.M. Kleinberg. Overview of the 2003 KDD cup. *SIGKDD Explorations*, 5:149–151, 2003.

[Gelman et al. 2003] A. Gelman, J.B. Carlin, H.S. Stern, and D.B. Rubin. *Bayesian Data Analysis, Second Edition.* London: Chapman & Hall, 2003.

[Hammack 2009] R.H. Hammack. Proof of a conjecture concerning the direct product of bipartite graphs. *European Journal of Combinatorics*, 30:1114–1118, 2009.

[Holme 2005] P. Holme. Core-periphery organization of complex networks. *Physical Review E*, 72:046111, 2005.

[Imrich 1998] W. Imrich. Factoring cardinal product graphs in polynomial time. *Discrete Mathematics*, 192:119–144, 1998.

[Imrich & Klavžar 2000] W. Imrich and S. Klavžar. *Product Graphs: Structure and Recognition.* New York: John Wiley & Sons, 2000.

[Kleinberg 1999] J.M. Kleinberg. The small-world phenomenon: An algorithmic perspective. Technical Report 99-1776, Department of Computer Science, Cornell University, 1999.

[Kleinberg et al. 1999] J.M. Kleinberg, S.R. Kumar, P. Raghavan, S. Rajagopalan, and A. Tomkins. The web as a graph: Measurements, models and methods. In *COCOON '99: Proceedings of the International Conference on Combinatorics and Computing*, 1999.

[Klemm & Eguíluz 2002] K. Klemm and V.M. Eguíluz. Highly clustered scale-free networks. *Physical Review E*, 65:036123, 2002.

[Kullback & Leibler 1951] S. Kullback and R.A. Leibler. On information and sufficiency. *Annals of Mathematical Statistics*, 22:79–86, 1951.

[Kumar et al. 2006] R. Kumar, J. Novak, and A. Tomkins. Structure and evolution of online social networks. In *KDD '06: Proceedings of the 12th ACM SIGKDD International Conference on Knowledge Discovery and Data Mining*, 611–617, 2006.

[Kumar et al. 2000] R. Kumar, P. Raghavan, S. Rajagopalan, D. Sivakumar, A. Tomkins, and E. Upfal. Stochastic models for the web graph. In *FOCS '00: Proceedings of the 41st Annual Symposium on Foundations of Computer Science*, 57, 2000.

[Kumar et al. 1999] S.R. Kumar, P. Raghavan, S. Rajagopalan, and A. Tomkins. Extracting large-scale knowledge bases from the web. In *Proceedings of the 25th VLDB Conference*, Edinburgh, Scotland, 1999.

[Langville & Stewart 2004] A.N. Langville and W.J. Stewart. The Kronecker product and stochastic automata networks. *Journal of Computation and Applied Mathematics*, 167:429–447, 2004.

[Leskovec 2009] J. Leskovec. Networks, communities and Kronecker products. In *CNIKM '09: Complex Networks in Information and Knowledge Management*, 2009.

[Leskovec et al. 2005a] J. Leskovec, D. Chakrabarti, J.M. Kleinberg, and C. Faloutsos. Realistic, mathematically tractable graph generation and evolution, using Kronecker multiplication. In *PKDD '05: Proceedings of the 9th European Conference on Principles and Practice of Knowledge Discovery in Databases*, 133–145, 2005.

[Leskovec & Faloutsos 2007] J. Leskovec and C. Faloutsos. Scalable modeling of real graphs using Kronecker multiplication. In *ICML '07: Proceedings of the 24th International Conference on Machine Learning*, 2007.

[Leskovec et al. 2005b] J. Leskovec, J.M. Kleinberg, and C. Faloutsos. Graphs over time: Densification laws, shrinking diameters and possible explanations. In *KDD '05: Proceedings of the 11th ACM SIGKDD International Conference on Knowledge Discovery in Data Mining*, 177–187, 2005.

[Leskovec et al. 2007a] J. Leskovec, J.M. Kleinberg, and C. Faloutsos. Graph evolution: Densification and shrinking diameters. *ACM Transactions on Knowledge Discovery from Data (TKDD)*, 1:2, 2007.

[Leskovec et al. 2008a] J. Leskovec, K.J. Lang, A. Dasgupta, and M.W. Mahoney. Community structure in large networks: Natural cluster sizes and the absence of large well-defined clusters. *ArXiv*, http://arXiv.org/abs/0810.1355, 2008.

[Leskovec et al. 2008b] J. Leskovec, K.J. Lang, A. Dasgupta, and M.W. Mahoney. Statistical properties of community structure in large social and information networks. In *WWW '08: Proceedings of the 17th International World Wide Web Conference*, 2008.

[Leskovec et al. 2007b] J. Leskovec, M. McGlohon, C. Faloutsos, N. Glance, and M. Hurst. Patterns of cascading behavior in large blog graphs. In *SDM07: Proceedings of the SIAM Conference on Data Mining*, 551–556, 2007.

[Mahdian & Xu 2007] M. Mahdian and Y. Xu. Stochastic kronecker graphs. In *WAW '07: Proceedings of the 5th Workshop on Algorithms and Models for the Web-Graph*, 179–186, 2007.

[Mihail & Papadimitriou 2002] M. Mihail and C.H. Papadimitriou. On the eigenvalue power law. In *RANDOM '02: Proceedings of the 6th International Workshop on Randomization*, London: Springer-Verlag, 254–262, 2002.

[Milgram 1967] S. Milgram. The small-world problem. *Psychology Today*, 2:60–67, 1967.

[Nickel 2008] C.L.M. Nickel. Random dot product graphs: A model for social networks. PhD. thesis, Dept. of Applied Mathematics and Statistics, Johns Hopkins University, 2008.

[Palmer et al. 2002] C.R. Palmer, P.B. Gibbons, and C. Faloutsos. ANF: A fast and scalable tool for data mining in massive graphs. In *KDD '02: Proceedings of the 8th ACM SIGKDD International Conference on Knowledge Discovery and Data Mining*, 81–90, 2002.

[Pennock et al. 2002] D.M. Pennock, G.W. Flake, S. Lawrence, E.J. Glover, and C.L. Giles. Winners don't take all: Characterizing the competition for links on the Web. *Proceedings of the National Academy of Sciences*, 99:5207–5211, 2002.

[Petrov 1995] V.V. Petrov. *Limit Theorems of Probability Theory.* Oxford, UK: Oxford University Press, 1995.

[Ravasz & Barabási 2003] E. Ravasz and A.-L. Barabási. Hierarchical organization in complex networks. *Physical Review E*, 67:026112, 2003.

[Ravasz et al. 2002] E. Ravasz, A.L. Somera, D.A. Mongru, Z.N. Oltvai, and A.-L. Barabási. Hierarchical organization of modularity in metabolic networks. *Science*, 297:1551–1555, 2002.

[Redner 1998] S. Redner. How popular is your paper? An empirical study of the citation distribution. *European Physical Journal B*, 4:131–134, 1998.

[Richardson et al. 2003] M. Richardson, R. Agrawal, and P. Domingos. Trust management for the semantic web. In *Proceedings of the International Semantic Web Conference (ISWC)*, Lecture Notes in Comput. Sci. 2870, New York, Springer, 2003, pp. 351–368.

[Ripeanu et al. 2002] M. Ripeanu, I. Foster, and A. Iamnitchi. Mapping the gnutella network: Properties of large-scale peer-to-peer systems and implications for system design. *IEEE Internet Computing Journal*, 6:50–57, 2002.

[Rissanen 1978] J. Rissanen. Modelling by the shortest data description. *Automatica*, 14:465–471, 1978.

[RouteViews 1997] RouteViews. University of Oregon Route Views Project. Online data and reports. http://www.routeviews.org, 1997.

[Sales-Pardo et al. 2007] M. Sales-Pardo, R. Guimera, A.A. Moreira, and L.A. Amaral. Extracting the hierarchical organization of complex systems. *Proceedings of the National Academy of Sciences*, 104:15224–15229, 2007.

[Schwarz 1978] G. Schwarz. Estimating the dimension of a model. *The Annals of Statistics*, 6:461–464, 1978.

[Siganos et al. 2006] G. Siganos, S.L. Tauro, and M. Faloutsos. Jellyfish: A conceptual model for the as internet topology. *Journal of Communications and Networks*, 8:339–350, 2006.

[Sole & Goodwin 2000] R. Sole and B. Goodwin. *Signs of Life: How Complexity Pervades Biology.* New York: Perseus Books Group, 2000.

[Tauro et al. 2001] S.L. Tauro, C. Palmer, G. Siganos, and M. Faloutsos. A simple conceptual model for the internet topology. In *GLOBECOM '01: Global Telecommunications Conference*, 3:1667–1671, 2001.

[Tsourakakis 2008] C.E. Tsourakakis. Fast counting of triangles in large real networks, without counting: algorithms and laws. In *ICDM '08 : IEEE International Conference on Data Mining*, 2008.

[Van Loan 2000] C.F. Van Loan. The ubiquitous Kronecker product. *Journal of Computation and Applied Mathematics*, 123:85–100, 2000.

[Vázquez 2003] A. Vázquez. Growing network with local rules: preferential attachment, clustering hierarchy, and degree correlations. *Physical Review E*, 67:056104, 2003.

[Wasserman & Pattison 1996] S. Wasserman and P. Pattison. Logit models and logistic regressions for social networks. *Psychometrika*, 60:401–425, 1996.

[Watts & Strogatz 1998] D.J. Watts and S.H. Strogatz. Collective dynamics of 'small-world' networks. *Nature*, 393:440–442, 1998.

[Waxman 1988] B.M. Waxman. Routing of multipoint connections. *IEEE Journal on Selected Areas in Communications*, 6:1617–1622, 1988.

[Weichsel 1962] P.M. Weichsel. The Kronecker product of graphs. *American Mathematical Society*, 13:37–52, 1962.

[Winick & Jamin 2002] J. Winick and S. Jamin. Inet-3.0: Internet Topology Generator. Technical Report CSE-TR-456-02, University of Michigan, Ann Arbor, 2002.

[Wiuf et al. 2006] C. Wiuf, M. Brameier, O. Hagberg, and M.P. Stumpf. A likelihood approach to analysis of network data. *Proceedings of the National Academy of Sciences*, 103:7566–7570, 2006.

[Young & Scheinerman 2007] S.J. Young and E.R. Scheinerman. Random dot product graph models for social networks. In *WAW '07: Proceedings of the 5th Workshop on Algorithms and Models for the Web-Graph*, 138–149, 2007.

[Zheleva et al. 2009] E. Zheleva, H. Sharara, and L. Getoor. Co-evolution of social and affiliation networks. In *Proceedings of the 15th SIGKDD Conference on Knowledge Discovery and Data Mining (KDD)*, 1007–1016, 2009.

Chapter 10

The Kronecker Theory of Power Law Graphs

Jeremy Kepner[*]

Abstract

An analytical theory of power law graphs is presented based on the Kronecker graph generation technique. Explicit, stochastic, and instance Kronecker graphs are used to highlight different properties. The analysis uses Kronecker exponentials of complete bipartite graphs to formulate the substructure of such graphs. The Kronecker theory allows various high-level quantities (e.g., degree distribution, betweenness centrality, diameter, eigenvalues, and iso-parametric ratio) to be computed directly from the model parameters.

10.1 Introduction

Power law graphs are ubiquitous and arise in the Internet [Faloutsos 1999], the web [Broder 2000], citation graphs [Redner 1998], and online social networks (see [Chakrabarti 2004]). Power law graphs have the general property that the histograms of their degree distribution $Deg()$ fall off with a power law and are approximately linear in a log-log representation. Mathematically this observation can be stated as

$$Slope[\log(Count[Deg(G)])] \approx -\text{constant}$$

*MIT Lincoln Laboratory, 244 Wood Street, Lexington, MA 02420 (kepner@ll.mit.edu).

This work is sponsored by the Department of the Air Force under Air Force Contract FA8721-05-C-0002. Opinions, interpretations, conclusions, and recommendations are those of the author and are not necessarily endorsed by the United States Government.

The analytical theory describing the specific structure of these graphs is just beginning to be developed. In this work, an analytical framework is proposed and used to illustrate the precise substructures that may exist within power law graphs. In addition, this analytical framework will also allow many higher level statistical quantities to be computed directly.

There are a variety of techniques for generating power law graphs. The simulations produced by these techniques are used to test out techniques for analyzing real-world graphs. This work focuses on the innovative Kronecker product technique [Leskovec 2005, Chakrabarti 2004], which generates power law graphs and with enough tunable parameters to provide detailed fits to real-world graphs. The Kronecker product approach reproduces a large number of statistical properties of real graphs: degree distribution, hop plot, singular value distribution, diameter, densification, etc. In addition, and perhaps more importantly, it is based on an adjacency matrix representation of graphs, which provides a rich set of theoretical tools for more detailed analysis [Gilbert 2006]. To review, for a graph $G = (V, E)$ with N vertices and M edges, the $N \times N$ adjacency matrix $\mathbf{A}$ has the property $\mathbf{A}(i,j) = 1$ if there is an edge e_{ij} from vertex v_i to vertex v_j and is zero otherwise.

The outline of this work is as follows. First, an overview of some results is presented. Second, the Kronecker graph generation algorithm is reviewed. Next are some basic results on simplified Kronecker graphs based on fully connected bipartite graphs $\mathbf{B}(n,m)$ of sets with n and m vertices. The fifth section presents the fundamental analytic constructs necessary for a more sophisticated analysis of Kronecker graphs. Section 10.6 gives results for a more complex model of Kronecker graphs. Finally Section 10.7 discusses the implications of this work.

10.2 Overview of results

The principal results presented in this work are

1. The basic properties of Kronecker products of graphs, which include

   (a) The Kronecker product of two bipartite graphs

   $$P_{\mathbf{B}}(\mathbf{B}(n_1,m_1) \otimes \mathbf{B}(n_2,m_2)) = \mathbf{B}(n_1 n_2, m_1 m_2) \cup \mathbf{B}(n_2 m_1, n_1 m_2)$$

   (b) The permutation functions for manipulating Kronecker graphs. These include the bipartite $P_{\mathbf{B}}$, recursive bipartite $P_{\bar{k}}$, and pop P_{Pop} permutations.

2. The construction of a useful analytic model of Kronecker graphs of the form $\mathbf{G}(n,m)^{\otimes k}$ where $\mathbf{G}(n,m) = \beta\mathbf{B}(n,m) + \alpha\mathbf{I} + \gamma\mathbf{1}$, and a useful approximation to this model is given by

$$\mathbf{G}(n,m)^{\otimes k} \overset{P}{\approx} \beta^k \mathbf{B}^{\otimes k} + \text{“}k\text{”}(\alpha\mathbf{I} + \gamma\mathbf{1}) \otimes \beta^{k-1}\mathbf{B}^{\otimes k-1}$$

3. The graph substructures generated by $\mathbf{B}^{\otimes k}$ are the union of many smaller bipartite graphs

$$P_{\bar{k}}(\mathbf{B}(n,m)^{\otimes k}) = \bigcup_{i=1}^{2^k} \mathbf{B}(n_k(i), m_k(i))$$

or equivalently

$$\mathbf{B}(n,m)^{\otimes k} \stackrel{P}{=} \bigcup_{r=0}^{k-1} \bigcup^{\binom{k-1}{r}} \mathbf{B}(n^{k-r}m^r, n^r m^{k-r})$$

4. The graph substructures generated by $(\mathbf{B}+\mathbf{I})^{\otimes k}$ are

$$(\mathbf{B}+\mathbf{I})^{\otimes k} \stackrel{P}{=} \sum_{r=1}^{k} \text{“}\binom{k}{r}\text{”} \bigcup^{N^{k-1}} \mathbf{B}^{\otimes k}$$

The first and second order terms are related by

$$P_{\bar{k}}(\mathbf{B}^{\otimes k-1-l} \otimes \mathbf{I} \otimes \mathbf{B}^{\otimes l}) = \chi_l^k \otimes \chi_{i_1 i_2}^k$$

where χ_l^k is the connection between blocks of vertices and $\chi_{i_1 i_2}^k$ is the strength of the connection.

5. The above graph substructures can be used to compute higher level statistics. For example, given $n > m$ and $r = 0, \ldots, k$

   (a) The degree distribution for $Deg(\mathbf{B}(n,m)^{\otimes k})$ is

   $$Count[Deg = n^r m^{k-r}] = \binom{k}{r} n^{k-r} m^r$$

   (b) The betweenness centrality distribution for $C_b(\mathbf{B}(n,m)^{\otimes k})$ is

   $$Count[C_b = (n/m)^{2r-k}(n^{k-r}m^r - 1)] = \binom{k}{r} n^{k-r} m^r$$

   (c) The degree distribution of $(\mathbf{B}+\mathbf{I})^{\otimes k}$ is given by

   $$Count[Deg = (n+1)^r (m+1)^{k-r}] = \binom{k}{r} n^{k-r} m^r$$

6. Various additional results

   (a) A fast, space efficient Kronecker graph generation algorithm

   (b) The degree distribution of an arbitrary Kronecker matrix based on Poisson statistics.

10.3 Kronecker graph generation algorithm

The Kronecker product graph generation algorithm ([Leskovec 2005, Chakrabarti 2004] is quite elegant and can be described as follows. First, let $\mathbf{A} \in \mathbb{R}^{M_B M_C \times N_B N_C}$, $\mathbf{B} \in \mathbb{R}^{M_B \times N_B}$, and $\mathbf{C} \in \mathbb{R}^{M_C \times N_C}$, then the Kronecker product is defined as follows [Van Loan 2000]:

$$\mathbf{A} = \mathbf{B} \otimes \mathbf{C} = \begin{pmatrix} \mathbf{B}(1,1)\mathbf{C} & \mathbf{B}(1,2)\mathbf{C} & \cdots & \mathbf{B}(1,M_B)\mathbf{C} \\ \mathbf{B}(2,1)\mathbf{C} & \mathbf{B}(2,2)\mathbf{C} & \cdots & \mathbf{B}(2,M_B)\mathbf{C} \\ \vdots & \vdots & & \vdots \\ \mathbf{B}(N_B,1)\mathbf{C} & \mathbf{B}(N_B,2)\mathbf{C} & \cdots & \mathbf{B}(N_B,M_B)\mathbf{C} \end{pmatrix}$$

Now let $\mathbf{G} \in \mathbb{R}^{N \times N}$ be an adjacency matrix. The Kronecker exponent to the power k is as follows

$$\mathbf{G}^{\otimes k} = \mathbf{G}^{\otimes k-1} \otimes \mathbf{G}$$

which generates an $N^k \times N^k$ adjacency matrix. This simple model naturally produces self-similar graphs, yet even a small $\mathbf{G}$ matrix provides enough parameters to fine-tune the generator to a particular real-world data set.

At this point, it is worth noting that the model $\mathbf{G}^{\otimes k}$ will be used in multiple contexts. If $\mathbf{G}$ is an explicit adjacency matrix consisting of only 1's and 0's, then likewise $\mathbf{G}^{\otimes k}$ will also be such an "explicit" adjacency matrix. Most of the subsequent theoretical analysis will be on these matrices, which reveal the graph substructures and how the higher level statistical quantities vary with the parameters of the model. Of course, such a precisely structured adjacency matrix does not correspond to any real-world, organically generated graph.

To produce more realistic statistical models, let $0 \leq \mathbf{G}(i,j) \leq 1$ be a matrix of probabilities. The "stochastic" adjacency matrix generated by $\mathbf{G}^{\otimes k}$ then contains at each (i,j) entry the probability that an edge exists from vertex i to j. A limited amount of theoretical analysis will be done on these matrices.

To create a specific "instance" of this matrix, edges are randomly selected from the stochastic matrix. One of the powerful features of the Kronecker product generation technique is that an arbitrarily large instance of the graph can be created efficiently without ever having to form the full stochastic adjacency matrix.

10.3.1 Explicit adjacency matrix

An example worth closer consideration is when $\mathbf{G}$ represents a star graph with n spokes ($N = n+1$ total vertices). The Kronecker product of such a graph naturally leads to graphs with many of the properties of real-world graphs. Denote the $N \times N$ adjacency matrix for a star graph as

$$\mathbf{S}(n+1) = \mathbf{S}(N) = \begin{pmatrix} 0 & 1 & \cdots & 1 \\ 1 & 0 & \cdots & 0 \\ \vdots & \vdots & & \vdots \\ 1 & 0 & \cdots & 0 \end{pmatrix}$$

Likewise, let $\mathbf{I}(N)$ denote the $N \times N$ identity matrix

$$\mathbf{I}(N) = \begin{pmatrix} 1 & 0 & \cdots & 0 \\ 0 & 1 & \cdots & 0 \\ \vdots & \vdots & & \vdots \\ 0 & 0 & \cdots & 1 \end{pmatrix}$$

Then set

$$\mathbf{G}(N) = \mathbf{S}(N) + \mathbf{I}(N)$$

which implies

$$\mathbf{G}^{\otimes k} = (\mathbf{S} + \mathbf{I})^{\otimes k}$$

Figure 10.1 shows the explicit adjacency matrix formed by $\mathbf{G}^{\otimes 3}$ for $N = 4$.

10.3.2 Stochastic adjacency matrix

A stochastic adjacency matrix sets the probability of edges between any two vertices. In the case of $N = 4$ the total number of free parameters is 16. To simplify this situation, we separate the entries into two values consisting of a foreground (β,α) and a background (γ) contribution. This results in a model of the form

$$\mathbf{G}(N) = \beta\mathbf{S}(N) + \alpha\mathbf{I}(N) + \gamma\mathbf{1}(N)$$

where $1 > \beta,\ \alpha \gg \gamma > 0$, and $\mathbf{1}(N)$ is $N \times N$ matrix of all ones. Figure 10.1 shows the stochastic adjacency matrix, formed by $\mathbf{G}^{\otimes 3}$ for $N = 4$.

The stochastic representation does lend itself to some analytical treatment. For instance, define the following probability density functions for the rows and columns of $\mathbf{G}$

$$\mathbf{g}_\sigma^{col}(1,j) = \sum_{i=1}^{N} \mathbf{G}(i,j) / \sum \mathbf{G}$$

$$\mathbf{g}_\sigma^{row}(i,1) = \sum_{j=1}^{N} \mathbf{G}(i,j) / \sum \mathbf{G}$$

The probability of there being an in/out edge for a particular vertex is given by ρ_i where

$$\rho = \mathbf{g}_\sigma^{\otimes k}$$

where the "row" and "col" superscripts have been dropped. The expected number of in/out edges for a particular vertex n_i is given by a Poisson distribution with an expected value of λ_i

$$Prob(n_i) = \lambda_i^n e^{-\lambda_i} / n!$$

where $\lambda = N_e \rho$ and N_e is the number of total edges in the graph. The degree distribution Deg of the adjacency matrix is obtained by summing these over all the vertices

$$Count[Deg = n] = \sum_{i=1}^{N^k} \lambda_i^n e^{-\lambda_i} / n!$$

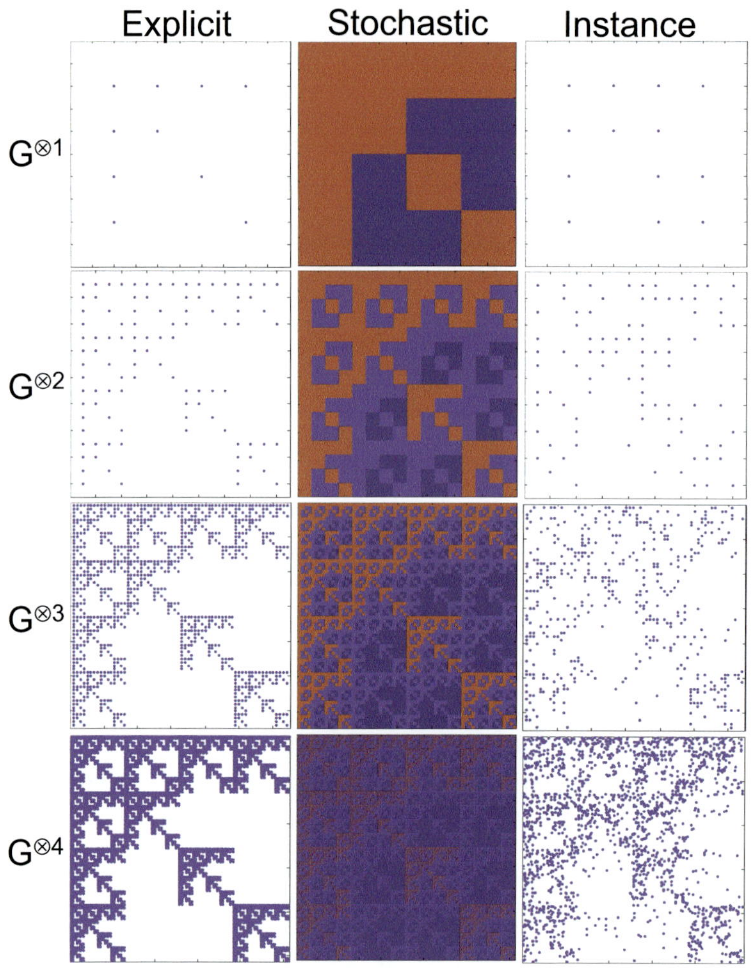

Figure 10.1. Kronecker adjacency matrices.
Explicit, stochastic, and instance matrices for a star plus identity matrix: $G = \mathbf{S}(4) + \mathbf{I}(4)$, where $N_e = 8N_v$.

The above sum requires a high amount of numerical precision to compute, so the following more practical formula should be used

$$Count[Deg = n] = \sum_{i=1}^{N^k} \exp\left[n(\ln(\rho_i) + \ln(N_e)) - \lambda_i - \sum_{n'=1}^{n} \ln(n')\right]$$

This example illustrates some of the analysis we can perform on the stochastic adjacency matrix. However, while it is possible to compute some higher level statistical quantities, it is more difficult to get at the detailed substructure of the graph.

10.3.3 Instance adjacency matrix

One of the advantages of the Kronecker method is that a specific instance of the graph can be generated efficiently without ever having to form the entire adjacency matrix. To create a single edge e_{ij} from the stochastic adjacency matrix created by $\mathbf{G}^{\otimes k}$ requires generating a vertex pair i and j. Let i_k^N and j_k^N be k-bit representations of i and j where each bit can take N values. Furthermore, let $i_k^N(k')$ and $j_k^N(k')$ be the k'-bit of the representation. In short

$$i_k^N(k'), j_k^N(k') \in \{0, \ldots, N-1\} \quad \forall k' \in \{1, \ldots, k\}$$

Each bit is randomly set using the following formulas

$$i_k^N(k') = \arg_{i'}(\mathbf{G}_\sigma^{row}(i', 1) < r_i < \mathbf{G}_\sigma^{row}(i'+1, 1))$$

$$j_k^N(k') = \arg_{j'}(\hat{\mathbf{G}}_\sigma^{col}(i_k^N(k'), j') < r_j < \hat{\mathbf{G}}_\sigma^{col}(i_k^N(k'), j'+1))$$

where random numbers $r_i, r_j \leftarrow U[0,1]$ are drawn from a uniform distribution. $\mathbf{G}_\sigma^{row}$ is the cumulative distribution function that there is an edge in a given row. $\hat{\mathbf{G}}_\sigma^{row}$ is the joint cumulative distribution function that there is an edge in a given column. The row cumulative distribution function is computed as follows

$$\mathbf{G}_\sigma^{row}(i, 1) = \mathbf{G}_\sigma^{row}(i-1, 1) + \mathbf{g}_\sigma^{row}(i-1, 1)$$

where $i = 2, \ldots, N+1$, and $\mathbf{G}_\sigma^{row}(1,1) = 0$. Likewise, the column cumulative joint distribution function is

$$\hat{\mathbf{G}}_\sigma^{col}(i, j) = \mathbf{G}_\sigma^{row}(i, j-1) + \hat{\mathbf{g}}_\sigma^{row}(i, j-1)$$

where $i = 1, \ldots, N$, $j = 2, \ldots, N+1$, $\mathbf{G}_\sigma^{col}(i, 1) = 0$, and

$$\hat{\mathbf{g}}_\sigma^{row}(i, j) = \frac{\mathbf{G}(i,j)}{\sum \mathbf{G}} \frac{1}{\mathbf{g}_\sigma^{row}(i, 1)}$$

The above formulas for i_k^N and j_k^N are repeated k times to fill in all the bits, after which it is a simple matter to compute i and j from their bit representations. The procedure culminates by setting the corresponding value in the adjacency matrix $\mathbf{A}(i,j) = 1$. The whole procedure is repeated for every edge to be created. Edges can be created completely independently provided the random number generators are seeded differently. Figure 10.1 shows an instance adjacency matrix formed by $\mathbf{G}^{\otimes 3}$ for $N = 4$ with $N_e = 8N$. Figure 10.2 shows the degree distribution for a stochastic Kronecker graph with a million vertices and the corresponding predicted distribution obtained from the stochastic adjacency matrix. The peak structure is a reflection of the underlying analytic structure of the Kronecker graph (see next section).

10.4 A simple bipartite model of Kronecker graphs

Consider the model generated by taking the Kronecker product of a complete bipartite graph

$$\mathbf{G} = \mathbf{B}(n, m)$$

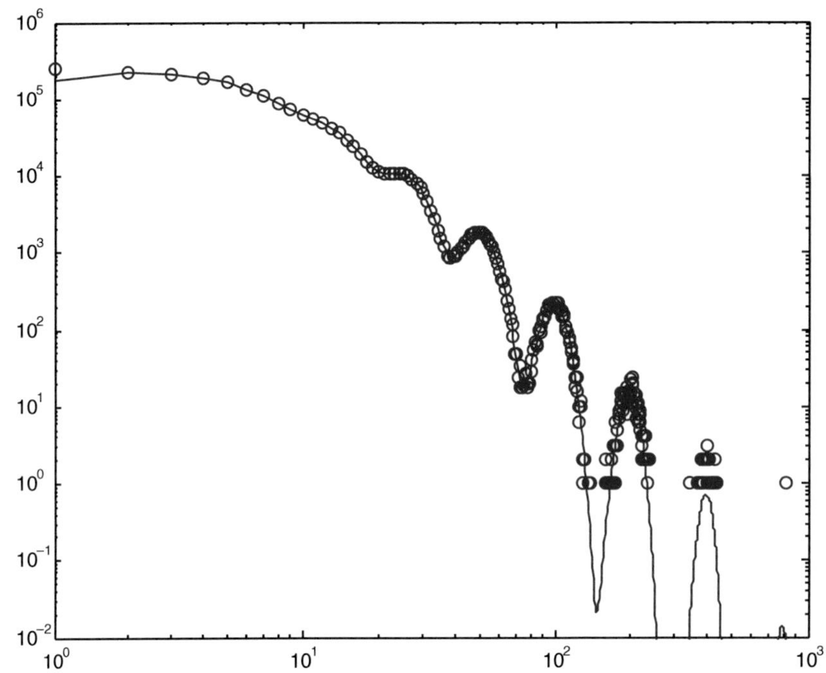

Figure 10.2. Stochastic and instance degree distribution. The degree distribution for a stochastic Kronecker graph with a million vertices (circles) and the corresponding predicted distribution obtained from the stochastic adjacency matrix (solid line). The peak structure is a reflection of the underlying analytic structure of the Kronecker graph.

This will be a reasonable model of certain simple Kronecker graphs and will be a useful building block for modeling more complex Kronecker graphs. This work builds on the original work of Weischel [Weischel 1962], who first looked at the relation between Kronecker products and graphs.

10.4.1 Bipartite product

Let the adjacency matrix of a bipartite graph be denoted by $\mathbf{B}(n, m)$

$$\mathbf{B}(n,m) = \begin{pmatrix} 0 & \mathbf{1}^{m \times n} \\ \mathbf{1}^{n \times m} & 0 \end{pmatrix}$$

Likewise, if the arguments of $\mathbf{B}()$ are matrices $\mathbf{A}$ and C, then

$$\mathbf{B}(\mathbf{A}, \mathbf{C}) = \begin{pmatrix} 0 & \mathbf{C} \\ \mathbf{A} & 0 \end{pmatrix}$$

Furthermore, let

$$\mathbf{B}(\mathbf{A}) = \begin{pmatrix} 0 & \mathbf{A}^T \\ \mathbf{A} & 0 \end{pmatrix}$$

and it is clearly the case that

$$\mathbf{B}(n,m) = \mathbf{B}(\mathbf{1}^{m\times n})$$

The adjacency matrix of a star graph is denoted by $\mathbf{S}(n+1)$ and is related to $\mathbf{B}$ by

$$\mathbf{S}(n+1) = \mathbf{B}(n,1)$$

The Kronecker product of two bipartite graphs is given by

$$\boxed{P_{\mathbf{B}}(\mathbf{B}(n_1,m_1) \otimes \mathbf{B}(n_2,m_2)) = \mathbf{B}(n_1n_2,m_1m_2) \cup \mathbf{B}(n_2m_1,n_1m_2)}$$

where $P_{\mathbf{B}} = P_{\mathbf{B}(n_1,m_1,n_2,m_2)}$ is the bipartite permutation function (see next section). The union notation is defined as follows

$$\mathbf{A} \cup \mathbf{C} = \begin{pmatrix} \mathbf{A} & 0 \\ 0 & \mathbf{C} \end{pmatrix}$$

and has the additional form

$$\bigcup^{N} \mathbf{A} = \begin{pmatrix} \mathbf{A} & & 0 \\ & \ddots & \\ 0 & & \mathbf{A} \end{pmatrix}$$

For convenience, where it is not necessary to keep track of the precise permutations, the $\stackrel{P}{=}$ notation is used. For the bipartite product this would be written as

$$\mathbf{B}(n_1,m_1) \otimes \mathbf{B}(n_2,m_2) \stackrel{P}{=} \mathbf{B}(n_1n_2,m_1m_2) \cup \mathbf{B}(n_2m_1,n_1m_2)$$

The different representations of the Kronecker product (graphical, matrix, and algebraic) are shown in Figure 10.3.

10.4.2 Bipartite Kronecker exponents

Ignoring the required permutations for the time being, the Kronecker powers of a bipartite graph are as follows

$$\mathbf{B}(n,m)^{\otimes 2} \stackrel{P}{=} \mathbf{B}(n^2,m^2) \bigcup \mathbf{B}(nm,nm)$$

and

$$\mathbf{B}(n,m)^{\otimes 3} \stackrel{P}{=} \mathbf{B}(n^3,m^3) \bigcup^{2} \mathbf{B}(n^2m,nm^2) \bigcup \mathbf{B}(nm^2,n^2m)$$

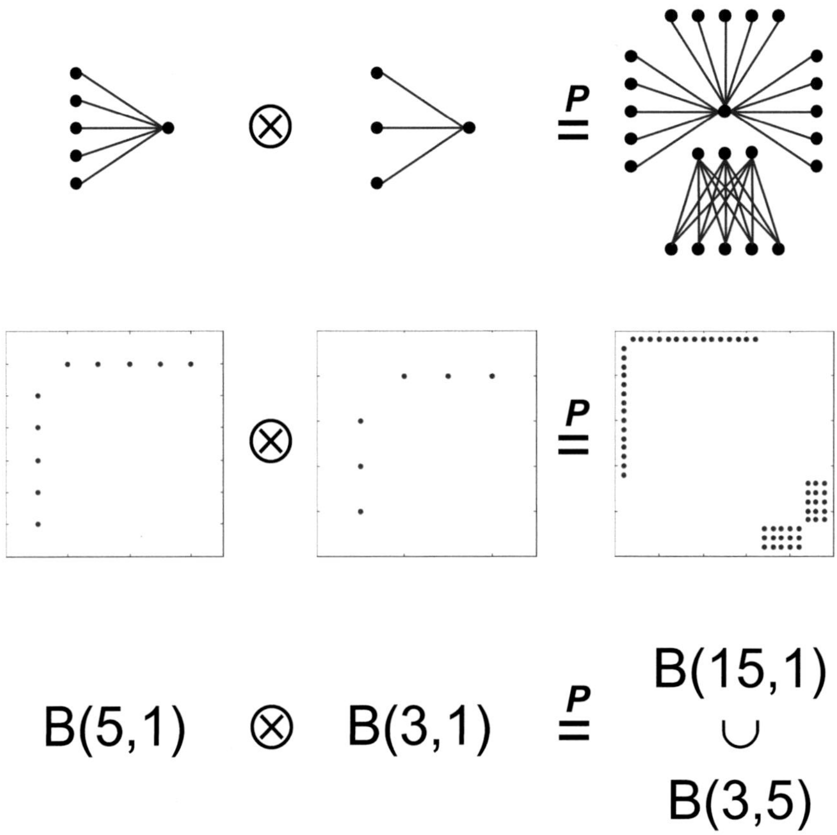

Figure 10.3. Graph Kronecker product.
Complementary representations of the Kronecker product of two graphs.

which generalizes to

$$\boxed{\mathbf{B}(n,m)^{\otimes k} \overset{P}{=} \bigcup_{r=0}^{k-1} \bigcup^{\binom{k-1}{r}} \mathbf{B}(n^{k-r}m^r, n^r m^{k-r})}$$

where

$$\binom{k}{r} = \frac{k!}{r!(k-r)!}$$

In short, the Kronecker exponential of a bipartite graph is the union of many smaller bipartite graphs. Note: to keep the expression for the coefficients a simple binomial coefficient, the fact that some terms can be combined further is ignored, i.e.,

$$\bigcup^{2} \mathbf{B}(n^2m, nm^2) \bigcup \mathbf{B}(nm^2, n^2m) = \bigcup^{3} \mathbf{B}(n^2m, nm^2)$$

The above formula clearly shows the substructures and how they grow with k. In particular, it is worth mentioning that going from k to $k+1$ results in mostly new structures that did not exist before.

The above general expression has the following special cases

$$\mathbf{B}(n,n)^{\otimes k} \stackrel{P}{=} \bigcup^{2^{k-1}} \mathbf{B}(n^k, n^k)$$

and

$$\mathbf{B}(n,1)^{\otimes k} \stackrel{P}{=} \bigcup_{r=0}^{k-1} \bigcup^{\binom{k-1}{r}} \mathbf{B}(n^{k-r}, n^r)$$

10.4.3 Degree distribution

The degree distribution of the above expression is

$$Deg(\mathbf{B}(n,m)^{\otimes k}) \stackrel{P}{=} \bigcup_{r=0}^{k-1} \bigcup^{\binom{k-1}{r}} Deg(\mathbf{B}(n^{k-r}m^r, n^r m^{k-r}))$$

where

$$Deg(\mathbf{B}(n,m)) = \{ \overbrace{\ldots\ m\ \ldots}^{n}, \overbrace{\ldots\ n\ \ldots}^{m} \}$$

and so

$$Deg(\mathbf{B}(n^{k-r}m^r, n^r m^{k-r})) = \{ \overbrace{\ldots\ n^r m^{k-r}\ \ldots}^{n^{k-r}m^r}, \overbrace{\ldots\ n^{k-r} m^r\ \ldots}^{n^r m^{k-r}} \}$$

The histogram of the degree distribution of the right set can be computed as follows

$$Count[Deg = n^{k-r}m^r] = \binom{k-1}{r} n^r m^{k-r}$$

Substituting $\tilde{r} = k - r$ gives

$$Count[Deg = n^{\tilde{r}}m^{k-\tilde{r}}] = \binom{k-1}{k-\tilde{r}} n^{k-\tilde{r}} m^{\tilde{r}}$$

where $\tilde{r} = 1, \ldots, k$. The histogram of the degree distribution of the left set can be computed as follows

$$Count[Deg = n^r m^{k-r}] = \binom{k-1}{r} n^{k-r} m^r$$

Substituting $\tilde{r} = r$ gives

$$Count[Deg = n^{\tilde{r}}m^{k-\tilde{r}}] = \binom{k-1}{\tilde{r}} n^{k-\tilde{r}} m^{\tilde{r}}$$

where $\tilde{r} = 0, \ldots, k-1$. Combining the formulas for the left and right sets and setting $\tilde{r} \to r$ produce

$$Count[Deg = n^r m^{k-r}] = [\overbrace{\binom{k-1}{k-r}}^{r=1,\ldots,k} + \overbrace{\binom{k-1}{r}}^{r=0,\ldots,k-1}] n^{k-r} m^r$$

Finally, we can simplify the binomial coefficients to obtain the following elegant expression

$$\boxed{Count[Deg = n^r m^{k-r}] = \binom{k}{r} n^{k-r} m^r}$$

where $r = 0, \ldots, k$.

For the case where $\mathbf{B} = \mathbf{B}(n, n)$, this yields the trivial distribution of all vertices having the exact same degree (which is clearly not a power law). For the more interesting case where $\mathbf{B} = \mathbf{B}(n, 1)$, the distribution is

$$Count[Deg = n^r] = \binom{k}{r} n^{k-r}$$

Under the assumption $n > m$, the above degree distribution is ordered in r and the slope of this degree distribution can be readily computed

$$Slope[n^r m^{k-r} \to n^{r+1} m^{k-r-1}] = -1 + \frac{\log_n[(k-r)/(r+1)]}{\log_n[n/m]}$$

which for the first and last points gives

$$Slope[n^0 m^k \to n^1 m^{k-1}] = -1 + \log_n(k)/\log_n(n/m)$$

$$Slope[n^{k-1} \to n^k] = -1 - \log_n(k)/\log_n(n/m)$$

More interestingly, this distribution has the property that over any symmetric interval $r \to k - r$

$$Slope[n^r m^{k-r} \to n^{k-r} m^r] = -1$$

which is why it can be closely associated with a power law graph (see Figure 10.4).

10.4.4 Betweenness centrality

The same technique can be applied to any other statistic of interest. For example, consider the betweenness centrality metric that is frequently used in graph analysis [Freeman 1977, Brandes 2001, Bader 2006]. This metric is defined as

$$C_b(v) = \sum_{s \neq v \neq t \in V} \frac{\sigma_{st}(v)}{\sigma_{st}}$$

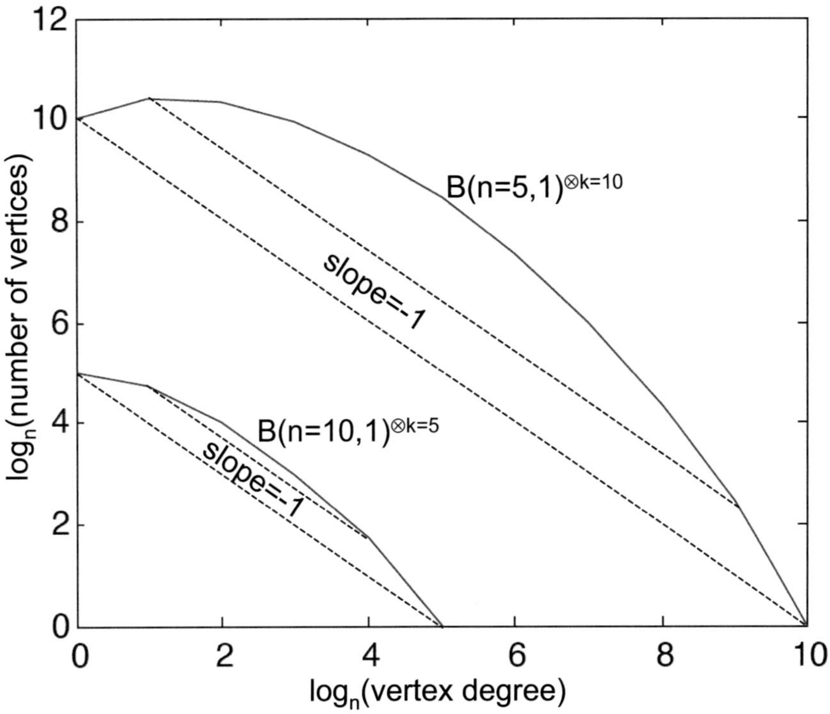

Figure 10.4. Theoretical degree distribution.
Degree distribution derived from the Kronecker product of bipartite graphs. The slope over any symmetric interval is always -1.

where σ_{st} is the number of shortest paths between vertex s and vertex t, and $\sigma_{st}(v)$ is the number of shortest paths between vertex s and vertex t that pass through vertex v. Applying C_b leads to

$$C_b(\mathbf{B}(n,m)^{\otimes k}) \stackrel{P}{=} \bigcup_{r=0}^{k-1} \bigcup^{\binom{k-1}{r}} C_b(\mathbf{B}(n^{k-r}m^r, n^r m^{k-r}))$$

where

$$C_b(\mathbf{B}(n,m)) = \{ \overbrace{\ldots\ m(m-1)/n\ \ldots}^{n}, \overbrace{\ldots\ n(n-1)/m\ \ldots}^{m} \}$$

for $n, m > 1$. For $\mathbf{B}(n,1)$, this becomes

$$C_b(\mathbf{B}(n,1)) = \{ \overbrace{\ldots\ 0\ \ldots}^{n}, n(n-1)\}$$

In the general case, the above can be written as

$$C_b(\mathbf{B}(n^{k-r}m^r, n^r m^{k-r}))$$

$$= \{ \overbrace{\dots \frac{n^r m^{k-r}(n^r m^{k-r}-1)}{n^{k-r}m^r} \dots}^{n^{k-r}m^r}, \overbrace{\dots \frac{n^{k-r}m^r(n^{k-r}m^r-1)}{n^r m^{k-r}} \dots}^{n^r m^{k-r}} \}$$

$$= \{ \overbrace{\dots (m/n)^{k-2r}(n^r m^{k-r}-1) \dots}^{n^{k-r}m^r}, \overbrace{\dots (n/m)^{k-2r}(n^{k-r}m^r-1) \dots}^{n^r m^{k-r}} \}$$

The histogram of the betweenness centrality distribution of the right-hand set can be computed as follows

$$Count[C_b = (n/m)^{k-2r}(n^{k-r}m^r - 1)] = \binom{k-1}{r} n^r m^{k-r}$$

Likewise, the betweenness centrality distribution of the left-hand set is

$$Count[C_b = (n/m)^{2r-k}(n^r m^{k-r} - 1)] = \binom{k-1}{r} n^r m^{k-r}$$

Combining the above two formulas by using the same steps that were used in the previous section to compute the degree distributions results in the following

$$\boxed{Count[C_b = (n/m)^{2r-k}(n^r m^{k-r} - 1)] = \binom{k}{r} n^{k-r} m^r}$$

where $r = 0, \dots, k$.

Selected values of the histogram for the degree distribution and the betweenness centrality of $\mathbf{B}(n,1)^{\otimes k}$ are as follows

Deg	C_b	**Count**
1	n^{-k}	n^k
n^1	n^{2-k}	kn^{k-1}
n^{k-1}	n^{k-2}	kn
n^k	n^k	1

10.4.5 Graph diameter and eigenvalues

The diameter of a graph $Diam()$ is the maximum of the shortest path between any two vertices (assuming all vertices are reachable). Clearly

$$Diam(\mathbf{B}) = 2$$

Likewise

$$Diam(\mathbf{A} \cup \mathbf{C}) = \infty$$

since there are unreachable nodes. By this result, the diameter of the Kronecker exponential of a bipartite graph is

$$Diam(\mathbf{B}^{\otimes k}) = \infty$$

A bipartite graph has just two nonzero eigenvalues

$$eig(\mathbf{B}) = \{(nm)^{1/2}, -(nm)^{1/2}\}$$

The eigenvalues of $\mathbf{B}^{\otimes k}$ fall directly from Theorem 2 of [Leskovec 2005] which states that the eigenvalues of a Kronecker product are just the Kronecker product of the eigenvalues. Thus, $\mathbf{B}^{\otimes k}$ has 2^k nonzero eigenvalues

$$eig(\mathbf{B}(n,m)^{\otimes k}) = \{\overbrace{(nm)^{k/2}, \ldots, (nm)^{k/2}}^{2^{k-1}}, \overbrace{-(nm)^{k/2}, \ldots, -(nm)^{k/2}}^{2^{k-1}}\}$$

10.4.6 Iso-parametric ratio

Another property of interest is the "surface" to "volume" ratio of subsets of the graph [Chung 2005], which can be measured using various iso-parametric ratios. One such ratio is given by

$$IsoPar(X) = \frac{\sum_{i \in X, j \in \bar{X}} \mathbf{A}(i,j) + \sum_{i \in \bar{X}, j \in X} \mathbf{A}(i,j)}{\sum_{i,j \in X} \mathbf{A}(i,j)}$$

where X is a subset of vertices with set complement $\bar{X} = V \setminus X$, and $\mathbf{A}$ is the adjacency matrix of the graph. For an unweighted symmetric graph, this can also be written as

$$IsoPar(X) = \frac{2\sum_{i \in X, j \in V} \mathbf{A}(i,j)}{\sum_{i,j \in X} \mathbf{A}(i,j)} - 2$$

Applying this to the graph $\mathbf{B}(n,m)^{\otimes k}$ results in a few interesting cases.

Consider the case when the subset consists of only those vertices in $n_k(i)$ (or $m_k(i)$)

$$IsoPar(n_k(i)) = \frac{4n_k(i)m_k(i)}{0} - 2 = \frac{2n^k m^k}{0} - 2 = \infty$$

In other words, since $n_k(i)$ is part of a complete bipartite subgraph, there are no edges between the vertices in $n_k(i)$. In general, this will also be true of a small random subset of vertices (i.e., most vertices are not connected to each other).

Next, consider the case when the subset consists of all the vertices in the subgraph $\mathbf{B}(n_k(i), m_k(i))$:

$$IsoPar(n_k(i) \cup m_k(i)) = \frac{4n_k(i)m_k(i)}{2n_k(i)m_k(i)} - 2 = \frac{4n^k m^k}{2n^k m^k} - 2 = 0$$

In other words, there are no connections to any other vertices outside this subgraph. Likewise, for any random subgraphs $\mathbf{B}(n_k(i), m_k(i))$ and $\mathbf{B}(n_k(i'), m_k(i'))$

$$IsoPar(n_k(i) \cup m_k(i) \cup n_k(i') \cup m_k(i')) = \frac{8n^k m^k}{4n^k m^k} - 2 = 0$$

10.5 Kronecker products and useful permutations

This section gives additional basic results on Kronecker products that will be useful in exploring more complex Kronecker graphs.

Other Kronecker product identities are as follows. $\mathbf{I}(N)$ is the $N \times N$ identity matrix. The Kronecker product of the identity with itself is

$$\mathbf{I}(N_1) \otimes \mathbf{I}(N_2) = \mathbf{I}(N_1 N_2)$$

The Kronecker product of $\mathbf{I}$ with another matrix $\mathbf{C}$ is

$$\mathbf{I}(N) \otimes \mathbf{C} = \bigcup^{N} \mathbf{C}$$

Let $\mathbf{1}(N)$ denote the $N \times N$ matrix with all values set equal to 1. The Kronecker product of $\mathbf{1}$ with itself is

$$\mathbf{1}(N_1) \otimes \mathbf{1}(N_2) = \mathbf{1}(N_1 N_2)$$

10.5.1 Sparsity

The sparsity denotes the fraction of the entries in the matrix that are nonzeros. For the identity matrix the sparsity is

$$\sigma(\mathbf{I}) = \sigma_{\mathbf{I}} = \frac{N}{N^2} = N^{-1}$$

Likewise, for a bipartite matrix

$$\sigma(\mathbf{B}(n,m)) = \sigma_{\mathbf{B}} = \frac{2nm}{N^2}$$

The sparsity is related to the Kronecker product as follows

$$\sigma(\mathbf{A}) = \sigma(\mathbf{B})\sigma(\mathbf{C})$$

where $\mathbf{A} = \mathbf{B} \otimes \mathbf{C}$. The sparsity is related to the Kronecker exponent as follows

$$\sigma(\mathbf{G}^{\otimes k}) = \sigma(\mathbf{G})^k$$

10.5.2 Permutations

The permutation function P will be used in the following overloaded manner (always in the context of square matrices)

$i' = P(i)$,
$i = P^{-1}(i')$,
$P(\mathbf{A})$ permutes $\mathbf{A}$,
$\mathbf{PAP}^T$ means that $\mathbf{P}$ is a permutation matrix.

P also has the identities $P(\mathbf{I}) = I$, $P(\mathbf{1}) = \mathbf{1}$, and $P(\mathbf{B} + \mathbf{C}) = P(\mathbf{B}) + P(\mathbf{C})$.

10.5.3 Pop permutation

If $\mathbf{B} \in R^{N_B \times N_B}$, $\mathbf{C} \in R^{N_C \times N_C}$, and

$$\mathbf{A} = \mathbf{B} \otimes \mathbf{C}$$

then there exists a permutation with

$$P_{\mathbf{A}}(\mathbf{A}) = P_{\mathbf{B}}(\mathbf{B}) \otimes P_{\mathbf{C}}(\mathbf{C})$$

where $P_{\mathbf{B}}$ and $P_{\mathbf{C}}$ are arbitrary permutations of B and C. Furthermore, $P_{\mathbf{A}}$ can be computed from $P_{\mathbf{B}}$ and $P_{\mathbf{C}}$ as follows

$$\begin{aligned} P_{\mathbf{A}}(i_A) &= (P_{\mathbf{B}}(\lfloor i_A/N_C \rfloor + 1) - 1)N_C + P_{\mathbf{C}}(((i_A - 1) \bmod N_C) + 1) \\ &= (P_{\mathbf{B}}(i^2_{N_C}(i_A)) - 1)N_C + P_{\mathbf{C}}(i^1_{N_C}(i_A)) \end{aligned}$$

where

$$\begin{aligned} i^1_{N_B}(i) &= (i - 1) \bmod N_B \\ i^2_{N_C}(i) &= \lfloor (i - 1)/N_C \rfloor + 1 \end{aligned}$$

In general, $P_{\mathbf{A}}$ in the above context will be denoted as

$$P_{Pop(P_{\mathbf{B}}, P_{\mathbf{C}})}(\mathbf{B} \otimes \mathbf{C}) = P_{\mathbf{B}}(\mathbf{B}) \otimes P_{\mathbf{C}}(\mathbf{C})$$

because it allows "popping" the permutations outside the Kronecker product.

10.5.4 Bipartite permutation

The bipartite permutation $P_{\mathbf{B}}$ is the key permutation that allows reorganizing the Kronecker product of two bipartite graphs into separate disconnected graphs. The formula for $P_{\mathbf{B}} = P_{\mathbf{B}(n_1, m_1, n_2, m_2)}$ is

$P_{\mathbf{B}}(\cdots)$	$= \cdots$	i_1	i_2
$N_2 i_1 + i_2$	$n_2 i_1 + i_2$	$0, \ldots, n_1 - 1$	$1, \ldots, n_2$
$N_2 i_1 + n_2 + i_2$	$(n_1 n_2 + m_1 m_2) + n_2 i_1 + i_2$	$0, \ldots, n_1 - 1$	$1, \ldots, m_2$
$N_2(n_1 + i_1) + i_2$	$(n_1 n_2 + m_1 m_2) + n_2 i_1 + i_2$	$0, \ldots, m_1 - 1$	$1, \ldots, n_2$
$N_2(n_1 + i) + n2 + j$	$n_1 n_2 + m2 i_1 + i_2$	$0, \ldots, m_1 - 1$	$1, \ldots, m_2$

where $N_1 = n_1 + m_1$ and $N_2 = n_2 + m_2$.

10.5.5 Recursive bipartite permutation

The recursive bipartite permutation allows the precise construction of $\mathbf{B}^{\otimes k}$ as a union of 2^k bipartite graphs

$$P_{\bar{k}}(\mathbf{B}(n, m)^{\otimes k}) = \bigcup_{i=1}^{2^k} \mathbf{B}(n_k(i), m_k(i))$$

where $n_k(i)$ and $m_k(i)$ are defined by the recursive relations

$$n_k(2i-1) = n_{k-1}(i)n$$
$$n_k(2i) = m_{k-1}(i)n$$
$$m_k(2i-1) = m_{k-1}(i)m$$
$$m_k(2i) = n_{k-1}(i)m$$

with

$$n_1(1) = n \quad \text{and} \quad m_1(1) = 1$$

The permutation P function can be computed as follows. Let

$$P_{\bar{k}}(\mathbf{B}(n,m)^{\otimes k}) = P_k(P_{\overline{k-1}}(\mathbf{B}(n,m)^{\otimes k-1}) \otimes \mathbf{B}(n,m))$$

where the above permutations are applied recursively until

$$P_2(\mathbf{B}(n,m)^{\otimes 2}) = \mathbf{B}(n^2,m^2) \bigcup \mathbf{B}(nm,nm)$$

which implies

$$P_{\bar{2}} = P_2 = P_{\mathbf{B}}(n,m,n,m)$$

In other words, $P_{\bar{k}}$ is the recursive composition of each of the constituent permutations P_k. At level $k-1$, there are the following terms

$$P_{\overline{k-1}}(\mathbf{B}(n,m)^{\otimes k-1}) = \bigcup_{i=1}^{2^{k-1}} \mathbf{B}(n_{k-1}(i), m_{k-1}(i))$$

multiplying by $\mathbf{B}(n,m)$ to get $\mathbf{B}(n,m)^{\otimes k}$ implies each of the above terms will be permuted by the bipartite permutation function

$$P_{k,i} = P_{\mathbf{B}(n_{k-1}(i), m_{k-1}(i), n, m)}$$

These functions are then concatenated together with the union function to produce all of P_k

$$P_k = \bigcup_{i=1}^{2^{k-1}} P_{k,i}$$

The aggregate permutation $P_{\bar{k}}$ is then computed using P_{Pop}

$$P_{\bar{k}}\mathbf{I}(i_k) = P_k((P_{\overline{k-1}}(\lfloor i_k/N \rfloor + 1) - 1)N + ((i_k - 1) \bmod N) + 1)$$

The effect of this permutation can be seen in Figure 10.5.

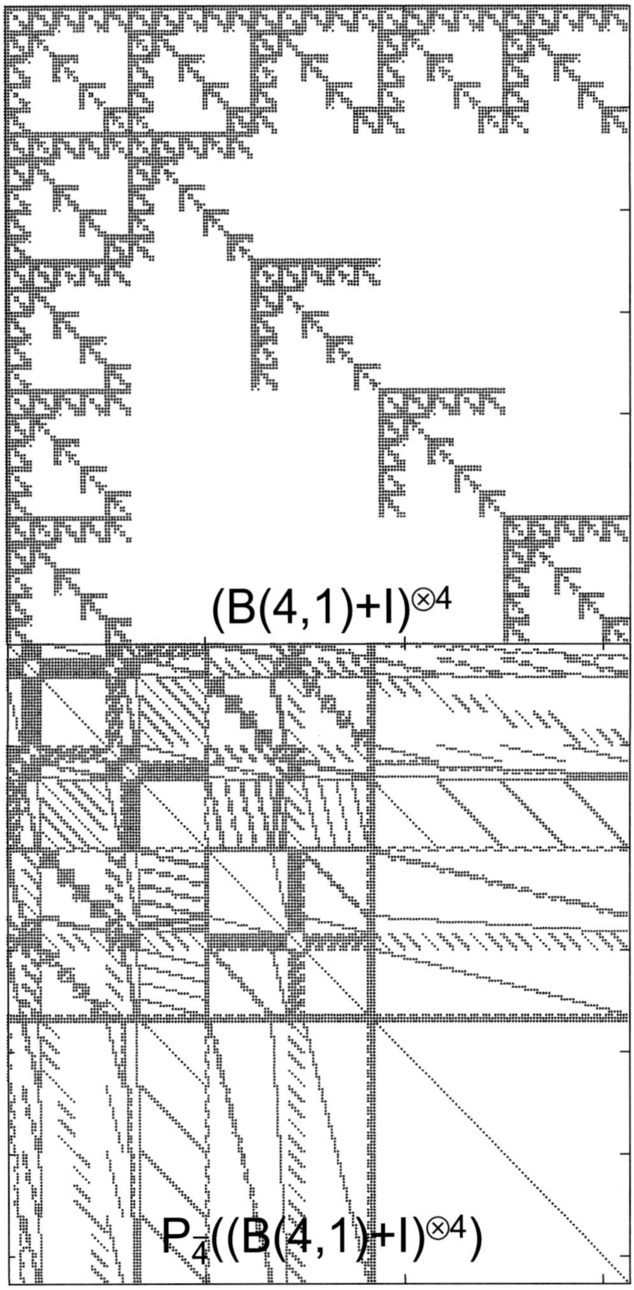

Figure 10.5. Recursive bipartite permutation.
The top figure shows the unpermuted $(\mathbf{B}+\mathbf{I})^{\otimes k}$ for $n = 4$, $m = 1$, and $k = 4$. The bottom figure shows the same adjacency matrix after applying the recursive bipartite permutation $P_{\bar{k}}$.

10.5.6 Bipartite index tree

The functions $n_k(i)$ and $m_k(i)$ form a tree which defines the overall structure of $\mathbf{B}(n,m)^{\otimes k}$. These values are powers of n and m. It is possible to define complex exponents ν and μ such that

$$n_k(i) = n^{Re(\nu_k(i))} n^{Im(\nu_k(i))}$$
$$m_k(i) = n^{Re(\mu_k(i))} n^{Im(\mu_k(i))}$$

and the recursion relations become

$$\nu_k(2i-1) = \nu_{k-1}(i) + 1$$
$$\nu_k(2i) = \mu_{k-1}(i) + 1$$
$$\mu_k(2i-1) = \mu_{k-1}(i) + \sqrt{-1}$$
$$\mu_k(2i) = \nu_{k-1}(i) + \sqrt{-1}$$

where

$$\nu_1(1) = 1 \quad \text{and} \quad \mu_1(1) = \sqrt{-1}$$

Observe that

$$Re(\nu) = Im(\mu) \quad \text{and} \quad Im(\nu) = Re(\mu)$$

and

$$Re(\nu) = k - Im(\nu) \quad \text{and} \quad Re(\mu) = k - Im(\nu)$$

Thus

$$n_k(i) = n^{Re(\nu_k(i))} m^{k-Re(\nu_k(i))}$$
$$m_k(i) = n^{k-Re(\nu_k(i))} m^{Re(\nu_k(i))}$$

So the entire structure can be determined by the single sequence $Re(\nu_k(i))$. $Re(\nu_6)$ in a stacked form is

```
1
2 1
3 1 2 2
4 1 2 3 3 2 3 2
5 1 2 4 3 3 4 2 4 2 3 3 4 2 3 3
6 1 2 5 3 4 5 2 4 3 4 3 5 2 3 4 5 2 3 4 4 3 4 3 5 2 3 4 4 3 4 3
```

Likewise, $Re(\nu_6)$ in a tree depiction is

```
                               1
               2                               1
       3               1               2               2
   4       1       2       3       3       2       3       2
 5   1   2   4   3   3   4   2   4   2   3   3   4   2   3   3
6 1 2 5 3 4 5 2 4 3 4 3 5 2 3 4 5 2 3 4 4 3 4 3 5 2 3 4 4 3 4 3
```

which reveals

$$Re(\nu_k(i)) + Re(\nu_k(i+1)) = k + 1, \quad i = 1, 3, \ldots$$

In addition, it is apparent that the higher/lower values (e.g., 6 and 1) are more closely associated with lower values of i while the more intermediate values (e.g., 4 and 3) become more prevalent as i increases.

10.6 A more general model of Kronecker graphs

A useful generating graph that can be used to model a wide range of power law graphs has the following form

$$\mathbf{G}(n+1) = \beta\mathbf{S}(n+1) + \alpha\mathbf{I}(n+1) + \gamma\mathbf{1}(n+1)$$

Noting that $\mathbf{S}(n+1) = \mathbf{B}(n,1)$ allows the more generalized model

$$\mathbf{G}(n,m) = \beta\mathbf{B}(n,m) + \alpha\mathbf{I}(n+m) + \gamma\mathbf{1}(n+m)$$

In addition to the aforementioned specialization, several special cases are also of interest. The trivial cases are

$$\alpha > 0, \quad \beta = 0, \quad \gamma = 0$$

which implies

$$\mathbf{G}(n,m)^{\otimes k} = \alpha^k \mathbf{I}((n+m)^k) = \alpha^k \mathbf{I}(N^k)$$

and

$$\alpha = 0, \quad \beta = 0, \quad \gamma > 0$$

which implies

$$\mathbf{G}(n,m)^{\otimes k} = \gamma^k \mathbf{1}((n+m)^k) = \gamma^k \mathbf{1}(N^k)$$

The less trivial case of

$$\alpha = 0, \quad \beta > 0, \quad \gamma = 0$$

has already been worked out in the previous section

$$\mathbf{G}(n,m)^{\otimes k} = \beta^k \mathbf{B}(n,m)^{\otimes k}$$

Finally, the case where

$$\alpha = \gamma > 1, \quad \beta = 0$$

deals with the expression

$$\mathbf{G}(n,m)^{\otimes k} = \alpha^k (\mathbf{I}(n+m) + \mathbf{1}((n+m))^{\otimes k}$$

In general, the most interesting case is where

$$1 > \beta > \alpha \gg \gamma > 0$$

This model corresponds to a foreground bipartite plus identity graph with a background probability of any vertex connecting to any other vertex

$$\begin{aligned}\mathbf{G}(n,m) &= \beta\mathbf{B}(n,m) + \alpha\mathbf{I}(n,m) + \gamma\mathbf{1}(n+m) \\ &= \beta\mathbf{B} + \alpha\mathbf{I} + \gamma\mathbf{1}\end{aligned}$$

However, because $\gamma \ll \alpha$, the situation can be approximated as follows

$$\begin{aligned}\mathbf{G}(n,m)^{\otimes k} &\overset{P}{\approx} (\beta\mathbf{B} + \alpha I)^{\otimes k} + \gamma\text{``}\binom{k}{k-1}\text{''}\mathbf{1}(N) \otimes (\beta\mathbf{B} + \alpha I)^{\otimes k-1} \\ &\quad + \text{lower order terms} \\ &\overset{P}{\approx} (\beta\mathbf{B} + \alpha I)^{\otimes k} + \gamma\text{``}k\text{''}\mathbf{1}(N) \otimes (\beta\mathbf{B} + \alpha I)^{\otimes k-1}\end{aligned}$$

For further theoretical analysis, consider the explicit form with the understanding that the appropriate α, β, and γ terms can be reinstated when needed for a stochastic representation. Under these conditions, the explicit form for the above formula is simply

$$\mathbf{G}(n,m)^{\otimes k} \overset{P}{\approx} (\mathbf{B}+\mathbf{I})^{\otimes k} + \text{“}k\text{”} \mathbf{1} \otimes (\mathbf{B}+\mathbf{I})^{\otimes k-1}$$

10.6.1 Sparsity analysis

The structure of $(\mathbf{B}+\mathbf{I})^{\otimes k}$ is given by

$$\boxed{(\mathbf{B}+\mathbf{I})^{\otimes k} \overset{P}{=} \sum_{r=1}^{k} \text{“}\binom{k}{r}\text{”} I^{\otimes k-r} \otimes \mathbf{B}^{\otimes r} \overset{P}{=} \sum_{r=1}^{k} \text{“}\binom{k}{r}\text{”} \bigcup^{N^{k-r}} \mathbf{B}^{\otimes r}}$$

where “ ” denotes that permutations are required to actually combine these terms. The sparsity structure of the above term is

$$\sigma((\mathbf{B}+\mathbf{I})^{\otimes k}) = \sum_{r=0}^{k} \binom{k}{r} \sigma_{\mathbf{I}}^{k-r} \sigma_{\mathbf{B}}^{r} = (\sigma_{\mathbf{B}} + \sigma_{\mathbf{I}})^k$$

Excluding the background ($\gamma \mathbf{1}$), the sparsity contribution of the rth term is

$$\sigma_r((\beta\mathbf{B}+\alpha\mathbf{I})^{\otimes k}) = \binom{k}{r} (\alpha\sigma_{\mathbf{I}})^{k-r} (\beta\sigma_{\mathbf{B}})^{r}$$

Computing the sparsity of the equations of previous sections for the case where $\beta\sigma_{\mathbf{B}} \gg \alpha\sigma_{\mathbf{I}}$ gives

$$\begin{aligned}\sigma(\mathbf{G}(n,m)^{\otimes k}) &\approx (\beta\sigma_{\mathbf{B}} + \alpha\sigma_{\mathbf{I}})^k + \gamma k \sigma_{\mathbf{1}} (\alpha\sigma_{\mathbf{I}} + \beta\sigma_{\mathbf{B}})^{k-1} \\ &\approx \beta^k \sigma_{\mathbf{B}} + \alpha\beta^k k \sigma_{\mathbf{I}} \sigma_{\mathbf{B}}^{k-1} + \gamma\beta^{k-1} k \sigma_{\mathbf{1}} \sigma_{\mathbf{B}}^{k-1}\end{aligned}$$

Applying this approximation back to the above equations gives

$$\begin{aligned}\mathbf{G}(n,m)^{\otimes k} &\overset{P}{\approx} \beta^k \mathbf{B}^{\otimes k} + \alpha\beta^k \text{“}k\text{”} \mathbf{I} \otimes \mathbf{B}^{\otimes k-1} + \gamma\beta^{k-1} \text{“}k\text{”} \mathbf{1} \otimes \mathbf{B}^{\otimes k-1} \\ &\overset{P}{\approx} \beta^k \mathbf{B}^{\otimes k} + \text{“}k\text{”} (\alpha\mathbf{I} + \gamma\mathbf{1}) \otimes (\beta\mathbf{B})^{\otimes k-1}\end{aligned}$$

Excluding the background, the explicit form is

$$\mathbf{G}(n,m)^{\otimes k} \overset{P}{\approx} \mathbf{B}^{\otimes k} + \text{“}k\text{”} \mathbf{B}^{\otimes k-1} \otimes \mathbf{I}$$

The relative contributions to the overall sparsity can be estimated as follows. In the case where $\mathbf{B} = \mathbf{B}(n,1)$, we have

$$\left(\frac{\sigma_{\mathbf{B}}}{\sigma_{\mathbf{I}} + \sigma_{\mathbf{B}}}\right)^k = \left(\frac{2(N-1)/N^2}{(3N-2)/N^2}\right)^k \approx \left(\frac{2}{3}\right)^k$$

In the case where $\mathbf{B} = \mathbf{B}(n, n)$, we have

$$\left(\frac{\sigma_{\mathbf{B}}}{\sigma_{\mathbf{I}} + \sigma_{\mathbf{B}}}\right)^k = \left(\frac{(N^2/2)/N^2}{(N + N^2/2)/N^2}\right)^k = \left(\frac{N}{N+2}\right)^k$$

In either case, as k increases, the $\mathbf{B}^{\otimes k}$ will go from dominating the overall sparsity to being a lesser component. That said, this term will always be the dominant single term and dominates the structure of the overall graph since more terms simply add a more "diffuse" component (see next section).

10.6.2 Second order terms

We can make the above expression more precise by inserting the full expression for the second term

$$\text{“}k\text{”}\mathbf{B}^{\otimes k-1} \otimes \mathbf{I} = \sum_{l=0}^{k-1} \mathbf{B}^{\otimes k-1-l} \otimes \mathbf{I} \otimes \mathbf{B}^{\otimes l}$$

which removes the arbitrary permutation giving

$$\mathbf{G}(n, m)^{\otimes k} \approx \mathbf{B}^{\otimes k} + \sum_{l=0}^{k-1} \mathbf{B}^{\otimes k-1-l} \otimes \mathbf{I} \otimes \mathbf{B}^{\otimes l}$$

Applying the recursive bipartite permutation produces

$$P_{\bar{k}}(\mathbf{G}(n, m)^{\otimes k}) \approx P_{\bar{k}}(\mathbf{B}^{\otimes k}) + \sum_{l=0}^{k-1} P_{\bar{k}}(\mathbf{B}^{\otimes k-1-l} \otimes \mathbf{I} \otimes \mathbf{B}^{\otimes l})$$

The first term has already been discussed and it describes a sequence of disjoint bipartite graphs $\mathbf{B}(n_k(i), m_k(i))$. The second term describes the connections between these bipartite graphs (see Figure 10.6).

The second order terms can be broken into two components

$$\boxed{P_{\bar{k}}(\mathbf{B}^{\otimes k-1-l} \otimes \mathbf{I} \otimes \mathbf{B}^{\otimes l}) = \chi_l^k \otimes \chi_{i_1 i_2}^k}$$

where χ_l^k is a $2^k \times 2^k$ adjacency matrix that specifies the connections between the blocks in the bipartite graph (see Figure 10.7). For convenience, let

$$\tilde{n}_k = \{n_k(1) \;\; m_k(1) \;\; \ldots \;\; n_k(2^{k-1}) \;\; m_k(2^{k-1})\}$$

Then $\chi_l^k(i_1, i_2) = 1$ means that there is a connection between the block of vertices specified by $\tilde{n}_k(i_1)$ and the block of vertices specified by $\tilde{n}_k(i_2)$. χ_l^k is computed by simply setting $n = 1$ and $m = 1$, which gives

$$\chi_l^k = P_{\bar{k}}(\mathbf{B}(1, 1)^{\otimes k-1-l} \otimes \mathbf{I}(2) \otimes \mathbf{B}(1, 1)^{\otimes l})$$

Furthermore, since each row or column of χ_l^k has only one value, the connections between blocks can be expressed as follows

$$i_2 = i_1 + \Delta_l^k(i_1)$$

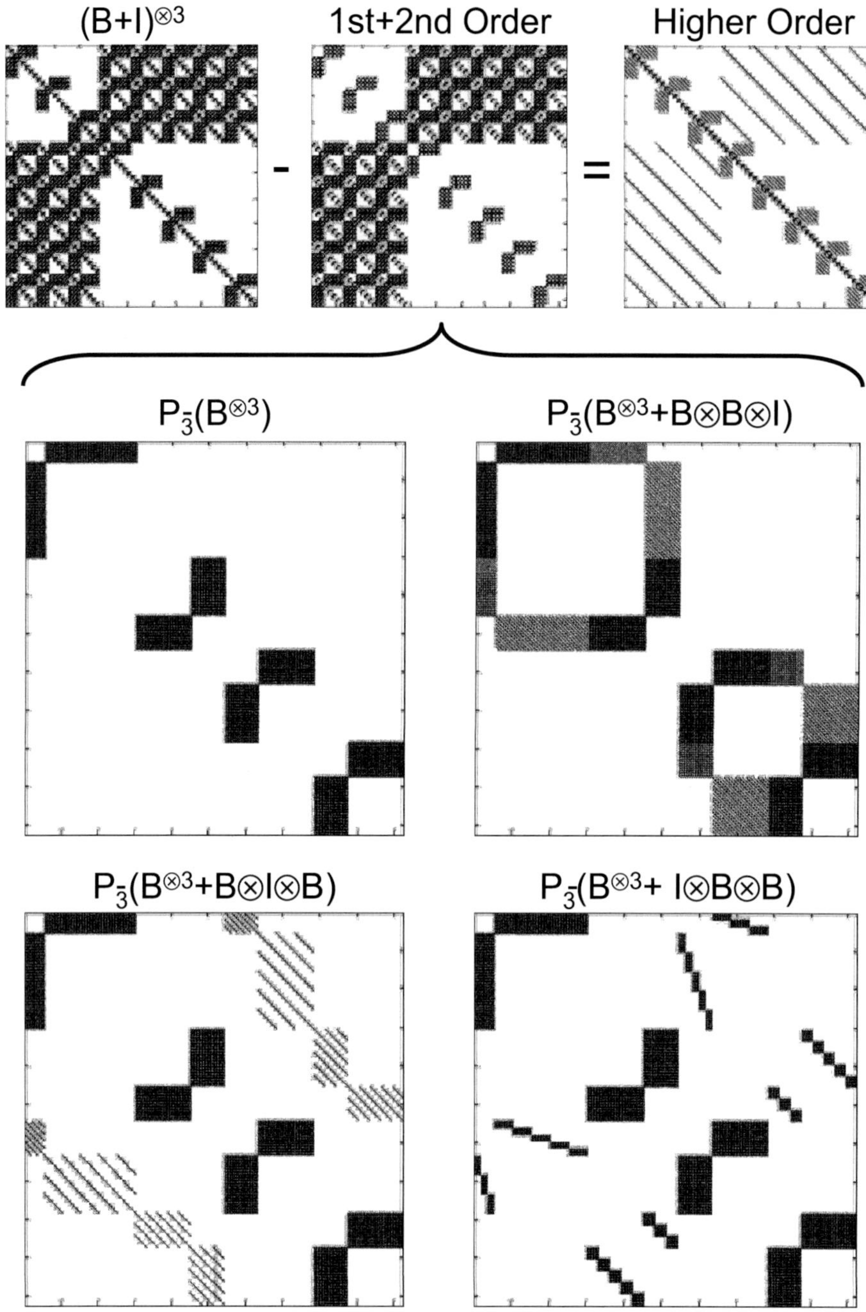

Figure 10.6. The revealed structure of $(\mathbf{B}+\mathbf{I})^{\otimes 3}$.
Upper part shows that a large fraction of the nonzero elements are from the first and second order terms. Lower part shows each of the second order terms after the recursive bipartite permutation has been applied.

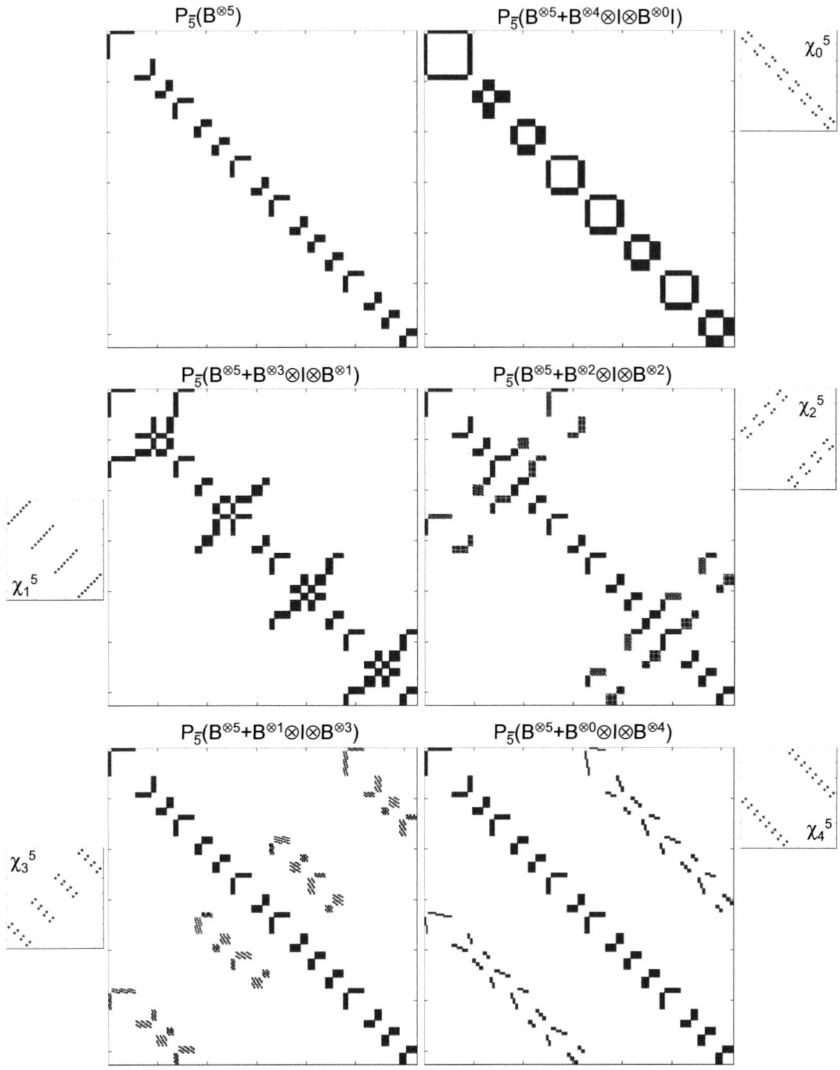

Figure 10.7. Structure of $(\mathbf{B}+\mathbf{I})^{\otimes 5}$ and corresponding χ_l^5.

These values are plotted in Figure 10.8 and show that the k blocks each block is connected to effectively span the set of 2^k blocks.

The strength of the connection between blocks of vertices is given by $\chi^k_{i_1 i_2}$ which is an $\tilde{n}_k(i_1) \times \tilde{n}_k(i_2)$ matrix with the following structure

$$\chi^k_{i_1 i_2} \stackrel{P}{=} \mathbf{1}(\tilde{n}_k(i_1)/n, \tilde{n}_k(i_2)/n) \otimes \mathbf{I}(n)$$

or

$$\chi^k_{i_1 i_2} \stackrel{P}{=} \mathbf{1}(\tilde{n}_k(i_1)/m, \tilde{n}_k(i_2)/m) \otimes \mathbf{I}(m)$$

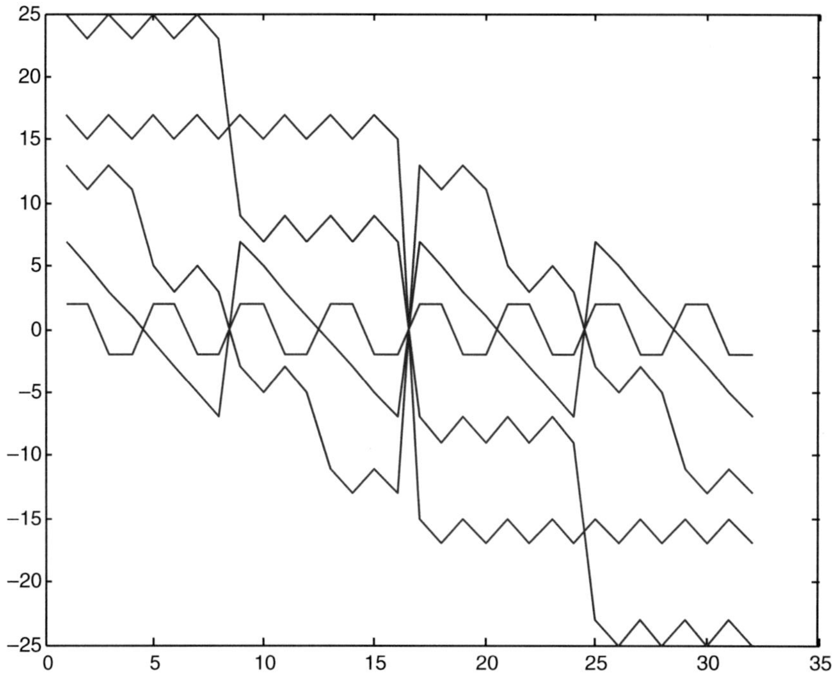

Figure 10.8. Block connections $\Delta_l^k(i)$.

Thus, although each block of vertices will be connected to k other blocks, the strength of these connections is weaker by a factor of n or m compared to the connections within each bipartite block.

10.6.3 Higher order terms

A similar analysis can be applied to the higher order terms. The overall effect is that the higher order terms are more numerous and more diffuse. In this respect, these higher order terms can be equated to a weakly structured background and obviate the need for the γ term.

The order with the largest contribution $\hat{r}$ is the solution to

$$\binom{k}{\hat{r}} \phi^{\hat{r}} = 0$$

where

$$\phi = \frac{\beta \sigma_{\mathbf{B}}}{\alpha \sigma_{\mathbf{I}}}$$

If $\phi = 1$, then the maximum is just the peak of the binomial coefficient $\hat{r} = k/2$. In general, $\phi > 0$ and so $\hat{r} < k/2$.

10.6.4 Degree distribution

To compute the degree distribution of the second order terms of a given vertex v requires first finding its corresponding block of vertices $\tilde{n}_k(i)$ by using the bipartite index tree. Then all the contributions from the different terms are summed

$$Deg(i(v)) = \sum_{l=0}^{k-1} \tilde{n}_k(i + \Delta_l^k(i))$$

Each of the above terms will tend to increase the number of unique values in the overall degree distribution. However, when summed with the other first order terms, they combine coherently. The result is that the degree distribution of the second and higher order terms is a simple horizontal offset from the degree distribution of the first order terms

$$n^r m^{k-r} \to n^r m^{k-r}[1 + r/n + (k-r)/m]$$

or

$$Count[Deg = n^r m^{k-r}[1 + r/n + (k-r)/m]] = \binom{k}{r} n^{k-r} m^r$$

where it is assumed $n > m$. Interestingly, the above expression can also be written in terms of the partial derivatives of n and m

$$[1 + \partial_n + \partial_m] n^r m^{k-r} = n^r m^{k-r}[1 + r/n + (k-r)/m]$$

For the case where $m = 1$, the above expression simplifies to

$$n^r \to n^r[1 + (k-r) + r/n]$$

Perhaps even more interesting is that the higher order terms also coherently sum in a similar fashion. The degree distribution of $(\mathbf{B}+\mathbf{I})^{\otimes k}$ is simply

$$\boxed{Count[Deg = (n+1)^r (m+1)^{k-r}] = \binom{k}{r} n^{k-r} m^r}$$

For the case where $m = 1$, the above expression simplifies to

$$Count[Deg = (n+1)^r] = \binom{k}{r} n^r$$

These offsets are illustrated in Figure 10.9. In each case, the effect of the higher order terms is to cause the slope of the degree distribution to become steeper.

10.6.5 Graph diameter and eigenvalues

The diameter of $(\mathbf{B}+\mathbf{I})^{\otimes k}$ can be readily deduced from Theorem 5 of [Leskovec 2005], which states that if $\mathbf{G}$ has a given diameter and contains $\mathbf{I}$, then $\mathbf{G}^{\otimes k}$ has the same diameter as $\mathbf{G}$. Thus

$$Diam((\mathbf{B}+\mathbf{I})^{\otimes k}) = 2$$

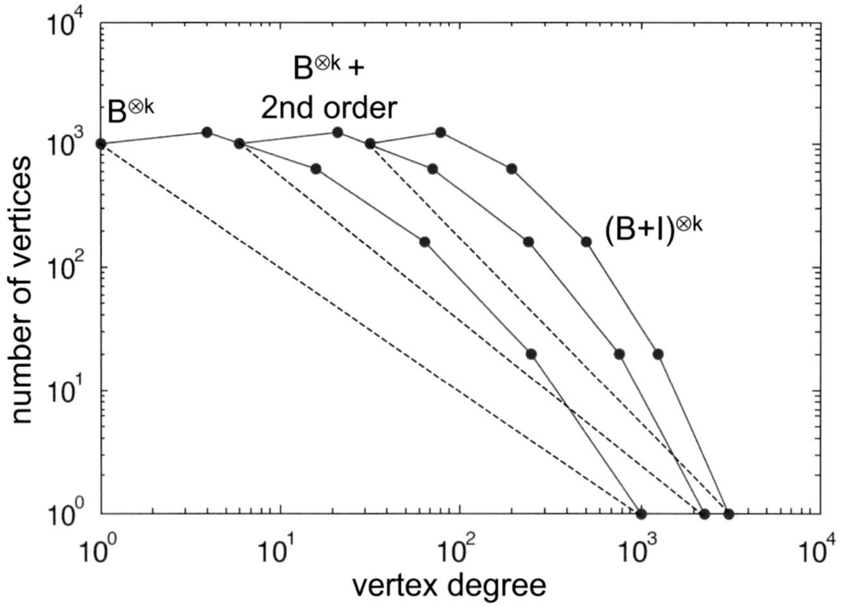

Figure 10.9. Degree distribution of higher orders. The degree distribution of the higher orders is a simple horizontal offset of the degree distribution of $\mathbf{B}^{\otimes k}$. Values shown are for $n = 4$, $m = 1$, and $k = 5$.

We can understand this by noting that even in the sparsest case ($\mathbf{B}(n, 1)$) the edges per vertex are

$$N_e/N_v \approx 3^k$$

Subtracting the set of these that are intrablock edges leaves (for reasonable values of n and k)

$$(\text{Interblock edges})/N_v \approx 3^k - n^{k/2} \approx 3^k$$

which is enough interblock edges to keep up with the 2^k growth rate in the number of blocks.

The more interesting situation occurs when an instance of the graph is created. In this situation, it is typical to hold the ratio N_e/N_v constant. Since, at best, each vertex is connected to N_e/N_v of the 2^k blocks, this implies

$$(N_e/N_v)^{Diam} \approx 2^k$$

or

$$Diam(Instance((\mathbf{B}+\mathbf{I})^{\otimes k})) \approx k/\lg(N_e/N_v) = O(\lg(N_e))$$

The eigenvalues of $\mathbf{B} + \mathbf{I}$ are given by

$$eig(\mathbf{B}+\mathbf{I}) = eig(\mathbf{B}) + 1 = \{(nm)^{1/2} + 1, \overbrace{1, \ldots, 1}^{N-2}, 1 - (nm)^{1/2}\}$$

The largest eigenvalues of $(\mathbf{B}+\mathbf{I})^{\otimes k}$ are thus

$$eig((\mathbf{B}+\mathbf{I})^{\otimes k}) = \{((nm)^{1/2}+1)^k, ((nm)^{1/2}+1)^{k-1}, ((nm)^{1/2}-1)^2((nm)^{1/2}+1)^{k-2}, \ldots\}$$

Interestingly, for these kinds of Kronecker graphs, there is not an obvious relation between the eigenvalues and the diameter.

10.6.6 Iso-parametric ratio

For $(\mathbf{B}+\mathbf{I})^{\otimes k}$ the iso-parametric ratio of only those vertices in $n_k(i)$ or $m_k(i)$ is

$$IsoPar(n_k(i)) = \frac{2\sum \mathbf{A}(n_k(i),:)}{n_k(i)} - 2$$

The denominator is from the $\mathbf{I}^{\otimes k}$ term, which implies that every vertex always has a self-loop. Letting $r = r(i)$ and using the formula for the degree distribution of $(\mathbf{B}+\mathbf{I})^{\otimes k}$ result in

$$\begin{aligned} IsoPar(n_k(i)) &= IsoPar(n^r m^{k-r}) \\ &= n^r m^{k-r} \frac{2(n+1)^{k-r}(m+1)^r}{n^r m^{k-r}} - 2 \\ &= 2(n+1)^{k-r}(m+1)^r - 2 \end{aligned}$$

For $m = 1$, the above expression simplifies to

$$IsoPar(n^r) = 2^{k+1}(n+1)^{k-r} - 2$$

which has a maximum value at $r = 0$ and decreases exponentially to a minimum value at $r = k$

$$\begin{aligned} IsoPar(n^0) &= 2^{k+1}(n+1)^k - 2 \\ &\approx 2^{k+1}(n+1)^k \\ IsoPar(n^k) &= 2^{k+1} - 2 \\ &\approx 2^{k+1} \end{aligned}$$

The next case, when the subset consists of all the vertices in subgraph $\mathbf{B}(n_k(i), m_k(i))$, is quite similar except there are additional terms in the denominator

$$\begin{aligned} IsoPar(n_k(i) \cup m_k(i)) &= 2\frac{\sum \mathbf{A}(n_k(i) \cup m_k(i),:)}{n_k(i)m_k(i) + n_k(i) + m_k(i) + [\chi \text{ terms}]} - 2 \\ &= \frac{n_i(k)\ \tilde{I}soPar(n_i(k)) + m_i(k)\ \tilde{I}soPar(m_i(k))}{n_k(i)m_k(i) + n_k(i) + m_k(i) + [\chi \text{ terms}]} - 2 \end{aligned}$$

where $\tilde{I}soPar = IsoPar + 2$. Substituting for $n_k(i)$ and $m_k(i)$ gives

$$\begin{aligned} &IsoPar(n_k(i) \cup m_k(i)) \\ &\quad = IsoPar(n^r m^{k-r} \cup n^{k-r} m^r) \\ &\quad = \frac{n^r m^{k-r}\ \tilde{I}soPar(n^r m^{k-r}) + n^{k-r} m^r\ \tilde{I}soPar(n^{k-r} m^r)}{2n^k m^k + n^r m^{k-r} + n^{k-r} m^r + [\chi \text{ terms}]} - 2 \\ &\quad = 2\frac{n^r m^{k-r}(n+1)^{k-r}(m+1)^r + n^{k-r} m^r (n+1)^r (m+1)^{k-r}}{2n^k m^k + n^r m^{k-r} + n^{k-r} m^r + [\chi \text{ terms}]} - 2 \end{aligned}$$

For $m = 1$, the above expression simplifies to

$$IsoPar(n^r \cup n^{k-r}) = 2\frac{(2n)^r(n+1)^{k-r} + (2n)^{k-r}(n+1)^r}{2n^k + n^r + n^{k-r} + [\chi \text{ terms}]} - 2$$

Ignoring the χ terms, the above expression has a maximum at $r = 0$ and $r = k$

$$\begin{aligned} IsoPar(n^0 \cup n^k) &= IsoPar(n^k \cup n^0) \\ &= 2\frac{(n+1)^k + (2n)^k}{3n^k + 1} - 2 \\ &\approx \left(\frac{2}{3}\right) 2^k \end{aligned}$$

and has a minimum at $r = k/2$

$$\begin{aligned} IsoPar(n^{k/2} \cup n^{k/2}) &= \frac{2(n+1)^{k/2}2^{k/2}}{n^{k/2} + 1} - 2 \\ &\approx 2^{k/2+1} \end{aligned}$$

It is interesting to note that iso-parametric ratios of the set $n_k(i) \cup m_k(i)$ are always much smaller than the iso-parametric ratio of either $n_k(i)$ or $m_k(i)$. In other words

$$\begin{aligned} IsoPar(n_k(i)) &\gg IsoPar(n_k(i) \cup m_k(i)) \\ IsoPar(m_k(i)) &\gg IsoPar(n_k(i) \cup m_k(i)) \end{aligned}$$

This is illustrated in Figure 10.10.

10.7 Implications of bipartite substructure

The previous sections have focused on analyzing the properties of the explicit graphs generated by Kronecker products. This section looks at some of the implications of these results to other types of graphs.

10.7.1 Relation between explicit and instance graphs

The properties of a particular graph instance randomly drawn from an explicit graph can be computed directly from the properties of the explicit graph. Let a particular instance graph have vertices N_v and edges N_e. Furthermore, let f_e denote the ratio of the number of edges in the instance graph to the number of edges in the explicit graph. For a $\mathbf{G}^{\otimes k}$, this is

$$N_e = f_e N_v^2 \sigma(\mathbf{G})^k$$

The degree distribution of an instance graph will then be the sum of the Poisson

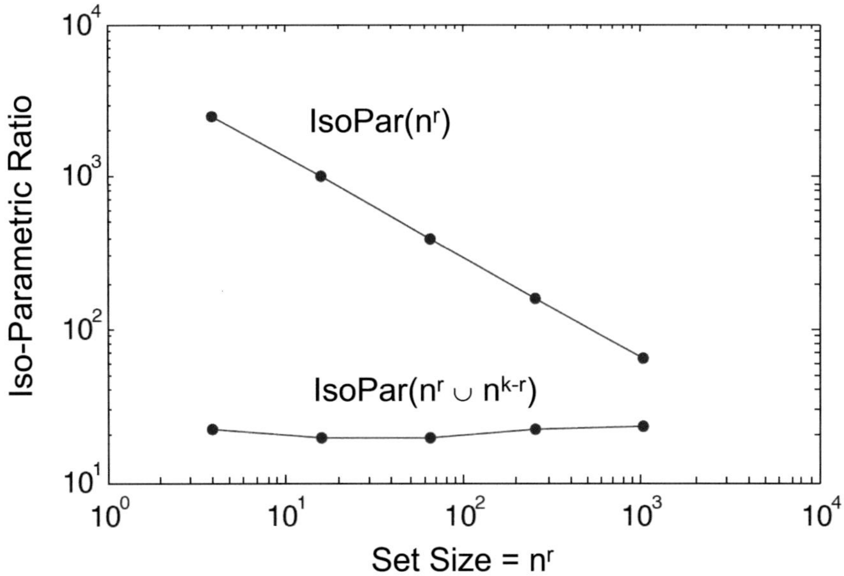

Figure 10.10. Iso-parametric ratios.
The iso-parametric ratios as a function of input set for $(\mathbf{B}+\mathbf{I})^{\otimes k}$. Values shown are for $n = 4$, $m = 1$, and $k = 5$.

distributions of the degree distribution of the explicit graph

$$Count[Deg(Instance(\mathbf{G}^{\otimes k})) = j] = \sum_{i=0}^{k} P_{\lambda_i}(j) \binom{k}{i} n^{k-i} m^i$$

where

$$P_{\lambda_i}(j) = \lambda_i^j e^{-\lambda_i} / j!$$

and λ_i is the expected number of edges per vertex for each block of vertices i. For $\mathbf{G} = \mathbf{B}(n, m)$, this is

$$\lambda_i = f_e n^i m^{k-i}$$

where

$$f_e = (N_e/N_v)/(2nm/N)^k$$

which for $m = 1$ is $f_e \approx (N_e/N_v)/2^k$. Likewise, for $\mathbf{G} = \mathbf{B} + \mathbf{I}$

$$\lambda_i = f_e n^i m^{k-i} \frac{\left(\frac{m+1}{m}\right)^k}{\left(1 + \frac{n-m}{m(m+1)}\right)^i}$$

where

$$f_e = (N_e/N_v)/((2nm + N)/N)^k$$

Figure 10.11 shows the degree distribution of 1,000,000 edge and 125,000 edge instance graphs taken from $\mathbf{B}(4, 1)^{\otimes 6}$.

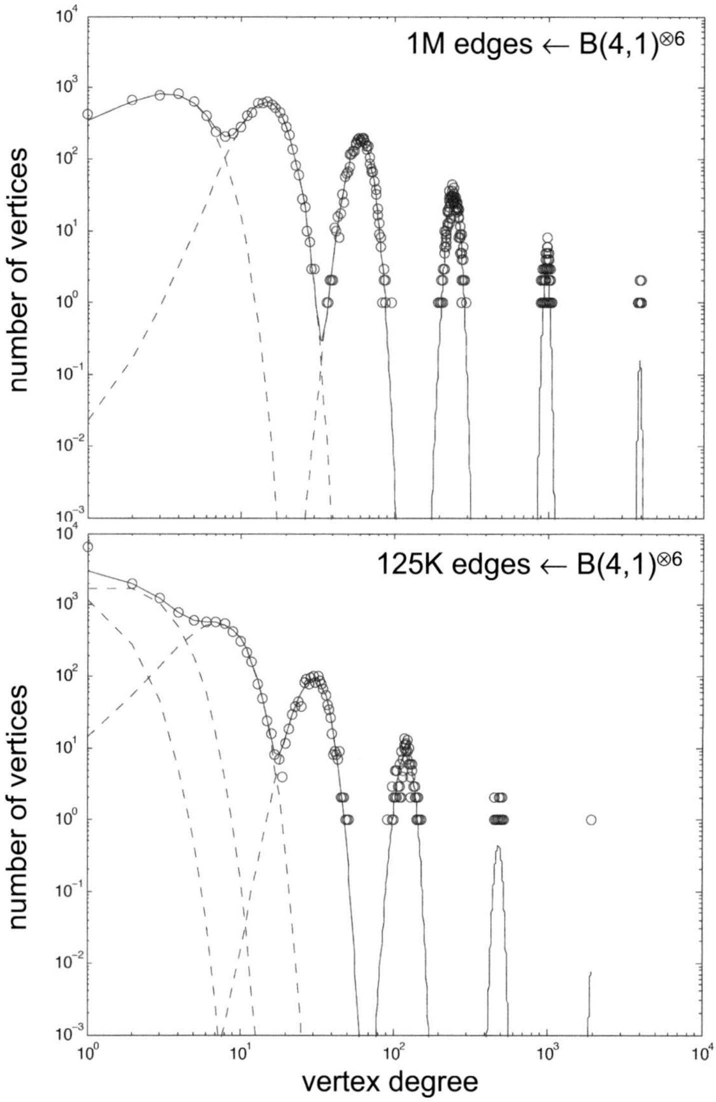

Figure 10.11. Instance degree distribution.
The degree distribution of 1,000,000 edge and 125,000 edge instance graphs taken from $\mathbf{B}(4,1)^{\otimes 6}$. Circles show the measured degree distribution of the instance graphs. Solid lines show the predicted degree distribution obtained by summing the Poisson distributions. Dashed lines show the Poisson distributions of each of the underlying terms in the explicit graph. The outlier points at high vertex degrees are because the vertex count can be less than one, and although the probability of one of these vertices occurring is low, the aggregate probability is enough to generate a small number of these high vertex degree points.

The peak-to-peak slope of the above distribution can be computed by assuming that λ_i is sufficiently large that the Poisson distributions can be approximated by a Gaussian distribution

$$P_{\lambda_i}(j) \approx \text{Gauss}_{\lambda_i,\lambda_i^{1/2}}(j) = (2\pi\lambda_i)^{-1/2} \exp[-(j-\lambda_i)^2/\lambda_i]$$

The peaks will occur at $j \approx \lambda_i$

$$P_{\lambda_i}(\lambda_i) \approx (2\pi\lambda_i)^{-1/2}$$

which results in the following formulas for the slopes

$$\mathit{Slope}[\lambda_i \to \lambda_{i+1}] \approx -\frac{3}{2} + \frac{\log_n((k-i)/(i+1))}{\log_n(n/m)}$$

and

$$\mathit{Slope}[\lambda_i \to \lambda_{k-i}] \approx -\frac{3}{2}$$

Thus, the peak-to-peak slope over any symmetric interval of an instance graph is also a power law, but with a slightly steeper slope.

10.7.2 Clustering power law graphs

A standard technique used for analyzing graphs is to attempt to cluster the graph. The basic clustering heuristic states the following [Radicchi 2004]

> Qualitatively, a community is defined as a subset of nodes within the graph such that connections between the nodes are denser than connections with the rest of the network.

Consider a graph that ideally suits the clustering heuristic: vertices inside clusters have many more connections within the cluster than outside the cluster. This is ideally represented by a series of N_1 loosely connected cliques of size N_2

$$\mathbf{I}(N_1) \otimes \mathbf{1}(N_2) + \gamma\mathbf{1}(N_1+N_2) = \bigcup^{N_1} 1(N_2) + \gamma\mathbf{1}(N_1+N_2)$$

If $\gamma = 0$, the degree distribution of such a graph is peaked around

$$\mathit{Count}[\mathit{Deg}(g) = N_2] = N_1 N_2$$

This will roughly correspond to a single Poisson peak (see Figure 10.11). If $\gamma > 0$, then the degree distribution will be broadened slightly around this peak. The distribution can be broadened further by varying N_2. Making this type of distribution consistent with a power law distribution is a challenge because it fundamentally consists of only one Poission distribution.

10.7.3 Dendragram and power law graphs

Another technique for analyzing graphs is to attempt to organize the graph into a dendragram or tree structure. The basic dendragram heuristic states the following [Radicchi 2004]

> The detection of the community structure in a network is generally intended as a procedure for mapping the network into a tree. In this tree (called a dendragram in the social sciences), the leaves are the nodes whereas the branches join nodes or (at higher level) groups of nodes, thus identifying a hierarchical structure of communities nested within each other.

Likewise, the ideal graph to apply a dendragram heuristic is a tree. A tree of degree k with l levels will also have a degree distribution that is peaked at

$$Count[Deg(G) = k + 1] = k^l$$

This will roughly correspond to a single Poisson peak (see Figure 10.11). The distribution can be broadened further by varying k with l, but making this type of distribution consistent with a power law distribution is a challenge because it fundamentally consists of only one Poission distribution.

10.8 Conclusions and future work

An analytical theory of power law graphs based on the Kronecker graph generation technique is presented. The analysis uses Kronecker exponentials of complete bipartite graphs to formulate the substructure of such graphs. This approach allows various high-level quantities (e.g., degree distribution, betweenness centrality, diameter, and eigenvalues) to be computed directly from the model parameters.

The analysis presented here on power law Kronecker matrices shows that the graph substructure is very different from those found in clusters or trees. Furthermore, the substructure changes qualitatively as a function of the degree distribution.

There are a number of avenues that can be pursued in subsequent work. These include

- Apply recursive bipartite permutation to various instance matrices.
- Examine if substructures produced in these generators are similar to those found in real data.
- Characterize vertices by their local bipartite adjacency matrix.
- Use models to predict exact computational complexity of algorithms on these graphs.

- Explore other "Kronecker" operations. For example

$$\mathbf{A} = \mathbf{B} \otimes^{+} \mathbf{C}$$

$$\mathbf{A} = \mathbf{B} \otimes^{max} \mathbf{C}$$

$$\mathbf{A} = \mathbf{B} \otimes^{xor} \mathbf{C}$$

- Block diagonal matrix as a generalization of I. For example

$$\mathbf{I}(N) \otimes \mathbf{1} = \bigcup^{N} \mathbf{1}$$

10.9 Acknowledgments

I would like to thank several individuals for their assistance with this work: John Gilbert for numerous helpful comments throughout the preparation of this manuscript, David Bader and Kamesh Madduri for introducing me to the R-MAT and Kronecker graph generators and for their assistance with betweenness centrality, and Jure Leskovec for his assistance with understanding the Kronecker graph generator. Finally, I would like to thank Bob Bond, Alan Edelman, Jeremy Fineman, Crystal Kahn, Mike Merrill, Nguyen Ngoc Huy, Raj Rao, Eric Robinson, and Viral Shah for many stimulating discussions I have with them on this topic.

References

[Brandes 2001] U. Brandes. A faster algorithms for betweenness centrality. *Journal of Mathematical Sociology,* 25:163–177, 2001.

[Broder 2000] A. Broder, R. Kumar, F. Maghoul1, P. Raghavan, S. Rajagopalan, R. Stata, A. Tomkins, and J. Wiener. Graph structure in the web: Experiments and models. In *Proceedings of World Wide Web Conference,* 2000.

[Bader 2006] D. Bader and K. Madduri. Efficient shared-memory algorithms and implementations for solving large-scale graph problems. In *SIAM Annual Meeting 2006.* Boston, Mass., 2006.

[Chakrabarti 2004] D. Chakrabarti, Y. Zhan, and C. Faloutsos. R-MAT: A recursive model for graph mining. In SDM04: Proceedings of the 4th SIAM International Conference on Data Mining, 442–446, 2004. http://www.siamorg/proceedings/datamining/2004/dm04.php.

[Chung 2005] F.R.K. Chung. Laplacians of graphs and Cheeger inequalities. *Annals of Combinatorics,* 9:1–19, 2005.

[Faloutsos 1999] M. Faloutsos, P. Faloutsos, and C. Faloutsos. On power-law relationships of the internet topology. In *Proceedings of the Conference on Applications, Technologies, Architectures, and Protocols for Computer Communication (SIGCOMM),* 251–262, 1999.

[Freeman 1977] L.C. Freeman. A set of measures of centrality based upon betweenness. *Sociometry,* 40:35–41, 1977.

[Gilbert 2006] J. Gilbert and V. Shah. An interactive environment for combinatorial supercomputing. In *SIAM Annual Meeting 2006*, Boston, Mass., 2006.

[Huy 2007] Nguyen Ngoc Huy, personal communication.

[Leskovec 2005] J. Leskovec, D. Chakrabarti, J. Kleinberg, and C. Faloutsos. Realistic, mathematically tractable graph generation and evolution, using Kronecker multiplication. In *European Conference on Principles and Practice of Knowledge Discovery in Databases (ECML/PKDD 2005)*, Porto, Portugal, 2005.

[Radicchi 2004] F. Radicchi, C. Castellano, F. Ceconi, V. Loreto, and D. Parisi. Defining and identifying communities in networks. Proceedings of the National Academy of Sciences (PNAS), 101:2658–2663, 2004.

[Redner 1998] S. Redner. How popular is your paper? An empirical study of the citation distribution. *European Physical Journal B,* 4:131–134, 1998.

[Van Loan 2000] C.F.V. Loan. The ubiquitous Kronecker product. *Journal of Computation and Applied Mathematics,* 123:85–100, 2000.

[Weischel 1962] P.M. Weischel. The Kronecker product of graphs. *Proceedings of the American Mathematical Society,* 13:47–52, 1962.

Chapter 11

Visualizing Large Kronecker Graphs

Huy Nguyen[*], *Jeremy Kepner*[†], *and Alan Edelman*[‡]

Abstract

Kronecker graphs have been shown to be one of the most promising models for real-world networks. Visualization of Kronecker graphs is an important challenge. This chapter describes an interactive framework to assist scientists and engineers in generating, analyzing, and visualizing Kronecker graphs with as little effort as possible.

11.1 Introduction

Kronecker graphs are of interest to the graph mining community because they possess many important patterns of realistic networks and are useful for theoretical analysis and proof (see [Leskovec et al. 2010] and Chapters 9 and 10). Once the model is fitted to the real networks, many applications can be built on top of Kronecker graphs, including graph compression, extrapolation, sampling, and anonymization. Moreover, there are efficient algorithms that can find Kronecker graphs that match important patterns of real networks [Leskovec et al. 2010]. Nevertheless, our understanding of Kronecker graphs is still limited. Chapter 9 (see also [Kepner 2008]) has shown that a simple combination of bipartite plus identity

*MIT CSAIL, 32 Vassar Street, Cambridge, MA 02139 (huy2n@mit.edu).
†MIT Lincoln Laboratory, 244 Wood Street, Lexington, MA 02420 (kepner@ll.mit.edu).
‡MIT Math Department, 77 Massachusetts Ave., Cambridge, MA 02139 (edelman@mit.edu).
This work is sponsored by the Department of the Air Force under Air Force Contract FA8721-05-C-0002. Opinions, interpretations, conclusions, and recommendations are those of the authors and are not necessarily endorsed by the United States Government.

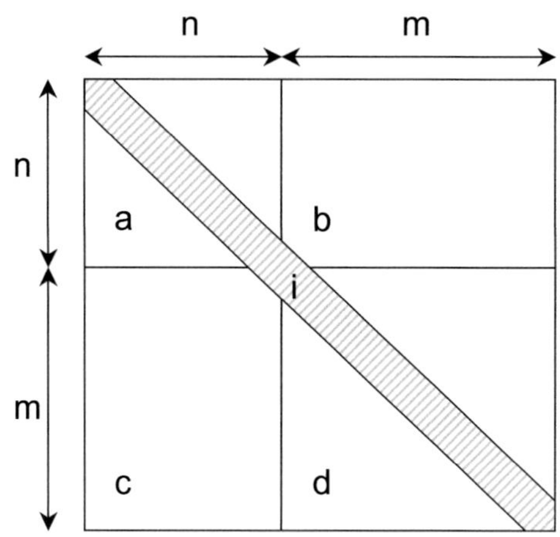

Figure 11.1. The seed matrix G.

graphs can generate a rich class of graphs that have many important patterns of a realistic network.

We have developed an interactive toolkit that assists scientists and algorithm designers in generating, analyzing, and visualizing Kronecker graphs. On a commodity workstation, the framework can interactively generate and analyze Kronecker graphs with millions of vertices. In order to visualize Kronecker graphs, we implement a visualizing algorithm on a parallel system that can then efficiently drive a large 250 megapixel display wall [Hill 2009]. This system allows for effective visualization of graphs of up to 100,000 vertices. Moreover, the framework is designed in an intuitive and interactive fashion that is easy to use and does not require much experience from the users in working with Kronecker graphs. We believe this tool can be a potential first step for anyone who wants to use Kronecker graphs as a model for their own networks.

11.2 Kronecker graph model

The working model of Kronecker graphs in our framework is a generalization of the bipartite stochastic model (see Chapter 9). In particular, let $\mathbf{G}$ be the seed matrix that is used to generate Kronecker graphs. $\mathbf{G}$ is a linear combination of a bipartite matrix $\mathbf{B}(n,m)$ and a diagonal matrix $\mathbf{I}$ (see Figure 11.1). $\mathbf{B}(n,m)$ is a four-quadrant matrix with size $(m+n) \times (m+n)$. The values of the entries in each of the quadrants are a, b, c, and d. In the diagonal matrix, all entries in the main diagonal have value i.

This simple model covers a range of Kronecker graphs. The stochastic model presented in [Leskovec et al. 2010] is a special case with $m = n = 1$ and $i = 0$. The model in [Kepner 2008] (where $\mathbf{G}$ is the union of a bipartite graph and an identity graph) is also a special case of our model with $a = d = 0$ and $b = c = i = 1$.

11.3 Kronecker graph generator

Let N and M be the desired number of vertices and edges of the Kronecker graph. If $N = (m+n)^k$, it is simple to generate a graph with the desired number of vertices. Specifically, we consider the $N \times N$ matrix $\mathbf{A} = \mathbf{G}^{\otimes k}$ where $\mathbf{G}$ is the seed matrix given above. For any pair of vertices i and j, a directed edge from i to j is created with probability $\mathbf{A}(i,j)$. In case the desired graph is sparse, $M = O(N)$, computing the whole matrix $\mathbf{A}$ is redundant. Instead, the algorithm can just generate the edge list directly from the seed matrix, with total running time $O(M \log N)$.

However, in case N is not a power of $m+n$, it is unclear how to generate Kronecker graphs from the algorithm above. A simple interpolation algorithm that creates a larger Kronecker graph (with $(m+n)^k$ vertices such that $(m+n)^{k-1} < N < (m+n)^k$) and then selects an appropriate subgraph does not work. Experiments show that simple interpolation produces large jumps (400%) on the edge/vertex (M/N) ratio in the resulting graph (see Figure 11.2). Sudden jumps in the edge/vertex ratio are not consistent with how real-world graphs grow.

The jumps in the edge/vertex ratio can be reduced by selecting the subgraph more intelligently using our "organic growth" interpolation algorithm. For a given desired N, our algorithm picks $(m+n)^k$—the smallest power of $m+n$ that is larger than N. Then, we generate an edge list (directly from the seed matrix) for the graph of size $(m+n)^k \times (m+n)^k$. Now, instead of taking a subgraph of size N, we randomly shuffle the labels of the vertices and then pick the N vertices with highest degree. As shown in Figure 11.2, generating Kronecker graphs in this way reduces the interpolation error of the edge/vertex ratio significantly.

11.4 Analyzing Kronecker graphs

Analyzing Kronecker graphs is the main feature of our framework. Once a graph is generated, it can be analyzed by three different methods: graph metrics, graph view, and graph organic growth. The graph metrics are a set of statistics about the structure of the graph that can be used to compare the similarity between the generated Kronecker graph and the target real network. The graph view helps users observe the generated graph from different perspectives and thereby identify important properties of the graph. Finally, the graph organic growth simulates the growing (or shrinking) process of a network graph using the previously described interpolation algorithm.

11.4.1 Graph metrics

We compute a set of statistics that can be used to derive many of the important metrics of a given graph.

- *Degree distribution power law exponent*: The degree distribution of a graph is a power law if the number of vertices with degree d in the graph is proportional to $d^{-\lambda}$ where $\lambda > 0$ is the *power law exponent*. Both the real networks and the Kronecker graphs can exhibit power law degree distribution.

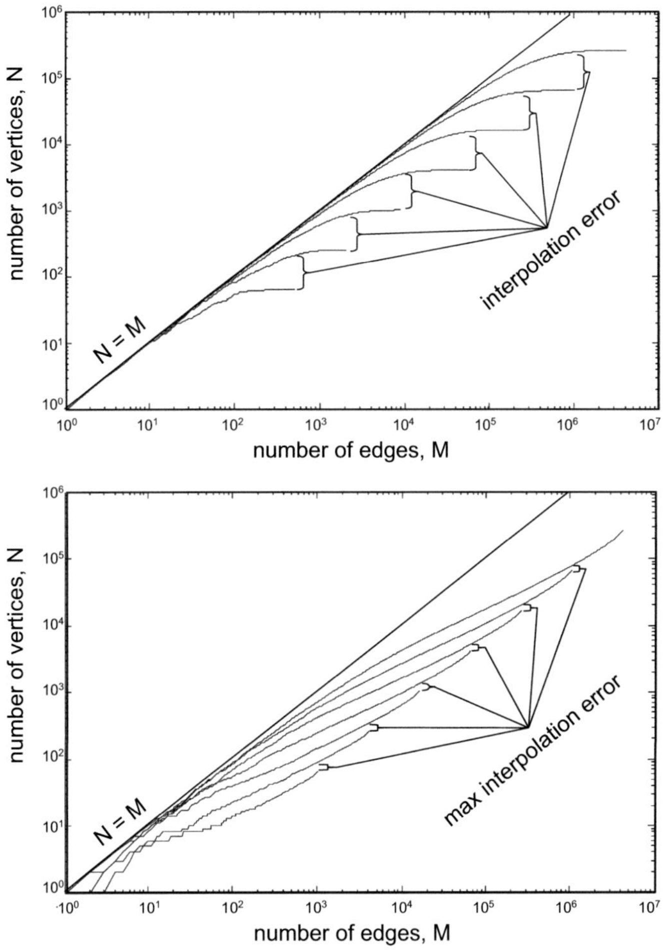

Figure 11.2. Interpolation algorithm comparison. Desired number of vertices versus the resulting number of edges for the simple interpolation algorithm (top) and our organic growth interpolation algorithm (bottom). The simple interpolation approach can produce a 400% jump in the number of edges when the desired number of vertices N is very close to $(n+m)^k$. The organic growth interpolation algorithm randomizes the vertex labels and selects the highest degree nodes to produce a much smoother curve with smaller jumps at these boundaries.

- *Densification power law exponent*: The number of edges M may be proportional to N^α where α is the densification power law exponent. Similar to the degree distribution power law exponent, this exponent can also be used as a metric of the graph [Leskovec et al. 2005].

- *Effective graph diameter*: The *effective graph diameter* was proposed in [Leskovec et al. 2005] as a robust measurement of the maximum distance between vertices in the graph. It is observed that many real-world networks have relatively small effective graph diameter (see [Leskovec et al. 2005, Albert 2001]). The diameter of the graph is not used here since it is susceptible to outliers.

- *Hop plot*: The function $f : k \to f(k)$, which is the fraction of pairs of vertices that have distance k in the graph [Palmer et al. 2002].

- *Node triangle participation*: The function $g : k \to g(k)$, which is the number of vertices that participate in k triangles in the graph [Tsourakakis 2008].

11.4.2 Graph view

Graph view is a visual way to study the structure of a Kronecker graph. The view is a 2D image of its 0/1 adjacency matrix under some permutation of the vertices. By adding more than one permutation to the framework, we hope that the images they create can give the users different perspectives about the graph structure and can help identify useful patterns. For example, if we view a Kronecker graph in the order it was generated, it is difficult to realize that its degree distribution obeys the power law. However, if we view the graph in degree sorted order, the power law property of its degree distribution can be easily noticed. Our framework allows a user to visualize the generated graph in four different permutations:

- Degree sorted: This view corresponds to vertices in nonincreasing order of degree.

- Randomized: This is the view taken from a random permutation.

- As generated: This is the view where the original order of vertices is used.

- Bipartite: This view exploits the inherent structure of a bipartite generating graph. The permutation takes advantage of the near-bipartiteness of the seed graph to efficiently detect the highly connected components in the graph and separate them from others. For more details on this permutation, see Chapter 10.

11.4.3 Organic growth simulation

In addition to the graph metrics and graph view, which only work with static Kronecker graphs, it is possible to use the organic growth interpolation algorithm to simulate how a graph might grow (or shrink). The simulation can show how a graph has been growing up from a single vertex to the current state and beyond. The main application of this feature is network extrapolation [Leskovec et al. 2010]. Given a real-world network and a Kronecker graph G that models that network, then by applying organic growth simulation on G, we can look into the future to see how the network might evolve. Similarly, organic growth simulation can also help us look into the past to see how the network might have looked in its early stages.

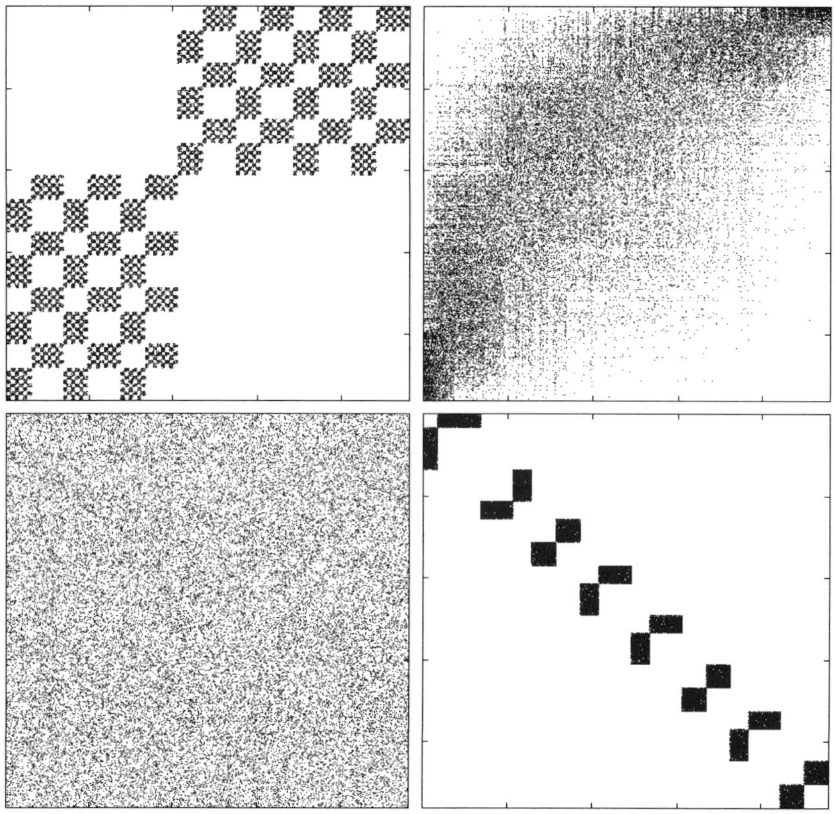

Figure 11.3. Graph permutations.
Views of different permutations of a Kronecker graph generated with a bipartite generating graph. Top left: As generated. Bottom left: Randomized. Top right: Degree sorted. Bottom right: Bipartite. The bipartite permutation clearly shows that the resulting graph consists of eight disconnected bipartite graphs.

11.5 Visualizing Kronecker graphs in 3D

Good visualization is an important part of an analytical framework. In this section, we describe our tool to visualize a Kronecker graph by projecting it onto the surface of a sphere. The idea of embedding a Kronecker graph onto a sphere was proposed and proved effective by Gilbert, Reinhardt, and Shah [Gilbert et al. 2007]. However, their embedding algorithm, which was designed for general graphs, did not take advantage of Kronecker graphs' structure. In our algorithm, we use the bipartite clustering method (as in the bipartite view) to partition the graph into well-connected components (see Figures 11.3 and 11.4) and embed them onto the sphere. As a result, the visualization quality has been improved significantly compared to the Fiedler mapping method (see Figure 11.5).

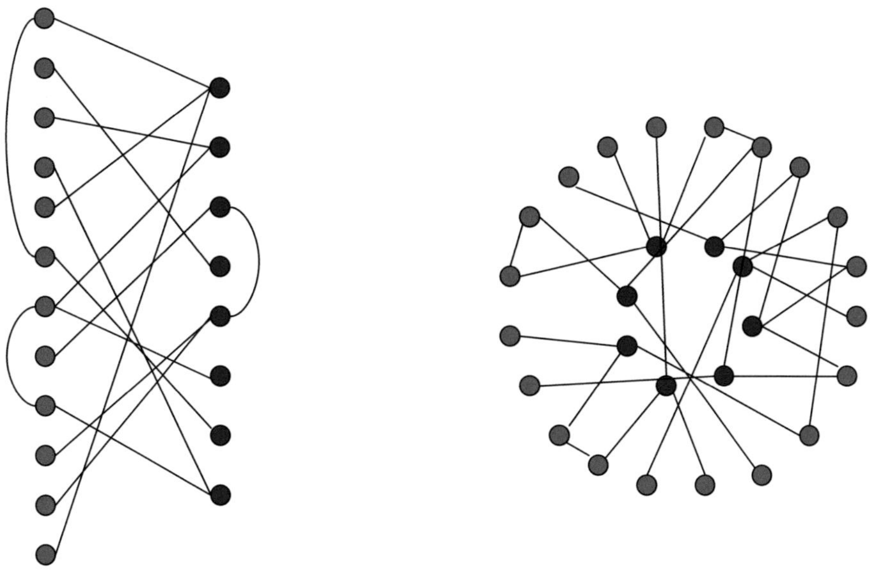

Figure 11.4. Concentric bipartite mapping.
Left: A near bipartite subgraph. Right: Mapping of subgraph on concentric circles.

11.5.1 Embedding Kronecker graphs onto a sphere surface

Given a Kronecker graph, our visualization first partitions such a graph into dense near-bipartite subgraphs using the bipartite permutation (see Chapter 10). Each subgraph is organized into a pair of concentric circles (see Figure 11.4), and these subgraphs are then placed onto a sphere so that they do not overlap. More specifically, for each such subgraph $\mathbf{B}(n, m)$, $\mathbf{B}$ comprises two nearly disjoint sets with n and m, where $n > m$. The subgraph is mapped onto two concentric circles such that n points are in the outer circle and m points are on the inner circle (see Figure 11.5). All the concentric circles are embedded on a sphere surface by using the Golden Section spiral method [Rusin 1998], which guarantees that the circles do not intersect and are evenly distributed. Because the majority of edges in the graph are internal to the subgraph, the visualization is pleasing to the eye and the overall structure of the Kronecker graph can be seen clearly.

11.5.2 Visualizing Kronecker graphs on parallel system

As the framework is designed to work with very large Kronecker graphs (up to 100,000 vertices), it is not practical to visualize them on a commodity workstation. Therefore, we implement our three-dimensional (3D) visualization algorithm on a parallel system with 60 display panels (2560×1600 pixel each) and 30 computational nodes (each node is responsible for 2 display panels), see Figure 11.6.

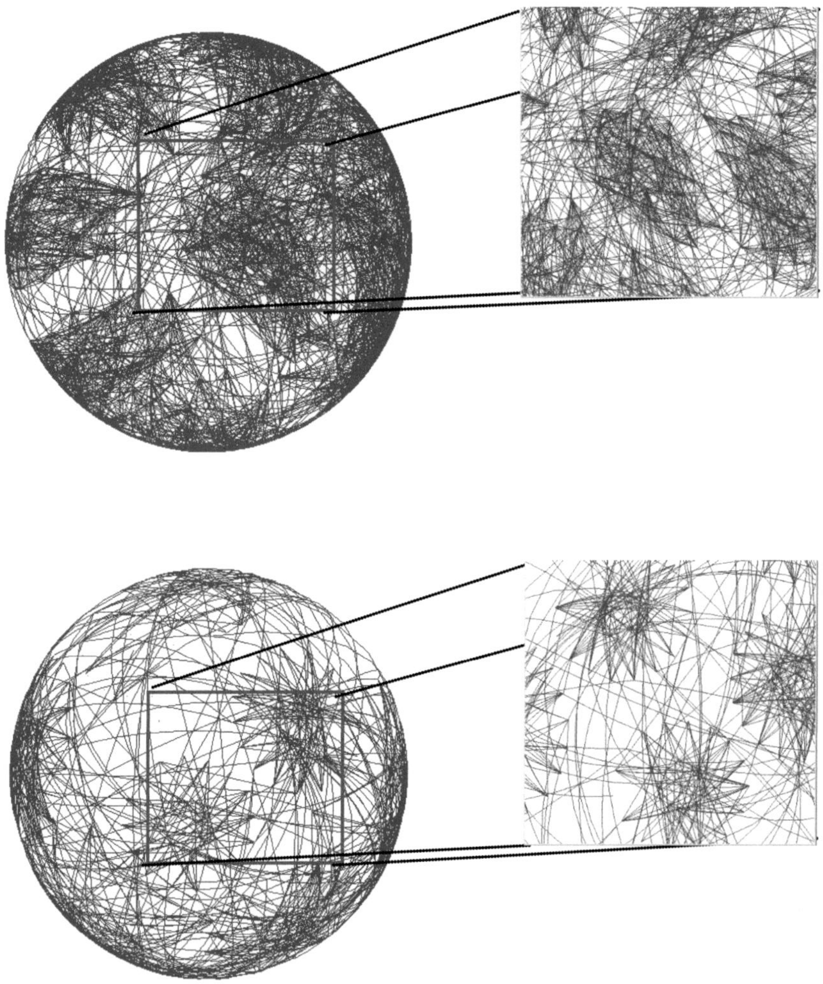

Figure 11.5. Kronecker graph visualizations.
Top: Fiedler mapping of a Kronecker graph onto a sphere. Bottom: Mapping of Kronecker graph with bipartite clustering method.

Figure 11.7 shows how the 3D visualization is designed on the parallel system. First, the graph is distributed to all computational nodes. Then, for each node, all vertices and edges of the graphs that are not visible on that node will be removed. Finally, visible edges and vertices are mapped onto a sphere surface using the algorithm above and rendered on the displaying panels.

Figure 11.6. Display wall.
Authors with 250 megapixel display wall showing a 100,000-vertex graph.

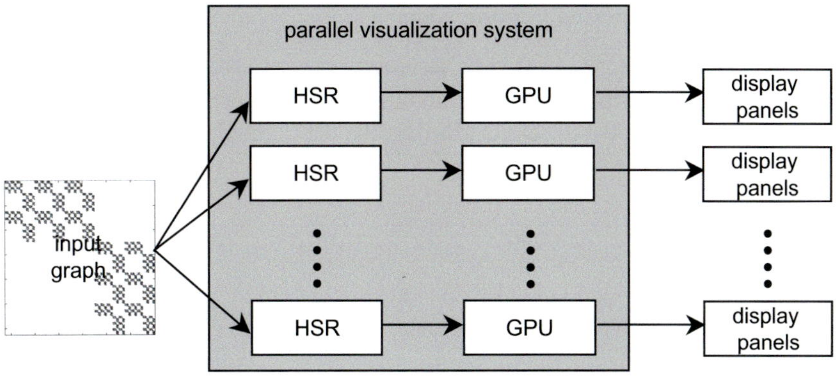

Figure 11.7. Parallel system to visualize a Kronecker graph in 3D. Components consist of input graph, hidden surface removal (HSR), graphics processing units (GPU), and video display panels.

References

[Albert 2001] R.Z. Albert and A.-L. Barabsi. Statistical mechanics of complex networks. *Reviews of Modern Physics,* 74:47–97, 2002.

[Gilbert et al. 2007] J.R. Gilbert, S. Reinhardt, and V. Shah. An interactive environment to manipulate large graphs. In *Proceedings of the 2007 IEEE International Conference on Acoustics, Speech, and Signal Processing,* 4:IV-1201–IV-1204 2007.

[Hill 2009] C. Hill. The Darwin Project. http://darwinproject.mit.edu/

[Kepner 2008] J. Kepner. Analytic theory of power law graphs. In *SIAM Parallel Processing 2008*, Minisymposium on HPC on Large Graphs, Atlanta, GA, 2008.

[Leskovec et al. 2010] J. Leskovec, D. Chakrabarti, J. Kleinberg, C. Faloutsos, and Z. Ghahramani. Kronecker graphs: An approach to modeling networks. *Journal of Machine Learning Research*, 11:985–1042, 2010.

[Leskovec et al. 2005] J. Leskovec, J. Kleinberg, and C. Faloutsos. Graphs over time: Densification laws, shrinking diameters and possible explanations. In *Proceedings of the Eleventh ACM SIGKDD International Conference on Knowledge Discovery in Data Mining (KDD '05)*, 177–187, 2005.

[Palmer et al. 2002] C.R. Palmer, P.B. Gibbons, and C. Faloutsos. ANF: A fast and scalable tool for data mining in massive graphs, In *Proceedings of the International Conference on Knowledge Discovery in Data Mining (KDD '02)*, 81–90, 2002.

[Rusin 1998] D. Rusin. Topics on sphere distributions. 1998. http://www.math.niu.edu/~rusin/known-math/95/sphere.faq.

[Tsourakakis 2008] C.E. Tsourakakis. Fast counting of triangles in large real networks without counting: Algorithms and laws. In *Proceedings of the IEEE International Conference on Data Mining*, 608–617, 2008.

Part III

Computation

Chapter 12
Large-Scale Network Analysis

*David A. Bader**, *Christine E. Heitsch*†, *and Kamesh Madduri*‡

Abstract

Centrality analysis deals with the identification of *critical* vertices and edges in real-world graph abstractions. Graph-theoretic centrality heuristics such as betweenness and closeness are widely used in application domains ranging from social network analysis to systems biology. In this chapter, we discuss several new results related to large-scale graph analysis using centrality indices. We present the *first parallel algorithms* and efficient implementations for evaluating these compute-intensive metrics. Our parallel algorithms are optimized for real-world networks, and they exploit topological properties such as the low graph diameter and unbalanced degree distributions. We evaluate centrality indices for several large-scale networks such as web crawls, protein-interaction networks

*College of Computing, Georgia Institute of Technology, Atlanta, GA 30332 (bader@cc.gatech.edu).

†School of Mathematics, Georgia Institute of Technology, Atlanta, GA 30332 (heitsch@math.gatech.edu).

‡Computational Research Division, Lawrence Berkeley National Laboratory, Berkeley, CA 94720 (KMadduri@lbl.gov).

Sections 12.2–12.4 based on "Parallel algorithms for evaluation centrality indices in real-world networks" by D.A. Bader and K. Madduri, which appeared in *Proceedings of the 35th International Conference on Parallel Processing* ©2006 IEEE.

Section 12.5 based on "A graph-theoretic analysis of the human protein-interaction network using multicore parallel algorithms" by D.A. Bader and K. Madduri, which appeared in *Proceedings of the 6th IEEE International Workshop on High-Performance Computational Biology* ©2007 IEEE.

Select text from Section 12.6 derived from "Analytic Betweenness Centrality for Meshes and Random Graphs," D.A. Bader, C. Heitsch, and K. Madduri, submitted to the *Journal of Graph Algorithms and Applications*, August 2010.

(PINs), movie-actor networks, and patent citation networks that are *three orders of magnitude larger* than instances that can be processed by current social network analysis packages. As an application to systems biology, we present the novel case study of betweenness centrality analysis applied to eukaryotic PINs. We make an important observation that proteins with high betweenness centrality, but low degree, are abundant in the human and yeast PINs.

12.1 Introduction

Graph abstractions are used to model interactions in a variety of application domains such as social networks (friendship circles, organizational networks), the Internet (network topologies, the web-graph, peer-to-peer networks), transportation networks, electrical circuits, genealogical research, and computational biology (protein-interaction networks, food webs). These networks seem to be entirely unrelated and indeed represent quite diverse relations, but experimental studies [Barabási & R. Albert 2007, Broder et al. 2000, Newman 2003] show that they share common traits such as a low average distance between the vertices (the *small-world* property), heavy-tailed degree distributions modeled by power laws, and high local densities. Modeling these networks based on experiments and measurements, and the study of interesting phenomena and observations [Callaway et al. 2000, Cohen et al. 2001, Pastor-Satorras & Vespignani 2001, Zanette 2001], continue to be active areas of research. Several models (see, e.g., [Guillaume & Latapy 2004, Newman et al. 2002, Palmer & Steffan 2000]) have been proposed to generate synthetic graph instances with these characteristics.

Complex network analysis traces its roots to the social sciences (see [Scott 2000, Wasserman & Faust 1994]), and seminal contributions in this field date back more than sixty years. There are several analytical tools (see, for instance, [Huisman & van Duijn 2005]) for visualizing social networks, determining empirical quantitative indices, and clustering. In most applications, graph abstractions and algorithms are frequently used to help capture the salient features. Thus, social network analysis (SNA) from a graph-theoretic perspective is about extracting interesting information, given a large graph constructed from a real-world data set.

Network modeling has received considerable attention in recent times, but algorithms are relatively less studied. Real-world graphs are typically characterized by low diameter, heavy-tailed degree distributions modeled by power laws, and self-similarity. They can be very large and sparse, with the number of vertices and edges ranging from several hundreds of thousands to billions. On current workstations, it is not possible to do exact in-core computations on these graphs because of the limited physical memory. In such cases, parallel computing techniques can be applied to obtain exact solutions for memory and compute-intensive graph problems quickly. For instance, recent experimental studies on breadth-first search for large-scale graphs show that a parallel in-core implementation is two orders of magnitude faster than an optimized external memory implementation [Ajwani et al. 2006, Bader & Madduri 2006a]. The design of efficient parallel graph algorithms is quite challenging [Lumsdaine et al. 2007] because massive graphs that occur in real-world applications are often not amenable to a balanced partitioning among processors of a parallel system [Lang 2005]. Algorithm design is simplified on parallel shared memory systems; they offer a higher memory bandwidth and lower latency than clusters, and the global shared memory obviates the need for partitioning the graph. However, the locality characteristics of parallel graph algorithms tend to be poorer than their sequential counterparts [Cong & Sbaraglia 2006], and so achieving good performance is still a challenge.

The key contributions of our work are as follows.

- We present the *first parallel algorithms* for efficiently computing the following centrality metrics: degree, closeness, stress, and betweenness (see, for instance, [Bader & Madduri 2006b]). We optimize the algorithms to exploit typical topological features of real-world graphs, and demonstrate the capability to process data sets that are *three orders of magnitude* larger than the networks that can be processed by existing social network analysis packages.

- We present the novel case study of betweenness centrality analysis applied to eukaryotic protein interaction networks (PINs). Jeong et al. [Jeong et al. 2001] empirically show that betweenness is positively correlated with a protein's essentiality and evolutionary age. We observe that proteins with *high betweenness centrality but low degree* are abundant in the human and yeast PINs, and that current small-world network models fail to model this feature [Bader & Madduri 2008].

- As a global shortest paths-based analysis metric, betweenness is highly correlated with routing and data congestion in information networks; see [Holme 2003, Singh & Gupte 2005]. We investigate the centrality of the integer torus, a popular interconnection network topology for supercomputers. We state and prove an empirical conjecture for betweenness centrality of all the vertices in this regular topology. This result is used as a validation technique in the HPCS Graph Analysis benchmark [Bader et al. 2006].

This chapter is organized as follows. Section 12.2 gives an overview of various centrality metrics and the sequential algorithms to compute them. We present our new parallel algorithms to compute centrality indices and optimizations for real-world networks in Section 12.3. Section 12.4 discusses implementation details and the performance of these algorithms on parallel shared memory and multithreaded architectures. We discuss the case study of betweenness applied to PINs in Section 12.5 and our result on betweenness for an integer torus in Section 12.6.

12.2 Centrality metrics

One of the fundamental problems in network analysis is to determine the *importance* or *criticality* of a particular vertex or an edge in a network. Quantifying *centrality* and *connectivity* helps us identify or isolate regions of the network that may play interesting roles. Researchers have been proposing metrics for centrality for the past fifty years, and there is no single accepted definition. The metric of choice is dependent on the application and the network topology. Almost all metrics are empirical and can be applied to element-level [Brin & Page 1998], group-level [Doreian & L.H. Albert 1989], or network-level [Bacon] analyses. We discuss several commonly used vertex centrality indices in this section. Edge centrality indices are similarly defined.

Preliminaries

Consider a graph $G = (V, E)$, where V is the set of vertices representing *actors* or *entities* in the complex network, and E is the set of edges representing relationships between the vertices. The number of vertices and edges is denoted by N and M, respectively. The graphs can be directed or undirected. We will assume that each edge $e \in E$ has a positive integer weight $w(e)$. For unweighted graphs, we use $w(e) = 1$. A *path* from vertex s to t is defined as a sequence of edges $\langle u_i, u_{i+1} \rangle$, $0 \leq i < l$, where $u_0 = s$ and $u_l = t$. The *length* of a path is the sum of the weights of edges. We use $d(s,t)$ to denote the distance between vertices s and t (the minimum length of any path connecting s and t in G). Let's denote the total number of shortest paths between vertices s and t by σ_{st}, and the number passing through vertex v by $\sigma_{st}(v)$.

Degree centrality

In an undirected network, the degree centrality of a vertex is simply the count of the number of adjacencies or neighbors it has. For directed graphs, we can define two variants: in-degree centrality and out-degree centrality. This is a simple local measure based on the notion of neighborhood and is straightforward to compute. In many networks, a high degree vertex is considered an important or central player, and this index quantifies that observation.

Closeness centrality

This index measures the closeness, in terms of *distance*, of a vertex to all other vertices in the network. Vertices with a smaller total distance are considered more important. Several closeness-based metrics [Bavelas 1950, Nieminen 1973] have been developed by the SNA community. A commonly used definition is the reciprocal of the total shortest path distance from a particular vertex to all other vertices

$$CC(v) = \frac{1}{\sum_{u \in V} d(v,u)}$$

Unlike degree centrality, this is a global metric. To calculate the closeness centrality of a vertex v, we may perform a breadth-first traversal (BFS, for unweighted graphs) or use a single-source shortest paths (SSSP, for weighted graphs) algorithm. The closeness centrality of a single vertex can be determined in linear time for unweighted networks.

Stress centrality

Stress centrality is a metric based on shortest paths counts, first presented in [Shimbel 1953]. It is defined as

$$SC(v) = \sum_{s \neq v \neq t \in V} \sigma_{st}(v)$$

Intuitively, this metric deals with the *communication work* done by each vertex in a network. The number of shortest paths that pass through a vertex v gives an estimate of the amount of stress a vertex v is under, assuming communication is carried out through shortest paths all the time. This index can be calculated by using a variant of the all-pairs shortest paths algorithm and by computing the number of shortest paths between every pair of vertices.

Betweenness centrality

Betweenness centrality is another shortest paths enumeration-based metric, introduced by Freeman in [Freeman 1977]. Let $\delta_{st}(v)$ denote the *pairwise dependency*, or the fraction of shortest paths between s and t that pass through v

$$\delta_{st}(v) = \frac{\sigma_{st}(v)}{\sigma_{st}}$$

Betweenness centrality of a vertex v is defined as

$$BC(v) = \sum_{s \neq v \neq t \in V} \delta_{st}(v) \tag{12.1}$$

This metric can be thought of as a normalized version of stress centrality. Betweenness centrality of a vertex measures the control a vertex has over communication in the network and can be used to identify critical vertices in the network. High centrality indices indicate that a vertex can reach other vertices on relatively short paths or that a vertex lies on a considerable fraction of shortest paths connecting pairs of other vertices.

This index has been extensively used in recent years for analysis of social as well as other large-scale complex networks. Some applications include the analysis of biological networks [del Sol et al. 2005, Jeong et al. 2001, Pinney et al. 2005], study of sexual networks and AIDS [Liljeros et al. 2001], identification of key actors in terrorist networks [Coffman et al. 2004, Krebs 2002], organizational behavior, supply chain management [Cisic et al. 2000], and transportation networks (see [Guimer et al. 2005]).

There are a number of commercial and research software packages for SNA (e.g., Pajek [Batagelj & A. Mrvar 1998], InFlow [Krebs 2005], UCINET [UCINET]) that can be used to determine these centrality metrics. However, they can only process comparatively small networks (in most cases, sparse graphs with less than 40,000 vertices). Our goal is to develop fast, high-performance implementations of these metrics to process large-scale real-world graphs with millions to billions of vertices.

Algorithms for computing betweenness centrality

We can evaluate the betweenness centrality of a vertex v (defined in equation (12.1)) by determining the number of shortest paths between every pair of vertices s and t and the number of shortest paths that pass through v. There is no known algorithm to compute the exact betweenness centrality score of a single vertex without solving an all-pairs shortest paths problem instance in the graph. In this chapter, we will constrain our discussion of parallel betweenness centrality algorithms to directed, unweighted graphs. To process undirected graphs, the network can be easily modified by replacing each edge by two oppositely directed edges. While the approach to parallelization for unweighted graphs [Bader & Madduri 2006b] works for weighted low-diameter graphs as well, the concurrency in each parallel phase is dependent on the weight distribution.

Earlier algorithms compute betweenness centrality in two steps: first, the number and length of shortest paths between all pairs of vertices are computed and stored, and second, the pairwise dependencies (the fractions $\frac{\sigma_{st}(v)}{\sigma_{st}}$) for each s-t pair are summed. The complexity of this approach is $O(N^3)$ time and $O(N^2)$ space. Exploiting the sparse nature of real-world networks, Brandes [Brandes 2001] presented an improved sequential algorithm to compute the betweenness centrality score for all vertices in an unweighted graph in $O(MN)$ time and $O(M+N)$ space. The main idea is to perform N breadth-first graph traversals and augment each traversal to compute the number of shortest paths passing through each vertex. The second key idea is that pairwise dependencies $\delta_{st}(v)$ can be aggregated without computing all of them explicitly. Define the *dependency* of a source vertex $s \in V$ on a vertex $v \in V$ as $\delta_s(v) = \sum_{t \in V} \delta_{st}(v)$. The betweenness centrality of a vertex v can be then expressed as $BC(v) = \sum_{s \neq v \in V} \delta_s(v)$. Brandes showed that the dependency values $\delta_s(v)$ satisfy the following recursive relation

$$\delta_s(v) = \sum_{w:d(s,w)=d(s,v)+1} \frac{\sigma_{sv}}{\sigma_{sw}}(1 + \delta_s(w)) \tag{12.2}$$

Thus, the sequential algorithm computes betweenness in $O(MN)$ time by iterating over all the vertices $s \in V$ and computing the dependency values $\delta_s(v)$ in two stages. First, the distance and shortest path counts from s to each vertex are determined. Second, the vertices are revisited starting with the farthest vertex from s first, and dependencies are accumulated according to equation (12.2).

12.3 Parallel centrality algorithms

In this section, we present novel parallel algorithms to compute the various centrality metrics, optimized for real-world networks. To the best of our knowledge, these are the first parallel algorithms for centrality analysis. We exploit the typical low-diameter (small-world) property to reveal an additional level of parallelism in graph traversal and take the unbalanced degree distribution into consideration while designing algorithms for the shortest path-based enumeration metrics. In addition to the exact algorithms, we also discuss approaches to approximate closeness and betweenness in [Bader et al. 2007].

We use a compact array representation for the graph that requires $M + N + 1$ machine words. We assume that the vertices are labeled with integer identifiers between 0 and $N - 1$. All the adjacencies of a vertex are stored contiguously in a block of memory, and the neighbors of vertex i are stored next to ones of vertex $i + 1$. The size of the adjacency array is M, and we require an array of pointers to this adjacency array, which is of size $N + 1$ words. This representation is motivated by the fact that all the adjacencies of a vertex are visited in the graph traversal step after the vertex is first discovered.

For parallel algorithm analysis, we use a complexity model similar to the one proposed by Helman and JáJá [Helman & JáJá 2001], which has been shown to work well in practice. This model takes into account both computational complexity and memory contention. We measure the overall complexity of an algorithm using $T_m(N, M, N_P)$, the maximum number of noncontiguous accesses made by any processor to memory, and $T_c(N, M, N_P)$, the upper bound on the local computational complexity. N_P denotes the number of processors or parallel threads of execution.

Degree centrality

We store the in- and out-degree of each vertex during construction of the graph abstraction. Thus, determining the degree centrality of a particular vertex is a constant-time lookup operation. As noted previously, degree centrality (also referred to as vertex connectivity in statistical physics literature) is a useful local metric and probably the most studied measure in complex network analysis.

Closeness centrality

Closeness centrality of a vertex v can be computed on a sequential processor by a breadth-first traversal from v (or single-source shortest paths in case of weighted graphs) and requires no auxiliary data structures. Thus, vertex centrality computation can be done in parallel by a straightforward parallelization of BFS. In a typical network analysis scenario, we would require centrality scores of all the vertices in the graph. We can then evaluate N BFS (shortest path) trees in parallel, one for each vertex $v \in V$. On N_P processors, this would yield $T_c = O(\frac{NM+N^2}{N_P})$ and $T_m = O(\frac{N^2}{N_P})$ for unweighted graphs. For weighted graphs, using a naive queue-based representation for the expanded frontier, we can compute all the centrality metrics in $T_c = O(\frac{NM+N^3}{N_P})$ and $T_m = O(\frac{N^2}{N_P})$. The bounds can be further improved with the use of efficient priority queue representations.

Evaluating closeness centrality of all the vertices in a graph is computationally intensive; hence, it is valuable to investigate approximate algorithms. Using a random sampling technique, Eppstein and Wang [Eppstein & Wang 2001] showed that the closeness centrality of all vertices in a weighted, undirected graph can be approximated with high probability in $O(\frac{\log N}{\epsilon^2}(N \log N + M))$ time and an additive error of at most $\epsilon\Delta_G$ (ϵ is a fixed constant, and Δ_G is the diameter of the graph). The algorithm proceeds as follows. Let k be the number of iterations needed to obtain the desired error bound. In iteration i, pick vertex v_i uniformly at random from V and solve the SSSP problem with v_i as the source. The estimated centrality is given by

$$\widetilde{CC}(v) = \frac{k}{N\Sigma_{i=1}^{k} d(v_i, u)}$$

The error bounds follow from a result by Hoeffding [Hoeffding 1963] on probability bounds for sums of independent random variables.

We design a parallel algorithm for approximate closeness centrality as follows. Each processor runs SSSP computations from $\frac{k}{N_P}$ vertices and stores the evaluated distance values. The cost of this step is given by $T_c = O(\frac{k(M+N)}{N_P})$ and $T_m = O(\frac{kM}{N_P})$ for unweighted graphs. For real-world graphs, the number of sample vertices k can be set to $\Theta(\frac{\log N}{\epsilon^2})$ to obtain the error bounds given above. The approximate closeness centrality value of each vertex can then be calculated in $O(k) = O(\frac{\log N}{\epsilon^2})$ time, and the summation for all N vertices would require $T_c = O(\frac{N \log N}{N_P \epsilon^2})$ and constant T_m.

Stress and betweenness centrality

Computing stress and betweenness centrality involves shortest path enumeration between every pair of vertices in the network, and there is no known algorithm to compute the betweenness/stress centrality score of a single vertex or an edge in linear time. We design two novel parallel algorithms for vertex betweenness centrality that retain the computational complexity of Brandes' sequential algorithm and that are particularly suited for real-world sparse networks.

We observe that parallelism can be exploited in the centrality computation at two levels:

- The BFS/SSSP computations from each vertex can be done concurrently, provided the centrality running sums are updated atomically.
- A single BFS/SSSP computation can be parallelized. Further, adjacencies of a vertex can be processed concurrently.

We will refer to the parallelization approach that concurrently computes the shortest path trees as the *coarse-grained* parallel betweenness centrality algorithm, and the latter approach, in which a single BFS/SSSP traversal is parallelized, as the *fine-grained* algorithm. Algorithm 12.1 gives the pseudocode for the fine-grained approach and describes the two main stages that are parallelized in each iteration. The loops that are executed in parallel (see lines 3, 13, 14, and 26) are indicated in the schematic. The coarse-grained algorithm uses the same data structures, but only the main loop (step 2) is parallelized. Note that accesses to shared data structures and updates to the distance and path counts need to be protected with appropriate synchronization constructs, which we do not indicate in Algorithm 12.1.

Algorithm 12.1. Synchronous betweenness centrality.
A level-synchronous parallel algorithm for computing betweenness centrality of vertices in unweighted graphs.

$\mathbf{b} : \mathbb{R}_+^N = \text{PARALLELBC}(G = (V, E))$

$\mathbf{b} = 0$
for $k \in V$
  do for $t \in V$ **in parallel**
    do $\mathbf{P}(t) : \mathbb{Z} =$ empty multiset
   $\boldsymbol{\sigma} : \mathbb{Z}^N = 0$
   $\mathbf{d} : \mathbb{Z}^N = -1$
   $\boldsymbol{\sigma}(k) = 1,\ \mathbf{d}(k) = 0$
   $i = 0,\ \mathbf{S}(i) : \mathbb{Z} =$ empty stack
   PUSH$(k, \mathbf{S}(i))$
   $c = 1$
   Graph traversal for shortest path discovery & counting
   while $c > 0$
    do $c = 0$
     for $v \in \mathbf{S}(i)$ **in parallel**
      do for each neighbor w of v **in parallel**
       do if $\mathbf{d}(w) < 0$
        do PUSH$(w, \mathbf{S}(i+1))$
         $c \mathrel{+}= 1$
         $\mathbf{d}(w) = \mathbf{d}(v) + 1$
       if $\mathbf{d}(w) = \mathbf{d}(v) + 1$
        do $\boldsymbol{\sigma}(w) \mathrel{+}= \boldsymbol{\sigma}(v)$
         APPEND$(v, \mathbf{P}(w))$
     $i \mathrel{+}= 1$
   $i \mathrel{-}= 1$
   Dependency accumulation by back-propagation
   $\boldsymbol{\delta} : \mathbb{R}^N = 0$
   while $i > 0$
    do for $w \in \mathbf{S}(i)$ **in parallel**
      do for $v \in \mathbf{P}(w)$
       do $\boldsymbol{\delta}(v) \mathrel{+}= \frac{\boldsymbol{\sigma}(v)}{\boldsymbol{\sigma}(w)}(1 + \boldsymbol{\delta}(w))$
      $\mathbf{b}(w) \mathrel{+}= \boldsymbol{\delta}(w)$
     $i \mathrel{-}= 1$

The fine-grained parallel algorithm proceeds as follows. Starting at the source vertex k, we successively expand the frontier of visited vertices and augment breadth-first graph traversal (we also refer to this as *level-synchronous* graph traversal) to count the number of shortest paths passing through each vertex. We maintain a multiset of *predecessors* $\mathbf{P}(w)$ associated with each vertex w. A vertex v belongs to the predecessor multiset of w if $\langle v, w \rangle \in E$ and $d(w) = d(v) + 1$. We implement each multiset as a dynamic array that can be resized at runtime, and so $\mathbf{P}$ is a two-dimensional data structure, with one array for each of the N vertices. The APPEND routine used in line 21 of the algorithm atomically adds a vertex to the multiset. Clearly, the size of a predecessor multiset for a vertex is bounded by its in-degree (or degree, in the case of an undirected graph). The predecessor information is used in the dependency accumulation step, which implements equation (12.2). The other key data structure used in both the stages is $\mathbf{S}$, the stack of visited vertices. $\mathbf{S}(i)$ stores all the vertices that are at a distance i from the source vertex, and the vertices are added atomically to the stack using the PUSH routine. We need $O(N)$ storage for $\mathbf{S}$. The rest of the data structures ($\mathbf{b}$, $\boldsymbol{\sigma}$, $\mathbf{d}$, and $\boldsymbol{\delta}$) are one-dimensional arrays. In an arbitrary network, there is a possibility that the path counts grow exponentially in the graph traversal, which might lead to an arithmetic overflow for some values of the path counts array $\boldsymbol{\sigma}$. We use 64-bit unsigned integers in our betweenness implementation and have so far never encountered a path count overflow occurrence for a real-world network instance. The final scores need to be divided by two (not shown in the algorithm) if the graph is undirected, as all the shortest paths are counted twice.

There are performance tradeoffs associated with both the coarse-grained and fine-grained algorithms, when implemented on parallel systems. The coarse-grained algorithm assigns each processor a fraction of the vertices from which to initiate graph traversals computations. The vertices can be assigned dynamically to processors, so that work is distributed as evenly as possible. For this approach, the graph traversal requires no synchronization, and the centrality metrics can be computed exactly, provided they are accumulated atomically. Alternately, each processor can store its partial sum of the centrality score for every vertex, and all the sums can be merged using an efficient global reduction operation. However, the problem with the coarse-grained algorithm is that the auxiliary data structures ($\mathbf{S}$, $\mathbf{P}$, $\boldsymbol{\sigma}$, $\mathbf{d}$, and $\boldsymbol{\delta}$) need to be replicated on each processor for doing concurrent traversals. The memory requirements scale as $O(N_P(M+N))$, and this approach becomes infeasible for large-scale graphs.

In the fine-grained algorithm, we parallelize each graph traversal, and so the memory requirement is just $O(M+N)$. We exploit concurrency in the two stages of each iteration, graph traversal for path discovery and counting (lines 11 to 23) and dependency accumulation by back-propagation (lines 24 to 30), from the fact the graph has a low diameter. In previous work, we presented multithreaded algorithms and efficient implementations for fine-grained parallel BFS (see [Bader & Madduri 2006a]) and SSSP (see [Crobak et al. 2007, Madduri et al. 2007]). Similar ideas can be applied to reduce the synchronization overhead in the access to shared data structures in the graph traversal phase.

12.3.1 Optimizations for real-world graphs

The unbalanced degree distribution is another important graph characteristic we need to consider while optimizing centrality algorithms. It has been observed that real networks tend to have highly skewed degree distributions that, in some cases, can be approximated by power laws [Barabási & R. Albert 2007, Newman 2003]. We see a significant number of low degree vertices in these networks and a comparatively smaller number of vertices of very high degree (can be as large as $O(N)$) [Dall'Asta et al. 2006]. In some networks, centrality is strongly correlated with degree [Jeong et al. 2001]: intuitively, high degree vertices may have high centrality scores as a significant number of shortest paths pass through them. The degree distribution has very little effect on the performance of coarse-grained parallel centrality algorithms, as we do a full graph traversal on every outer-loop iteration. Each iteration roughly takes the same time, and even a static distribution of work among processors is reasonably well balanced. However, while designing fine-grained centrality algorithms, we need to explicitly consider unbalanced degree distributions. In a level-synchronized parallel BFS in which vertices are statically assigned to processors without considering their degree, it is highly probable that there will be phases with severe work imbalance. For instance, consider the case in which one processor is assigned a group of low degree vertices and another processor has to expand the BFS frontier from a set of high degree vertices (say, degree $O(N)$). In the worst case, we will not achieve any parallel speedup using this approach. Our fine-grained BFS and shortest path algorithms [Bader & Madduri 2006a, Madduri et al. 2007] are designed to be independent of degree distribution, and we use these optimized algorithms as the inner routines for the fine-grained centrality algorithms.

There are several other optimizations that are applicable to real-world networks. If a network is composed of several large disjoint subgraphs, we can run the linear-time connected components algorithm to preprocess the network and identify the components. The centrality indices of the various components can then be evaluated concurrently. Similarly, we can decompose a directed network into its strongly connected components.

Observe that, by definition, the betweenness centrality score of a degree-1 vertex is zero (Figure 12.1). Also, we show that it is not necessary to traverse the graph from a degree-1 vertex if we already have the shortest path tree from its adjacent vertex. For undirected networks, Algorithm 12.2 gives the modified pseudocode for the dependency accumulation stage (lines 25–30 of Algorithm 12.1) from the adjacency of a degree-1 vertex. With this revision, we can just skip the iteration from a degree-1 vertex. Instead, we increment the dependency score of all other vertices in the traversal from its adjacency (see line 5). Each degree-1 vertex contributes to an increase of $\delta(w)$ to the centrality score, and hence we add the term $n_1\delta(w)$. In addition, we need to increment the centrality score of the source vertex (k in this case) in this traversal itself. Line 7 of the routine gives the required increment to be factored in. This optimization is particularly effective in networks such as the web-graph and protein-interaction networks, in which there are a significant percentage of degree-1 vertices (unannotated proteins with few interactions; web pages linking to a hub-site). In case of directed networks, the increment values (steps 5 and 7) differ depending on the directivity of the edges, but the same idea is applicable.

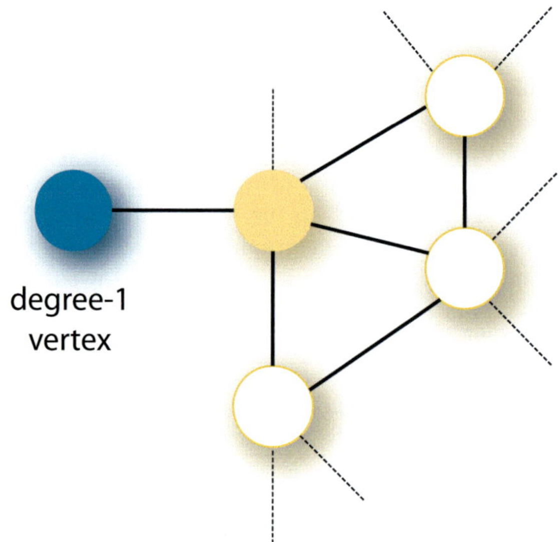

Figure 12.1. Betweenness centrality definition.
The betweenness centrality index of a degree-1 vertex is 0. We need not traverse the graph from a degree-1 vertex if we store the shortest path tree from its adjacency.

Algorithm 12.2. Betweenness centrality dependency accumulation.
Pseudocode for the parallel betweenness centrality algorithm dependency accumulation stage (replacement for lines 25–30 in Algorithm 12.1) for a graph traversal from vertex k. k has $n_1 > 0$ adjacencies whose degree is 1.

$\mathbf{b} : \mathbb{R}_+^N = \text{ParallelBCAccumDeg1Adj}(k, i, \mathbf{S}, \boldsymbol{\sigma}, \mathbf{P}, \boldsymbol{\delta})$

while $i > 0$
    do for $w \in \mathbf{S}(i)$ **in parallel**
        do for $v \in \mathbf{P}(w)$
            do $\boldsymbol{\delta}(v) \mathrel{+}= \frac{\boldsymbol{\sigma}(v)}{\boldsymbol{\sigma}(w)}(1 + \boldsymbol{\delta}(w))$
        $\mathbf{b}(w) \mathrel{+}= (n_1 + 1)\boldsymbol{\delta}(w)$
    $i \mathrel{-}= 1$
$\mathbf{b}(k) \mathrel{+}= n_1(\boldsymbol{\delta}(k) - 1)$

12.4 Performance results and analysis

12.4.1 Experimental setup

In this section, we discuss parallel performance results of closeness, stress, and betweenness centrality. We implement both the fine-grained and coarse-grained approaches for betweenness. In addition, we optimize our implementations for three different shared memory architectures: multicore systems, symmetric multiprocessors (SMPs), and massively multithreaded architectures. The multithreaded implementation is optimized for the Cray MTA-2 (see, for instance, [Bader & Madduri 2006a, Bader & Madduri 2006b]). The MTA-2 is a high-end shared memory system offering two unique features that aid considerably in the design of irregular algorithms: fine-grained parallelism and low-overhead word-level synchronization. It has no data cache; rather than using a memory hierarchy to reduce latency, the MTA-2 processors use hardware multithreading to tolerate the latency. The word-level synchronization support complements multithreading and makes performance primarily a function of parallelism. Since graph algorithms have an abundance of parallelism, yet often are not amenable to partitioning, the MTA-2 architectural features lead to superior performance and scalability.

For computing centrality metrics on weighted graphs, we use a fine-grained parallel SSSP algorithm [Madduri et al. 2007] as the inner routine for graph traversal. However, the centrality accumulation step cannot be easily parallelized, as the concurrency is dependent on the weight distribution. The coarse-grained algorithm is straightforward and just requires minor changes to Algorithm 12.1.

We report multithreaded performance results on a 40-processor Cray MTA-2 system with 160 GB uniform shared memory. Each processor has a clock speed of 220 MHz and support for 128 hardware threads. The code is written in C with MTA-2 specific pragmas and directives for parallelization. We compile the code using the MTA-2 C compiler (Cray Programming Environment [PE] 2.0.3). The MTA-2 code also compiles and runs on sequential processors without any modification.

Our test platform for the SMP implementations is an IBM Power 570 server. The IBM Power 570 is a symmetric multiprocessor with 16 1.9 GHz Power5 cores with simultaneous multithreading (SMT), 32 MB shared L3 cache, and 256 GB shared memory. The code is written in C with OpenMP directives for parallelization and compiled using the IBM XL C compiler v7.0.

We report multicore system performance results on the Sun Fire T2000 server, with the Sun UltraSPARC T1 (Niagara) processor. This system has eight cores running at 1.0 GHz, each of which is four-way multithreaded. There are eight integer units with a six-stage pipeline on chip, and four threads running on a core share the pipeline. The cores also share a 3 MB L2 cache, and the system has a main memory of 16 GB. There is only one floating-point unit (FPU) for all cores. We compile our code with the Sun C compiler v5.8.

Network data

We test our centrality metric implementations on a variety of real-world graphs, summarized in Table 12.1. We use the Recursive MATrix (R-MAT) (see [Chakrabarti et al. 2004]) random graph generation algorithm to generate synthetic input data that are representative of real-world networks with a small-world topology. The degree distributions of the Internet Movie Database (IMDB) test graph instance are shown in Figure 12.2. We observe that the degree distributions of most of the networks are unbalanced with heavy tails. This observation is in agreement with prior experimental studies.

Table 12.1. Networks used in the centrality analysis.

Data set	Source	Network description
ND-actor	[Barabási 2007]	An undirected graph of 392,400 vertices (movie actors) and 31,788,592 edges. An edge corresponds to a link between two actors, if they have acted together in a movie. The data set includes actor listings from 127,823 movies.
ND-web	[Barabási 2007]	A directed network with 325,729 vertices and 1,497,135 arcs. Each vertex represents a web page within the Univ. of Notre Dame *nd.edu* domain, and the arcs represent from → to links.
ND-yeast	[Barabási 2007]	Undirected network with 2114 vertices and 2277 edges. Vertices represent proteins, and the edges represent interactions between them in the yeast network.
UMD-human	[Batagelj & A. Mrvar 2006]	Undirected network with 18,669 vertices and 43,568 edges. Vertices represent proteins, and the edges represent interactions between them in the human interactome.
PAJ-patent	[Batagelj & A. Mrvar 2006]	A network of about 3 million U.S. patents granted between January 1963 and December 1999, and 16 million citations made among them between 1975 and 1999.
PAJ-cite	[Batagelj & A. Mrvar 2006]	The *Lederberg* citation data set, produced using HistCite, in PAJEK graph format with 8843 vertices and 41,609 edges.

12.4.2 Performance results

Figure 12.3 compares the single-processor execution time of closeness, betweenness, and stress centrality for three networks of different sizes, on the MTA-2 and the Power 570. All three metrics are of the same computational complexity and exhibit nearly similar running times practice.

The MTA-2 performance results are for the fine-grained centrality implementations. On SMPs, the coarse-grained version outperforms the fine-grained algorithm on current systems due to the parallelization and synchronization overhead involved

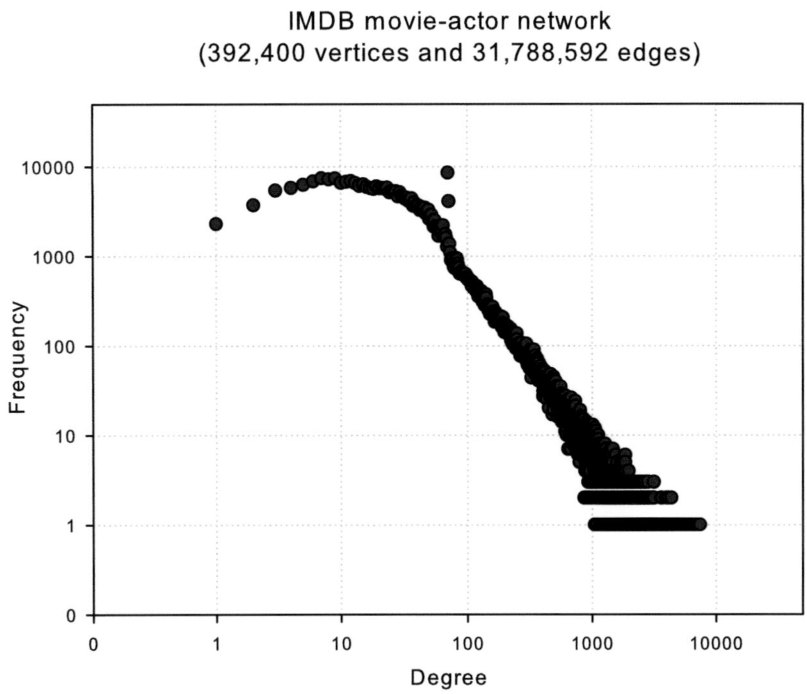

Figure 12.2. Vertex degree distribution of the IMDB movie-actor network.

in the fine-grained version. We only have a modest number of processors on current SMP systems, so each processor can run a concurrent shortest path computation and create auxiliary data structures. On our target system, the Power 570, we can compute centrality metrics for graphs with up to 100 million edges by using this coarse-grained implementation.

Figure 12.4 summarizes multiprocessor execution times for computing betweenness centrality on the Power 570 and the MTA-2. Figure 12.4(a) gives the running times for the ND-actor graph on the Power 570 and the MTA-2. As expected, the execution time scales nearly linearly with the number of processors. It is possible to evaluate the centrality metric for the entire ND-actor network in 42 minutes on 16 processors of the Power 570. We observe similar performance for the patents citation data. This includes the optimizations for undirected, unweighted real-world networks discussed in Section 12.3.1.

Figures 12.4(c) and 12.4(d) plot the execution time on the MTA-2 and the Power 570 for ND-web, and a synthetic graph instance of the same size generated using the R-MAT algorithm, respectively. Note that the actual execution time is dependent on the graph structure; for the same problem size, the synthetic graph instance takes much longer than the ND-web graph. The web crawl is a directed network, and splitting the network into its strongly connected components and the degree-1 optimization helps in significantly reducing the execution time.

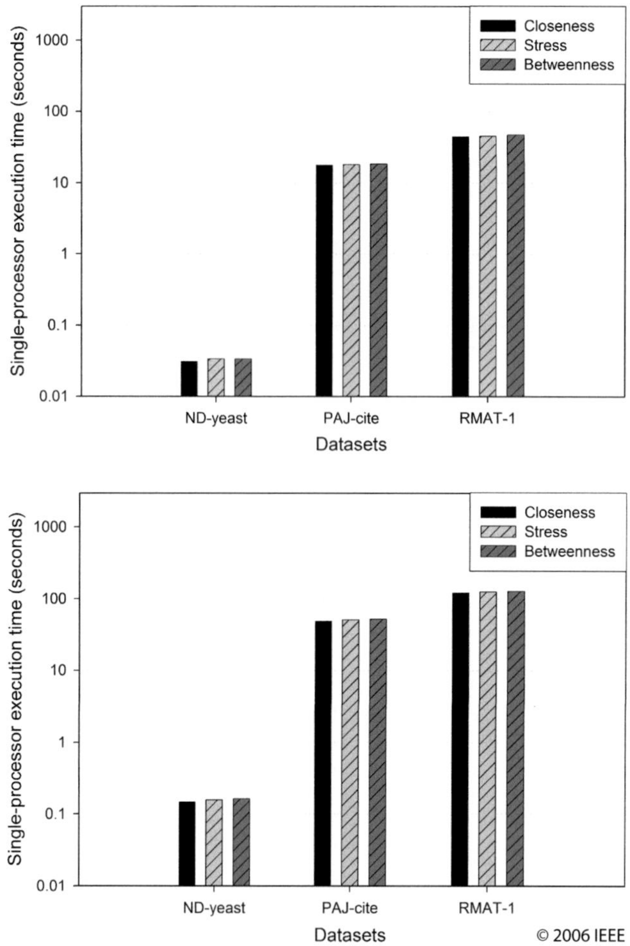

Figure 12.3. Single-processor comparison.
Single-processor execution time comparison of the centrality metric implementations on the IBM Power 570 (top) and the Cray MTA-2 (bottom).

12.5 Case study: Betweenness applied to protein-interaction networks

We will now apply the betweenness centrality metric to analyze the human interactome. Researchers have paid particular attention to the relation between centrality and *essentiality* or *lethality* of a protein (for instance, [Jeong et al. 2001]). A protein is said to be essential if the organism cannot survive without it. Essential proteins can only be determined experimentally, so alternate approaches to *predicting*

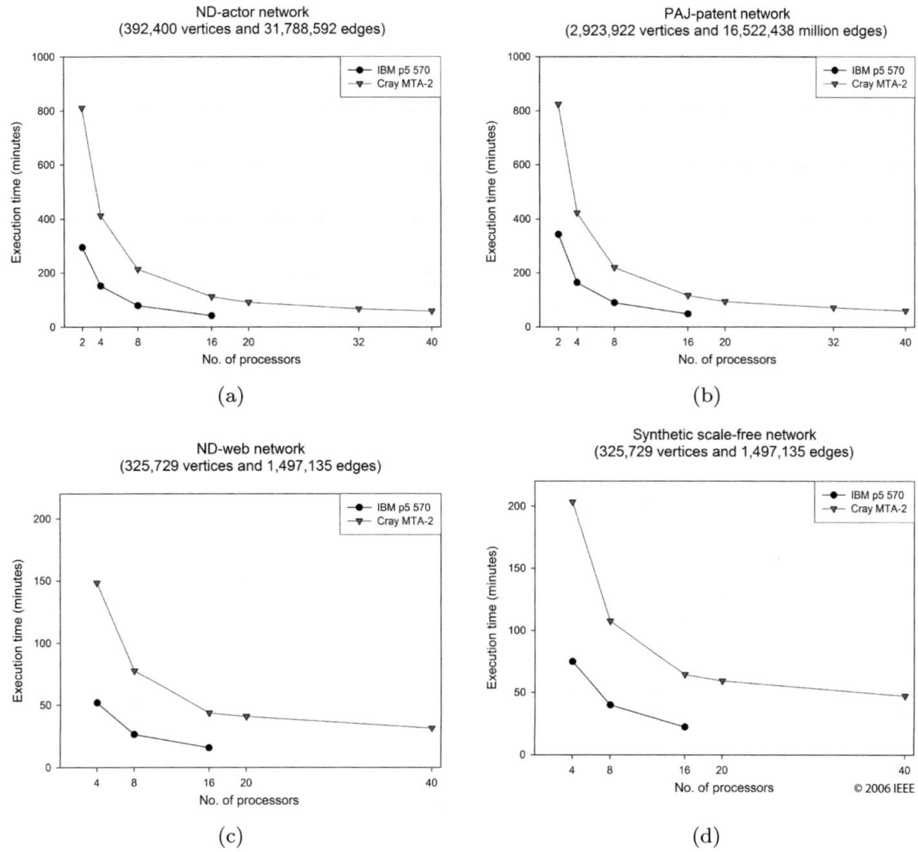

Figure 12.4. Parallel performance comparison.
Parallel performance of exact betweenness centrality computation for various graph instances on the Power 570 and the MTA-2.

essentiality are of great interest and have potentially significant applications, such as drug target identification [Jeong et al. 2003]. Previous studies on yeast have shown that proteins acting as hubs (or high degree vertices) are three times more likely to be essential. So we wish to analyze the interplay between degree and centrality scores for proteins in the human PIN. We derive our human protein-interaction map (referred to as HPIN throughout the chapter) by merging interactions from a human proteome analysis data set [Gandhi et al. 2006], a data snapshot from the Human Protein Reference Database [Peri et al. 2003], and IntAct [Hermjakob et al. 2004]. Each protein is represented by a vertex, and interactions between proteins are modeled as edges. This results in a high-confidence protein-interaction network of 18,869 proteins and 43,568 interactions.

Figure 12.5 plots the betweenness centrality scores of the top 1% (about 100) proteins in two lists, one ordered by degree and the other by the betweenness

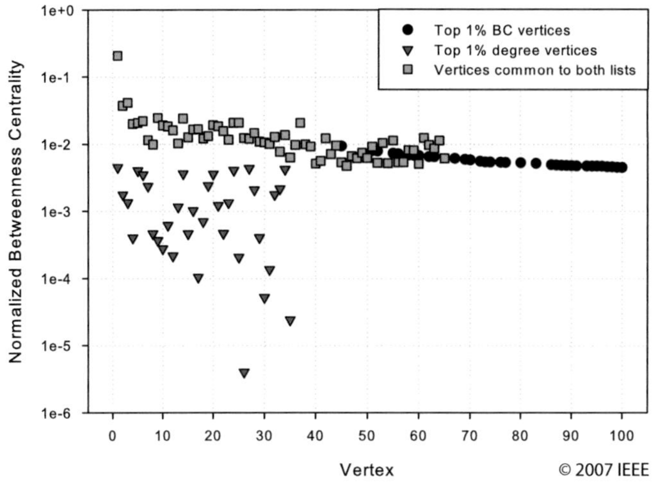

Figure 12.5. The top 1% proteins.
HPIN proteins sorted by betweenness centrality (BC) scores and the number of interactions.

centrality score. We observe that there is a strong correlation between the degree and betweenness centrality score: about 65% of the proteins are common to both lists. The protein with the highest degree in the graph also has the highest centrality score. This protein (*Solute carrier family 2 member 4*, Gene Symbol SLC2A4, HPRD ID 00688) belongs to the transport/cargo protein molecular class, and its primary biological function is transport. From Figure 12.5, it should also be noted that the top 1% proteins by degree show a significant variation in betweenness centrality scores. The scores vary by over four orders of magnitude, from 10^{-1} to 10^{-4}.

We next study the correlation of degree with betweenness centrality. Unlike degree, which ranges from 1 to 822, the values of betweenness centrality range over several orders of magnitude. The few highly connected (or high degree) vertices have high betweenness values, as there are many vertices directly and exclusively connected to these hubs. Thus, most of the shortest paths between these vertices go through these hubs. However, the low-connectivity vertices show a significant variation in betweenness values, as evidenced in Figure 12.6 (top). They exhibit a variation of betweenness of values up to four orders of magnitude. The high betweenness scores may suggest that these proteins are globally important. Interestingly, these vertices are completely absent in synthetically generated graphs designed to explain scale-free behavior (observe the variation of betweenness centrality scores among low degree vertices in Figure 12.6 (bottom).

Our observations are further corroborated by two recent results. As the yeast PIN has been comprehensively mapped, lethal proteins in the network have been identified. Gandhi et al. [Gandhi et al. 2006] demonstrated from an independent

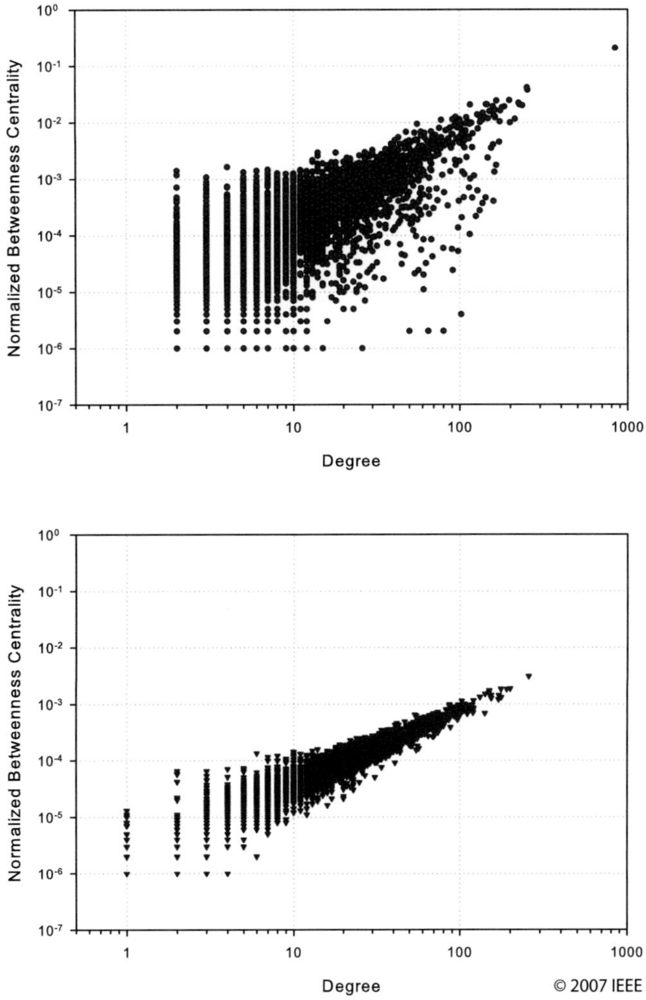

Figure 12.6. Normalized HPIN betweenness centrality. Normalized betweenness centrality scores as a function of the degree for HPIN (top) and a synthetic scale-free graph instance (bottom).

analysis that the relative frequency of a gene to occur as an essential one is higher in the yeast network than in the HPIN. They also observe that the lethality of a gene could not be confidently predicted on the basis of the number of interaction partners. Joy et al. [Joy et al. 2005] confirmed that proteins with high betweenness scores are more likely to be essential and that there are a significant number of high-betweenness, low-interaction proteins in the yeast PIN. For a detailed discussion of our analysis of HPIN, see [Bader & Madduri 2008].

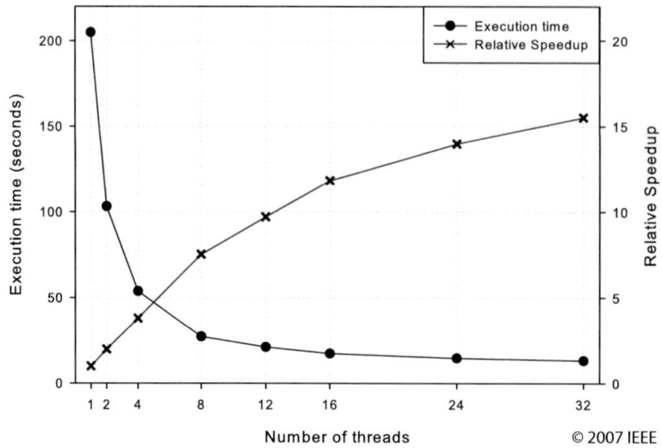

Figure 12.7. Betweenness centrality performance.
Betweenness centrality execution time and speedup on the Sun Fire T2000 system.

Figure 12.7 plots the execution time and relative speedup achieved on the Sun Fire T2000 for computing the betweenness centrality on HPIN. The performance scales nearly linearly up to 16 threads, but plateaus between 16 and 32 threads. This can be attributed to a saturation of memory bandwidth with 16 threads of execution, as well as the presence of only one floating-point unit on the entire chip. We use the floating-point unit for accumulating pair dependencies and centrality values.

12.6 Integer torus: Betweenness conjecture

Betweenness centrality is a key kernel in the DARPA HPCS SSCA#2 (see [Bader et al. 2006]), a benchmark extensively used to evaluate the performance of emerging high performance computing architectures for graph analytics. Parallel implementations of multithreaded graph algorithms are often prone to programming error and are computationally expensive to validate; since the integer torus is a regular network that is easy to generate, we propose using it as a test instance for the SSCA#2 benchmark. A simple validation check for the benchmark would be to compare the computed betweenness scores generated from a computational routine with the exact analytical expression derived in this section.

For $n \in \mathbb{N}$, let $\mathcal{T}_n$ denote an integer torus, that is the two-dimensional integer lattice mod n. Based on empirical evidence from extensive computational experimentation, we have the following.

Conjecture 1. *For* $v \in \mathcal{T}_n$

$$BC(v) = \frac{n^3}{2} - n^2 - \frac{n}{2} + 1 \text{ when } n \text{ is odd} \tag{12.3}$$

and

$$BC(v) = \frac{n^3}{2} - n^2 + 1 \text{ when } n \text{ is even} \tag{12.4}$$

There is a parity dependence because of the impact of geodesics whose horizontal and/or vertical distance is maximal when n is even. For $s, t \in \mathcal{T}_n$ with $s = (x, y)$ and $t = (x', y')$, let

$$d_H(s,t) = \min\{(x' - x) \bmod n, (x - x') \bmod n\}$$
$$d_V(s,t) = \min\{(y' - y) \bmod n, (y - y') \bmod n\}$$
$$d(s,t) = d_H(s,t) + d_V(s,t)$$

denote the horizontal, vertical, and total distance, respectively, from s to t. Note that $0 \le d_H(s,t), d_V(s,t) \le \frac{n}{2}$. When n is even and $d_H(s,t) = \frac{n}{2}$, then we say that s and t are horizontal *diameter achieving.* Similarly, when $d_V(s,t) = \frac{n}{2}$, then s and t are said to be vertical diameter achieving.

For $0 \le d_H(s,t), d_V(s,t) < \frac{n}{2}$, we have that

$$\sigma_{st} = \binom{d_H(s,t) + d_V(s,t)}{d_V(s,t)} = \binom{d_H(s,t) + d_V(s,t)}{d_H(s,t)}$$

If n is even and s and t are either horizontal or (exclusively) vertical diameter achieving, then

$$\sigma_{st} = 2\binom{d_H(s,t) + d_V(s,t)}{d_V(s,t)}$$

If n is even and s and t are both horizontal and vertical diameter achieving, then

$$\sigma_{st} = 4\binom{d_H(s,t) + d_V(s,t)}{d_V(s,t)}$$

Let $v = (p, q)$ for $p, q \in \mathbb{N}$ with $0 \le p, q < n$. Observe that there is a shortest path between s and t, which passes through v if and only if

$$d_H(s,t) = d_H(s,v) + d_H(v,t) \text{ and } d_V(s,t) = d_V(s,v) + d_V(v,t)$$

Since $\sigma_{st}(v) = \sigma_{sv}\sigma_{vt}$, we will calculate $BC(v)$ by counting all geodesics from s to v and from v to t where $\sigma_{st}(v)$ is nonzero. We exploit the symmetries of $\mathcal{T}_n$ and enumerate shortest paths through $v_0 = (0, 0)$ for particular subsets of $s, t \in \mathcal{T}_n$. For $S, T \subseteq \mathcal{T}_n$ and $v_0 = (0, 0)$, let

$$\Delta(S,T) = \sum_{s \in S, t \in T, s \ne v \ne t} \delta_{st}(v_0)$$

Let $m = \lfloor \frac{n}{2} \rfloor$ and $x, y \in \{-m, \ldots, 0, \ldots, m\}$ for $(x, y) \in \mathcal{T}_n$. We divide $\mathcal{T}_n$ into four quadrants, centered at v_0

$$Q_1 = \{(-x, -y) \in \mathcal{T}_n \mid 0 \le x, y \le m\}$$
$$Q_2 = \{(-x, y) \in \mathcal{T}_n \mid 0 \le x, y \le m\}$$
$$Q_3 = \{(x, y) \in \mathcal{T}_n \mid 0 \le x, y \le m\}$$
$$Q_4 = \{(x, -y) \in \mathcal{T}_n \mid 0 \le x, y \le m\}$$

Since exactly which subsets of $\mathcal{T}_n$ are used depends on the parity of n, we consider the two cases separately.

12.6.1 Proof of conjecture when *n* is odd

We first prove equation (12.3) of the conjecture, since there are no diameter achieving pairs of vertices when n is odd.

Lemma 12.1. *For $v_0 = (0,0) \in \mathcal{T}_n$ with n odd*

$$\begin{aligned} BC(v_0) &= \Delta(Q_1,Q_3) + \Delta(Q_3,Q_1) + \Delta(Q_2,Q_4) + \Delta(Q_4,Q_2) \\ &\quad -\Delta(Q_1,Q_2) - \Delta(Q_3,Q_4) - \Delta(Q_1,Q_4) - \Delta(Q_2,Q_3) \\ &= 4\cdot\Delta(Q_1,Q_3) - 4\cdot\Delta(Q_1,Q_2) \end{aligned}$$

Proof. In the calculation of $BC(v_0)$, we sum over all possible pairs s and t, where a path from s to t is counted as distinct from the reversed path that begins at t and ends at s. Thus, the first four terms count all possible paths from s to t along the four "diagonals" through v_0, with redundancy. By the symmetries of $\mathcal{T}_n$, these four values are all equal to $\Delta(Q_1,Q_3)$ which counts the fraction of shortest paths through v_0 from an $s \in Q_1$, the "lower-left" quadrant (modulo n) with respect to $(0,0)$, to t in the upper-right quadrant Q_3. The remaining additive factors then correct for the overcounting of geodesics along the vertical and horizontal, respectively, lines through v_0. Like before, these are all equal to $\Delta(Q_1,Q_2)$ where $\sigma_{st}(v_0) \neq 0$ if and only if $x = 0 = x'$ for $(x,y) \in Q_1$ and $(x',y') \in Q_2$.

Note that each of the paths where one of s and t lies on the horizontal line through v_0 and the other lies on the vertical line through v_0 is counted exactly twice (once in each direction) in $4\cdot\Delta(Q_1,Q_2)$. For instance, $\Delta(Q_1,Q_2)$ counts paths where s lies on the horizontal line to the left of v_0 (again modulo n) or on the vertical line below v_0 and where t lies on the horizontal line to the right of v_0 or on the vertical line above v_0. □

Theorem 12.2. *Let n be an odd integer. Suppose $v_0 = (0,0) \in \mathcal{T}_n$ and $m = \lfloor \frac{n}{2} \rfloor$, with*

$$\begin{aligned} Q_1 &= \{(-x,-y) \in \mathcal{T}_n \mid 0 \leq x,y \leq m\} \\ Q_2 &= \{(-x,y) \in \mathcal{T}_n \mid 0 \leq x,y \leq m\} \\ Q_3 &= \{(x,y) \in \mathcal{T}_n \mid 0 \leq x,y \leq m\} \end{aligned}$$

Then

$$\Delta(Q_1,Q_2) = \frac{m(m-1)}{2} \tag{12.5}$$

and

$$\Delta(Q_1,Q_3) = 1 + (m-1)(m+1)^2 \tag{12.6}$$

Proof. Let $v_0 = (0,0) \in \mathcal{T}_n$ for an odd integer n. Consider $s,t \in \mathcal{T}_n$ where $s \neq v_0 \neq t$ and v_0 lies on a shortest path from s to t. Let $m = \lfloor \frac{n}{2} \rfloor$ and

$$\begin{array}{ll} d_H(s,t) = h, & d_V(s,t) = k \\ d_H(s,v_0) = i, & d_V(s,v_0) = j \\ d_H(v_0,t) = i' = h-i, & d_V(v_0,t) = j' = k-j \end{array}$$

for $0 \le i \le h \le m < \frac{n}{2}$ and $0 \le j \le k \le m < \frac{n}{2}$ where not both $i = j = 0$ nor $i = h, j = k$ nor $h = k = 0$. In this notation, we have that

$$\sigma_{sv_0} = \binom{i+j}{i}, \ \sigma_{v_0t} = \binom{(h-i)+(k-j)}{(h-i)}, \text{ and } \sigma_{st} = \binom{h+k}{h}$$

Suppose further that $s \in Q_1$ and $t =\in Q_2$. Since $\sigma_{st}(v_0) = 0$ unless $x = 0 = x'$, we need only consider $i = 0$. Since $j = 0$ implies that $s = v_0$ and $j = k$ implies that $t = v_0$, we consider only $0 < j < k$. When $i = 0$ and $0 < j < k \le m < \frac{n}{2}$, then $\sigma_{st} = 1$ and $\sigma_{st}(v_0) = 1$. Thus, since $i = 0$ and $h = 0$

$$\begin{aligned} \Delta(Q_1,Q_2) &= \sum_{k=2}^{m} \sum_{j=1}^{k-1} \frac{\binom{0+j}{0}\binom{0+(k-j)}{0}}{\binom{0+k}{0}} \\ &= \frac{m(m-1)}{2} \end{aligned}$$

where we begin the outer summation at $k = 2$ since $h = 0$, $k = 0$ implies that $s = v_0 = t$ and $h = 0, k = 1$ implies that either $s = v_0$ or $t = v_0$. Hence, equation (12.5) in Theorem 12.2 holds.

Suppose now that $s \in Q_1$ and $t \in Q_3$. As with our previous equality, we enumerate $\Delta(Q_1,Q_3)$ by summing over the possible values of $0 \le i \le h \le m < \frac{n}{2}$ and $0 \le j \le k \le m < \frac{n}{2}$ where not both $i = j = 0$ nor $i = h, j = k$ nor $h = k = 0$. Also, as before, if $h = 0$, then $k > 1$ and vice versa. Thus

$$\Delta(Q_1,Q_3) = 1 + \sum_{h=0}^{m} \sum_{k=0}^{m} \left(-2 + \frac{1}{\binom{h+k}{k}} \sum_{i=0}^{h} \sum_{j=0}^{k} \binom{i+j}{i} \binom{(h-i)+(k-j)}{(h-i)} \right)$$

where we have corrected for $i = 0 = j$ and $h - i = 0 = k - j$ by subtracting out the terms $\binom{0+0}{0}\binom{h+k}{k}$ and $\binom{h+k}{k}\binom{0+0}{0}$. Likewise, we have corrected for $h = 0, k = 0$ by adding 1. Note that when $h = 0, k = 1$ and $h = 1, k = 0$, the expression inside the h and k summands is zero and no correction is needed. We know that

$$\sum_{j=0}^{k} \binom{i+j}{i} \binom{(h-i)+(k-j)}{(h-i)} = \binom{h+k+1}{k}$$

either as an application of equation (5.26) from [Graham et al. 1989] (found via identity # 3100005 on the Pascal's Triangle website* or proved directly via induction

*http://binomial.csuhayward.edu/.

and parallel summation. Hence, we have that

$$\begin{aligned}
\Delta(Q_1, Q_3) &= 1 + \sum_{h=0}^{m}\sum_{k=0}^{m}\left(-2 + \frac{1}{\binom{h+k}{k}}\sum_{i=0}^{h}\binom{h+k+1}{k}\right)\\
&= 1 + \sum_{h=0}^{m}\sum_{k=0}^{m}\left(-2 + \frac{1}{\binom{h+k}{k}}(h+1)\frac{h+k+1}{h+1}\binom{h+k}{k}\right)\\
&= 1 + \sum_{h=0}^{m}\sum_{k=0}^{m}(h+k-1)\\
&= 1 + (m-1)(m+1)^2
\end{aligned}$$

and equation (12.6) of Theorem 12.2 also holds. □

Since $m = \frac{n-1}{2}$ and

$$BC(v_0) = 4 \cdot \Delta(Q_1, Q_3) - 4 \cdot \Delta(Q_1, Q_2) = BC(v) \text{ for all } v \in \mathcal{T}_n$$

as an immediate consequence of Theorem 12.2, we have that

Corollary 12.3. *When n is odd*

$$BC(v) = \frac{n^3}{2} - n^2 - \frac{n}{2} + 1$$

12.6.2 Proof of conjecture when *n* is even

The proof of equation (12.4) of the conjecture is similar, although considerably more complicated when n is even and $m = \frac{n}{2}$. The complications are due to any diameter achieving pairs $s, t \in \mathcal{T}_n$, which double the number of geodesics when either $d_H(s,t)$ or $d_V(s,t) = \frac{n}{2}$ and quadruple it when s and t are both horizontal and vertical diameter achieving.

Again, we let

$$\begin{aligned}
Q_1 &= \{(-x,-y) \in \mathcal{T}_n \mid 0 \le x, y \le m\}\\
Q_2 &= \{(-x,y) \in \mathcal{T}_n \mid 0 \le x, y \le m\}\\
Q_3 &= \{(x,y) \in \mathcal{T}_n \mid 0 \le x, y \le m\}\\
Q_4 &= \{(x,-y) \in \mathcal{T}_n \mid 0 \le x, y \le m\}
\end{aligned}$$

Also, for $s, t, v_0 = (0,0) \in \mathcal{T}_n$, we will use the same notation of

$$\begin{aligned}
&d_H(s,t) = h, && d_V(s,t) = k\\
&d_H(s,v_0) = i, && d_V(s,v_0) = j\\
&d_H(v_0,t) = i' = h - i, && d_V(v_0,t) = j' = k - j
\end{aligned}$$

for $0 \leq i \leq h \leq m = \frac{n}{2}$ and $0 \leq j \leq k \leq m = \frac{n}{2}$ where not both $i = j = 0$ nor $i = h, j = k$ nor $h = k = 0$.

When n was odd, we were able to compute $BC(v_0)$ as a function only of $\Delta(Q_1, Q_3)$ and $\Delta(Q_1, Q_2)$. Now that n is even, the basic approach is the same except that we must consider a number of different subcases due to the impact of the diameter achieving pairs on the enumeration of

$$\delta_{st}(v_0) = \frac{\sigma_{st}(v_0)}{\sigma_{st}} = \frac{\sigma_{sv_0}\sigma_{v_0t}}{\sigma_{st}}$$

In our notation, we still have that

$$\sigma_{sv_0} = \binom{i+j}{i}, \ \sigma_{v_0t} = \binom{(h-i)+(k-j)}{(h-i)}, \text{ and } \sigma_{st} = \binom{h+k}{h}$$

when $0 \leq i \leq h < m = \frac{n}{2}$, $0 \leq j \leq k < m = \frac{n}{2}$. If instead $h = m$ but $i, h - i \neq m$, then we have

$$\sigma_{sv_0} = \binom{i+j}{i}, \ \sigma_{v_0t} = \binom{(h-i)+(k-j)}{(h-i)}, \text{ and } \sigma_{st} = 2\binom{h+k}{h}$$

By the symmetry between i and j, h and k, this is also true for $k = m$ and $j, k - j \neq m$. When $h = m$ and $i = 0$, $h = m$ and $i = m$, $k = m$ and $j = 0$, or $k = m$ and $j = m$, then

$$\delta_{st}(v_0) = \frac{2\binom{i+j}{i}\binom{(h-i)+(k-j)}{(h-i)}}{2\binom{h+k}{h}}$$

When $h = k = m$, then s and t are both horizontal and vertical diameter achieving. If $1 \leq i \leq m - 1$ and $1 \leq j \leq m - 1$, then

$$\sigma_{sv_0} = \binom{i+j}{i}, \ \sigma_{v_0t} = \binom{(h-i)+(k-j)}{(h-i)}, \text{ and } \sigma_{st} = 4\binom{h+k}{h}$$

If exactly one of i, $h - i$, j, or $k - j$ is also diameter achieving, then

$$\delta_{st}(v_0) = \frac{2\binom{i+j}{i}\binom{(h-i)+(k-j)}{(h-i)}}{4\binom{h+k}{h}}$$

while if both i and j or both $h-i$ and $k-j$ are diameter achieving, then

$$\delta_{st}(v_0) = \frac{4\binom{i+j}{i}\binom{(h-i)+(k-j)}{(h-i)}}{4\binom{h+k}{h}}$$

If we enumerate $\delta_{st}(v_0)$ for all these different cases, then we will be able to calculate $BC(v_0)$ as we did before.

Theorem 12.4. *Let n be an even integer. Suppose $v_0 = (0,0) \in \mathcal{T}_n$ and $m = \frac{n}{2}$. Then*

$$\begin{aligned} BC(v_0) &= 4 \cdot \Delta(S_1, T_1) \\ &+ 4 \cdot 2 \cdot \Delta(S_2, T_2) \\ &+ 2 \cdot 4 \cdot \Delta(S_3, T_3) \\ &+ 4 \cdot \Delta(S_4, T_4) \\ &+ 2 \cdot 4 \cdot \Delta(S_5, T_5) \\ &+ 2 \cdot \Delta(S_6, T_6) \\ &- 4 \cdot \Delta(S_7, T_7) \end{aligned}$$

where, for $d_H(s, v_0) = i$, $d_V(s, v_0) = j$, $d_H(s,t) = h$, and $d_V(s,t) = k$

$$\begin{aligned} S_1 &= \{s \in Q_1 \mid 0 \le i \le h \le m-1, 0 \le j \le k \le m-1\} \\ T_1 &= \{t \in Q_3 \mid 0 \le h-i \le h \le m-1, 0 \le k-j \le k \le m-1\} \end{aligned}$$

$$\begin{aligned} S_2 &= \{s \in Q_1 \mid 1 \le i < h = m, 0 \le j \le k \le m-1\} \\ T_2 &= \{t \in Q_3 \mid 1 \le h-i < h = m, 0 \le k-j \le k \le m-1\} \end{aligned}$$

$$\begin{aligned} S_3 &= \{s \in Q_1 \mid i = 0, 0 \le j \le k \le m-1\} \\ T_3 &= \{t \in Q_3 \mid h-i = m, 0 \le k-j \le k \le m-1\} \end{aligned}$$

$$\begin{aligned} S_4 &= \{s \in Q_1 \mid 1 \le i < h = m, 1 \le j < k = m\} \\ T_4 &= \{t \in Q_3 \mid 1 \le h-i < h = m, 1 \le k-j < k = m\} \end{aligned}$$

$$\begin{aligned} S_5 &= \{s \in Q_1 \mid i = 0, 1 \le j < k = m\} \\ T_5 &= \{t \in Q_3 \mid h-i = m, 1 \le k-j < k = m\} \end{aligned}$$

$$\begin{aligned} S_6 &= \{s \in Q_1 \mid i = 0, j = m\} \\ T_6 &= \{t \in Q_3 \mid h-i = m, k-j = 0\} \end{aligned}$$

$$\begin{aligned} S_7 &= \{s \in Q_1 \mid i = 0, 1 \le j < k = m\} \\ T_7 &= \{t \in Q_3 \mid h-i = 0, 1 \le k-j < k = m\} \end{aligned}$$

and in each case not both $i = j = 0$ nor $i = h, j = k$ nor $h = k = 0$.

Proof. The different pairs S_i, T_i for $1 \le i \le 7$ correspond to the following cases:

S_1, T_1	$0 \le h \le m-1$	$0 \le k \le m-1$	$0 \le i \le h$	$0 \le j \le k$
S_2, T_2	$h = m$	$0 \le k \le m-1$	$1 \le i \le h-1$	$0 \le j \le k$
	$0 \le h \le m-1$	$k = m$	$0 \le i \le h$	$1 \le j \le k-1$
S_3, T_3	$h = m$	$0 \le k \le m-1$	$i = 0$	$0 \le j \le k$
	$h = m$	$0 \le k \le m-1$	$i = m$	$0 \le j \le k$
	$0 \le h \le m-1$	$k = m$	$0 \le i \le h$	$j = 0$
	$0 \le h \le m-1$	$k = m$	$0 \le i \le h$	$j = m$
S_4, T_4	$h = m$	$k = m$	$1 \le i \le h-1$	$1 \le j \le k-1$
S_5, T_5	$h = m$	$k = m$	$i = 0$	$1 \le j \le k-1$
	$h = m$	$k = m$	$i = m$	$1 \le j \le k-1$
	$h = m$	$k = m$	$1 \le i \le h-1$	$j = 0$
	$h = m$	$k = m$	$1 \le i \le h-1$	$j = m$
S_6, T_6	$h = m$	$k = m$	$i = 0$	$j = m$
S_7, T_7	$h = 0$	$2 \le k \le m$	$i = 0$	$1 \le j < k = m$

For $S_1 \subset Q_1, T_1 \subset Q_3$ and $S_2 \subset Q_1, T_2 \subset Q_3$, there are corresponding distinct sets in each of Q_2, Q_4 and Q_3, Q_1 and Q_4, Q_2. This is also true for S_4, T_4. For S_3, T_3, however, we also have that $S_3 \subset Q_4$ and $T_3 \subset Q_2$. Thus, we only multiply $\Delta(S_3, T_3)$ by a factor of two to account for the reverse paths from $Q_3 \cap Q_2$ to $Q_1 \cap Q_4$. This is also the case for S_5, T_5. For S_6, T_6, we first note that $i = m, j = 0$ gives the same path, just in the opposite direction. Since this is the only such path, it is counted twice. Finally, S_7, T_7 correct for the overcounting along the horizontal and vertical lines through v_0, once in each direction, for a total factor of four. $\square$

Theorem 12.5. *Suppose the assumptions of Theorem* 12.4 *hold. Then*

$$\Delta(S_1, T_1) = 1 + (m-2)m^2$$

$$\Delta(S_2, T_2) = \frac{(m-1)m(3m+1)}{4(m+1)}$$

$$\Delta(S_3, T_3) = \frac{(m-1)m}{2(m+1)}$$

$$\Delta(S_4, T_4) = \frac{1}{4\binom{m+m}{m}}\left((m-3)\binom{2m+1}{m} + 2(1) + 2\binom{2m}{m}\right)$$

$$\Delta(S_5, T_5) = \frac{\binom{m+m+1}{m} - 1 - \binom{2m}{m}}{2\binom{2m}{m}}$$

$$\Delta(S_6, T_6) = \frac{1}{\binom{2m}{m}}$$

$$\Delta(S_7, T_7) = \frac{(m-1)^2}{2}$$

Proof. The result follows from the following summations, and extensive use of the equality on page 275

$$\Delta(S_1,T_1) = 1 + \sum_{h=0}^{m-1}\sum_{k=0}^{m-1}\left(-2 + \frac{1}{\binom{h+k}{k}}\sum_{i=0}^{h}\sum_{j=0}^{k}\binom{i+j}{i}\binom{(h-i)+(k-j)}{(h-i)}\right)$$

$$\Delta(S_2,T_2) = \sum_{k=0}^{m-1}\frac{1}{2\binom{m+k}{k}}\sum_{i=1}^{m-1}\sum_{j=0}^{k}\binom{i+j}{i}\binom{(m-i)+(k-j)}{(m-i)}$$

$$\Delta(S_3,T_3) = \sum_{k=0}^{m-1}\left(-1 + \frac{1}{\binom{m+k}{k}}\sum_{j=0}^{k}\binom{0+j}{0}\binom{(m-0)+(k-j)}{(m-0)}\right)$$

$$\Delta(S_4,T_4) = \frac{1}{4\binom{m+m}{m}}\sum_{i=1}^{m-1}\sum_{j=1}^{m-1}\binom{i+j}{i}\binom{(m-i)+(m-j)}{(m-i)}$$

$$\Delta(S_5,T_5) = \frac{1}{2\binom{m+m}{m}}\sum_{j=1}^{m-1}\binom{0+j}{0}\binom{(m-0)+(m-j)}{(m-0)}$$

$$\Delta(S_6,T_6) = \frac{1}{\binom{m+m}{m}}\binom{0+m}{0}\binom{(m-0)+(m-m)}{(m-0)}$$

$$\Delta(S_7,T_7) = \sum_{k=2}^{m-1}\sum_{j=1}^{k-1}\frac{\binom{0+j}{0}\binom{0+(k-j)}{0}}{\binom{0+k}{0}} + \sum_{j=1}^{m-1}\frac{\binom{0+j}{0}\binom{0+(m-j)}{0}}{2\binom{0+m}{0}} \qquad \square$$

As an immediate consequence of Theorems 12.4 and 12.5, we have the following.

Corollary 12.6. *When n is even,*

$$BC(v) = \frac{n^3}{2} - n^2 + 1$$

Acknowledgments

This work was supported in part by the National Science Foundation (NSF) Grant CNS-0614915 and a Career Award at the Scientific Interface from the Burroughs Wellcome Fund. We thank Cray, Inc. for access to the MTA-2. We acknowledge Sun Microsystems for their Academic Excellence Grant and donation of Niagara systems used in this work. We thank Jonathan Berry and Bruce Hendrickson for discussions on large-scale network analysis and betweenness centrality.

References

[Ajwani et al. 2006] D. Ajwani, R. Dementiev, and U. Meyer. A computational study of external-memory BFS algorithms. In *Proceedings of the 17th Annual ACM-SIAM Symposium on Discrete Algorithms (SODA-06)*, 601–610, Miami, FL, Philadelphia, 2006.

[Bacon] University of Virginia. Oracle of Bacon. www.oracleofbacon.org.

[Bader et al. 2007] D.A. Bader, S. Kintali, K. Madduri, and M. Mihail. Approximating betweenness centrality. In *Proc. 5th Workshop on Algorithms and Models for the Web-Graph (WAW2007)*, volume 4863 of *Lecture Notes in Computer Science*, 134–137, San Diego, CA, December 2007. Berlin, Heidelberg: Springer-Verlag,

[Bader & Madduri 2006a] D.A. Bader and K. Madduri. Designing multithreaded algorithms for breadth-first search and st-connectivity on the Cray MTA-2. In *Proc. 35th Int'l Conf. on Parallel Processing (ICPP)*, Columbus, OH, August 2006, IEEE Computer Society.

[Bader & Madduri 2006b] D.A. Bader and K. Madduri. Parallel algorithms for evaluating centrality indices in real-world networks. In *Proc. 35th Int'l Conf. on Parallel Processing (ICPP)*, Columbus, OH, August 2006, IEEE Computer Society.

[Bader & Madduri 2007] D.A. Bader and K. Madduri. A graph-theoretic analysis of the human protein interaction network using multicore parallel algorithms. In *Proc. 6th Workshop on High Performance Computational Biology (HiCOMB 2007)*, Long Beach, CA, 2007.

[Bader & Madduri 2008] D.A. Bader and K. Madduri. A graph-theoretic analysis of the human protein-interation network using multicore parallel algorithms. *Parallel Computing*, 34:627–639, 2008.

[Bader et al. 2006] D.A. Bader, K. Madduri, J.R. Gilbert, V. Shah, J. Kepner, T. Meuse, and A. Krishnamurthy. Designing scalable synthetic compact applications for benchmarking high productivity computing systems. *CTWatch Quarterly*, 2(4B), November 2006.

[Barabási 2007] A.-L. Barabási. Network databases. http://www.nd.edu/~networks/resources.htm, 2007.

[Barabási & R. Albert 2007] A.-L. Barabási and R. Albert. Emergence of scaling in random networks. *Science*, 286:509–512, 1999.

[Batagelj & A. Mrvar 1998] V. Batagelj and A. Mrvar. Pajek—program for large network analysis. *Connections*, 21:47–57, 1998.

[Batagelj & A. Mrvar 2006] V. Batagelj and A. Mrvar. Pajek datasets. http://vlado.fmf.uni-lj.si/pub/networks/data/, 2006.

[Bavelas 1950] A. Bavelas. Communication patterns in task oriented groups. *Journal of the Acoustical Society of America*, 22:271–282, 1950.

[Brandes 2001] U. Brandes. A faster algorithm for betweenness centrality. *Journal of Mathematical Sociology*, 25:163–177, 2001.

[Brandes & Erlebach 2005] U. Brandes and T. Erlebach, editors. *Network Analysis: Methodological Foundations*, volume 3418 of *Lecture Notes in Computer Science*. Berlin, Heidelberg: Springer-Verlag, 2005.

[Brin & Page 1998] S. Brin and L. Page. The anatomy of a large-scale hypertextual web search engine. *Computer Networks and ISDN Systems*, 30:107–117, 1998.

[Broder et al. 2000] A. Broder, R. Kumar, F. Maghoul, P. Raghavan, S. Rajagopalan, R. Stata, A. Tomkins, and J. Wiener. Graph structure in the web. *Computer Networks*, 33:309–320, 2000.

[Buckley & van Alstyne 2004] N. Buckley and M. van Alstyne. Does email make white collar workers more productive? Technical Report, University of Michigan, 2004.

[Callaway et al. 2000] D.S. Callaway, M.E.J. Newman, S.H. Strogatz, and D.J. Watts. Network robustness and fragility: Percolation on random graphs. *Physical Review Letters*, 85:5468–5471, 2000.

[Chakrabarti et al. 2004] D. Chakrabarti, Y. Zhan, and C. Faloutsos. R-MAT: A recursive model for graph mining. In *SDM04: Proceedings of the 4th SIAM International Conference on Datamining (SDM)*, 442–446, 2004. http://www.siam.org/proceedings/datamining/2004/dm04.php.

[Cisic et al. 2000] D. Cisic, B. Kesic, and L. Jakomin. Research of the power in the supply chain. International Trade, Economics Working Paper Archive EconWPA, April 2000.

[Coffman et al. 2004] T. Coffman, S. Greenblatt, and S. Marcus. Graph-based technologies for intelligence analysis. *Communications of the ACM*, 47:45–47, 2004.

[Cohen et al. 2001] R. Cohen, K. Erez, D. ben-Avraham, and S. Havlin. Breakdown of the Internet under intentional attack. *Physical Review Letters*, 86:3682–3685, 2001.

[Cong & Sbaraglia 2006] G. Cong and S. Sbaraglia. A study on the locality behavior of minimum spanning tree algorithms. In *Proc. 13th Int'l Conf. on High Performance Computing (HiPC 2006)*, Bangalore, India, Berlin Heidelberg: Springer-Verlag, 2006.

[Cormen et al. 1990] T. H. Cormen, C. E. Leiserson, and R. L. Rivest. *Introduction to Algorithms*. Cambridge, MA: MIT Press, 1990.

[Crobak et al. 2007] J.R. Crobak, J.W. Berry, K. Madduri, and D.A. Bader. Advanced shortest path algorithms on a massively-multithreaded architecture. In *Proc. Workshop on Multithreaded Architectures and Applications*, Long Beach, CA, 2007.

[Dall'Asta et al. 2006] L. Dall'Asta, I. Alvarez-Hamelin, A. Barrat, A. Vazquez, and A. Vespignani. Exploring networks with traceroute-like probes: Theory and simulations. *Theoretical Computer Science*, 355:6–24, 2006.

[del Sol et al. 2005] A. del Sol, H. Fujihashi, and P. O'Meara. Topology of small-world networks of protein-protein complex structures. *Bioinformatics*, 21:1311–1315, 2005.

[Doreian & L.H. Albert 1989] P. Doreian and L.H. Albert. Partitioning political actor networks: Some quantitative tools for analyzing qualitative networks. *Journal of Quantitative Anthropology*, 1:279–291, 1989.

[Eppstein & Wang 2001] D. Eppstein and J. Wang. Fast approximation of centrality. In *Proceedings of the 12th Annual Symposium on Discrete Algorithms (SODA '01)*, 228–229, Washington, DC, 2001.

[Faloutsos et al. 1999] M. Faloutsos, P. Faloutsos, and C. Faloutsos. On power-law relationships of the Internet topology. In *Proc. ACM SIGCOMM*, 251–262, Cambridge, MA, New York: ACM Press, 1999.

[Freeman 1977] L.C. Freeman. A set of measures of centrality based on betweenness. *Sociometry*, 40:35–41, 1977.

[Freeman 2004] L.C. Freeman. *The Development of Social Network Analysis: A Study in the Sociology of Science.* Vancouver: Empirical Press, 2004.

[Gandhi et al. 2006] T.K. Gandhi et al. Analysis of the human protein interactome and comparison with yeast, worm and fly interaction datasets. *Nature Genetics*, 38:285–293, 2006.

[Graham et al. 1989] R.L. Graham, D.E. Knuth, and O. Patashnik. *Concrete Mathematics: A Foundation for Computer Science.* Reading, MA: Addison–Wesley Publishing Advanced Book Program, 1989.

[Guillaume & Latapy 2004] J.-L. Guillaume and M. Latapy. Bipartite graphs as models of complex networks. In *Proc. 1st Int'l Workshop on Combinatorial and Algorithmic Aspects of Networking*, volume 3405 of *Lecture Notes in Computer Science*, 127–139, Alberta, Canada, Berlin Heidelberg: Springer-Verlag, 2004.

[Guimer et al. 2005] R. Guimerà, S. Mossa, A. Turtschi, and L.A.N. Amaral. The worldwide air transportation network: anomalous centrality, community structure, and cities' global roles. *Proceedings of the National Academy of Sciences USA*, 102:7794–7799, 2005.

[Helman & JáJá 2001] D. R. Helman and J. JáJá. Prefix computations on symmetric multiprocessors. *Journal of Parallel and Distributed Computing*, 61:265–278, 2001.

[Hermjakob et al. 2004] H. Hermjakob et al. IntAct: An open source molecular interaction database. *Nucleic Acids Research*, 32:D452–D455, 2004.

[Hoeffding 1963] W. Hoeffding. Probability inequalities for sums of bounded random variables. *Journal of the American Statistical Association*, 58:13–30, 1963.

[Holme 2003] P. Holme. Congestion and centrality in traffic flow on complex networks. *Advances in Complex Systems*, 6:163–176, 2003.

[Huisman & van Duijn 2005] M. Huisman and M.A.J. van Duijn. Software for social network analysis. In P.J. Carrington, J. Scott, and S. Wasserman, eds., *Models and Methods in Social Network Analysis*, Chapter 13, 270–316, Cambridge, UK: Cambridge University Press, 2005.

[Jeong et al. 2001] H. Jeong, S.P. Mason, A.-L. Barabási, and Z.N. Oltvai. Lethality and centrality in protein networks. *Nature*, 411:41–42, 2001.

[Jeong et al. 2003] H. Jeong, Z. Oltvai, and A.-L. Barabási. Prediction of protein essentiality based on genomic data. *ComPlexUs*, 1:19–28, 2003.

[Joy et al. 2005] M.P. Joy, A. Brock, D.E. Ingber, and S. Huang. High-betweenness proteins in the yeast protein interaction network. *Journal of Biomedicine and Biotechnology*, 2:96–103, 2005.

[Krebs 2005] V. Krebs. InFlow 3.1—Social network mapping software. http://www.orgnet.com/inflow3.html, 2005.

[Krebs 2002] V.E. Krebs. Mapping networks of terrorist cells. *Connections*, 24:43–52, 2002.

[Lang 2005] K. Lang. Fixing two weaknesses of the spectral method. In *Processing Advances in Neurals Information Processing Systems 18 (NIPS 2005)*, Vancouver, Canada, 2005.

[Liljeros et al. 2001] F. Liljeros, C.R. Edling, L.A.N. Amaral, H.E. Stanley, and Y. berg. The web of human sexual contacts. *Nature*, 411:907–908, 2001.

[Lumsdaine et al. 2007] A. Lumsdaine, D. Gregor, B. Hendrickson, and J. Berry. Challenges in parallel graph processing. *Parallel Processing Letters*, 17:5–20, 2007.

[Madduri et al. 2007] K. Madduri, D.A. Bader, J.W. Berry, and J.R. Crobak. An experimental study of a parallel shortest path algorithm for solving large-scale graph instances. In *Proc. The 9th Workshop on Algorithm Engineering and Experiments (ALENEX07)*, New Orleans, LA, 2007.

[Newman 2003] M.E.J. Newman. The structure and function of complex networks. *SIAM Review*, 45:167–256, 2003.

[Newman et al. 2002] M.E.J. Newman, S.H. Strogatz, and D.J. Watts. Random graph models of social networks. *Proceedings of the National Academy of Sciences USA*, 99:2566–2572, 2002.

[Nieminen 1973] U.J. Nieminen. On the centrality in a directed graph. *Social Science Research*, 2:371–378, 1973.

[Palmer & Steffan 2000] C.R. Palmer and J.G. Steffan. Generating network topologies that obey power laws. In *Proceedings of the IEEE Global Internet Symposium (GLOBECOM)*, 434–438, San Francisco, CA, 2000.

[Park et al. 2002] J. Park, M. Penner, and V.K. Prasanna. Optimizing graph algorithms for improved cache performance. In *Proceedings of the International Parallel and Distributed Processing Symposium (IPDPS 2002)*, Fort Lauderdale, FL, 2002. IEEE Computer Society.

[Pastor-Satorras & Vespignani 2001] R. Pastor-Satorras and A. Vespignani. Epidemic spreading in scale-free networks. *Physical Review Letters*, 86:3200–3203, 2001.

[Peri et al. 2003] S. Peri et al. Development of human protein reference database as an initial platform for approaching systems biology in humans. *Genome Research*, 13:2363–2371, 2003.

[Pinney et al. 2005] J.W. Pinney, G.A. McConkey, and D.R. Westhead. Decomposition of biological networks using betweenness centrality. In *Proc. 9th Ann. Int'l Conf. on Research in Computational Molecular Biology (RECOMB 2005)*, Cambridge, MA, 2005. Poster session.

[Scott 2000] J.P. Scott. *Social Network Analysis: A Handbook.* Newbury Park, CA: SAGE Publications, 2000.

[Shimbel 1953] A. Shimbel. Structural parameters of communication networks. *Bulletin of Mathematical Biophysics*, 15:501–507, 1953.

[Singh & Gupte 2005] B.K. Singh and N. Gupte. Congestion and decongestion in a communication network. *Physical Review E*, 71:055103, 2005.

[UCINET] Analytic Technologies. UCINET 6 social network analysis software. http://www.analytictech.com/ucinet.

[Wasserman & Faust 1994] S. Wasserman and K. Faust. *Social Network Analysis: Methods and Applications.* Cambridge, UK: Cambridge University Press, 1994.

[Watts & Strogatz 1998] D.J. Watts and S.H. Strogatz. Collective dynamics of small world networks. *Nature*, 393:440–442, 1998.

[Zanette 2001] D.H. Zanette. Critical behavior of propagation on small-world networks. *Physical Review E*, 64:050901, 2001.

Chapter 13

Implementing Sparse Matrices for Graph Algorithms

Aydın Buluç, John Gilbert†, and Viral B. Shah†*

Abstract

Sparse matrices are a key data structure for implementing graph algorithms using linear algebra. This chapter reviews and evaluates storage formats for sparse matrices and their impact on primitive operations. We present complexity results of these operations on different sparse storage formats both in the random access memory (RAM) model and in the input/output (I/O) model. RAM complexity results were known except for the analysis of sparse matrix indexing. On the other hand, most of the I/O complexity results presented are new. The chapter focuses on different variations of the triples (coordinates) format and the widely used compressed sparse row (CSR) and compressed sparse column (CSC) formats. For most primitives, we provide detailed pseudocodes for implementing them on triples and CSR/CSC.

13.1 Introduction

The choice of data structure is one of the most important steps in algorithm design and implementation. Sparse matrix algorithms are no exception. The representation of a sparse matrix not only determines the efficiency of the algorithm, but also influences the algorithm design process. Given this bidirectional relationship,

*High Performance Computing Research, Lawrence Berkeley National Laboratory, 1 Cyclotron Road, Berkeley, CA 94720 (abuluc@lbl.gov).

†Computer Science Department, University of California, Santa Barbara, CA 93106-5110 (gilbert@cs.ucsb.edu, viral@mayin.org).

this chapter reviews and evaluates sparse matrix data structures with key primitive operations in mind. In the case of array-based graph algorithms, these primitives are sparse matrix vector multiplication (SpMV), sparse general matrix matrix multiplication (SpGEMM), sparse matrix reference/assignment (SpRef/SpAsgn), and sparse matrix addition (SpAdd). The administrative overheads of different sparse matrix data structures, both in terms of storage and processing, are also important and are exposed throughout the chapter.

Let $\mathbf{A} \in \mathbb{S}^{M \times N}$ be a sparse rectangular matrix of elements from an arbitrary semiring $\mathbb{S}$. We use $nnz(\mathbf{A})$ to denote the number of nonzero elements in $\mathbf{A}$. When the matrix is clear from context, we drop the parentheses and simply use nnz. For sparse matrix indexing, we use the convenient MATLAB colon notation, where $\mathbf{A}(:, i)$ denotes the ith column, $\mathbf{A}(i, :)$ denotes the ith row, and $\mathbf{A}(i, j)$ denotes the element at the (i, j)th position of matrix $\mathbf{A}$. For one-dimensional arrays, $\mathbf{a}(i)$ denotes the ith component of the array. Indices are 1-based throughout the chapter. We use flops($\mathbf{A}\, op\, \mathbf{B}$) to denote the number of nonzero arithmetic operations required by the operation $\mathbf{A}\, op\, \mathbf{B}$. Again, when the operation and the operands are clear from the context, we simply use flops. To reduce notational overhead, we take each operation's complexity to be at least one, i.e., we say $O(\cdot)$ instead of $O(\max(\cdot, 1))$.

One of the traditional ways to analyze the computational complexity of a sparse matrix operation is by counting the number of floating-point operations performed. This is similar to analyzing algorithms according to their RAM complexities (see [Aho et al. 1974]). As memory hierarchies became dominant in computer architectures, the I/O complexity (also called the cache complexity) of a given algorithm became as important as its RAM complexity. Cache performance is especially important for sparse matrix computations because of their irregular nature and low ratio of flops to memory access. One approach to hiding the memory-processor speed gap is to use massively multithreaded architectures [Feo et al. 2005]. However, these architectures have limited availability at present.

In the I/O model, only two levels of memory are considered for simplicity: a fast memory and a slow memory. The fast memory is called cache and the slow memory is called disk, but the analysis is valid at different levels of memory hierarchy with appropriate parameter values. Both levels of memories are partitioned into blocks of size L, usually called the cache line size. The size of the fast memory is denoted by Z. If data needed by the CPU is not found in the fast memory, a *cache miss* occurs, and the memory block containing the needed data is fetched from the slow memory. The I/O complexity of an algorithm can be roughly defined as the number of memory transfers it makes between the fast and slow memories [Aggarwal & Vitter 1988]. The number of memory transfers does not necessarily mean the number of words moved between fast and slow memories, since the memory transfers happen in blocks of size L. For example, scanning an array of size N has I/O complexity N/L. In this chapter, $scan(\mathbf{A}) = nnz(\mathbf{A})/L$ is used as an abbreviation for the I/O complexity of examining all the nonzeros of matrix $\mathbf{A}$ in the order that they are stored. Figure 13.1 shows a simple memory hierarchy with some typical latency values as of today. Meyer, Sanders, and Sibeyn provided a contemporary treatment of algorithmic implications of memory hierarchies [Meyer et al. 2002].

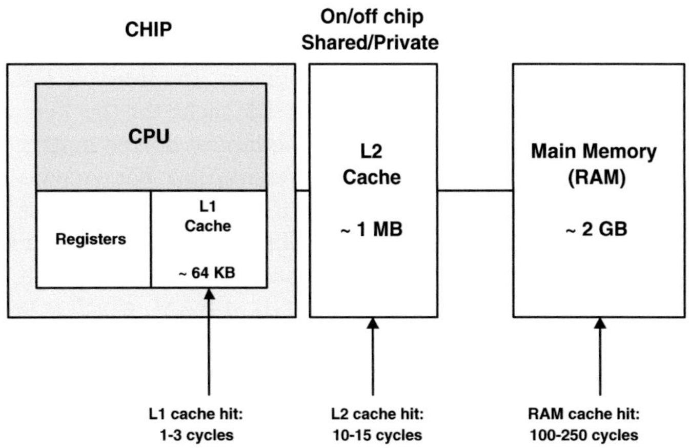

Figure 13.1. Typical memory hierarchy.
Approximate values of memory size and latency (partially adapted from Hennessy and Patterson [Hennessy & Patterson 2006]) assuming a 2 GHz processing core.

We present the computational complexities of algorithms in the RAM model as well as the I/O model. However, instead of trying to come up with the most I/O efficient implementations, we analyze the I/O complexities of the most widely used implementations, which are usually motivated by the RAM model. There are two reasons for this approach. First, I/O optimality is still an open problem for some of the key primitives presented in this chapter. Second, I/O efficient implementations of some key primitives turn out to be suboptimal in the RAM model with respect to the amount of work they do.

Now, we state two crucial assumptions that are used throughout this chapter.

Assumption 1. *A sparse matrix with dimensions $M \times N$ has $nnz \geq M, N$. More formally, $nnz = \Omega(N, M)$.*

Assumption 1 simplifies the asymptotic analysis of the algorithms presented in this chapter. It implies that when both the order of the matrix and its number of nonzeros are included as terms in the asymptotic complexity, only nnz is pronounced. While this assumption is common in numerical linear algebra (it is one of the requirements for nonsingularity), in some parallel graph computations it may not hold. In this chapter, however, we use this assumption in our analysis. In contrast, Buluç and Gilbert gave an SpGEMM algorithm specifically designed for *hypersparse* matrices (matrices with $nnz < N, M$) [Buluç & Gilbert 2008b].

Assumption 2. *The fast memory is not big enough to hold data structures of $O(N)$ size, where N is the matrix dimension.*

We argue that Assumption 2 is justified when the fast memory under consideration is either the L1 or L2 cache. Out-of-order CPUs can generally hide memory latencies from L1 cache misses, but not L2 cache misses [Hennessy & Patterson 2006]. Therefore, it is more reasonable to treat the L2 cache as the fast memory and RAM (main memory) as the slow memory. The largest sparse matrix that fills the whole machine RAM (assuming the triples representation that occupies 16 bytes per nonzero, has $2^{30}/16 = 2^{26}$ nonzeros per GB of RAM). A square sparse matrix, with an average of eight nonzeros per column, has dimensions $2^{23} \times 2^{23}$ per GB of RAM. A single dense vector of double-precision floating-point numbers with 2^{23} elements would require 64 MB of L2 cache memory per GB of RAM, which is typically much larger than the typical size of the L2 cache.

The goal of this chapter is to explain sparse matrix data structures progressively, starting from the least structured and most simple format (unordered triples) and ending with the most structured formats: compressed sparse row (CSR) and compressed sparse column (CSC). This way, we provide motivation on why the experts prefer to use CSR/CSC formats by comparing and contrasting them with simpler formats. For example, CSR, a dense collection of sparse row arrays, can also be viewed as an extension of the triples format enhanced with row indexing capabilities. Furthermore, many ideas and intermediate data structures that are used to implement key primitives on triples are also widely used with implementations on CSR/CSC formats.

A vast amount of literature exists on sparse matrix storage schemes. Some additional other specialized data structures that are worth mentioning include

- Blocked compressed stripe formats (BCSR and BCSC) use less bandwidth to accelerate bandwidth limited computations such as SpMV.

- Knuth storage allows fast access to both rows and columns at the same time, and it makes dynamic changes to the matrix possible. Therefore, it is very suitable for all kinds of SpRef and SpAsgn operations. Its drawback is its excessive memory usage ($5\ nnz + 2M$) and high cache miss ratio.

- Hierarchical storage schemes such as quadtrees [Samet 1984, Wise & Franco 1990] are theoretically attractive, but achieving good performance in practice requires careful algorithm engineering to avoid high cache miss ratios that would result from straightforward pointer-based implementations.

The rest of this chapter is organized as follows. Section 13.2 describes the key sparse matrix primitives. Section 13.3 reviews the triples/coordinates representation, which is natural and easy to understand. The triples representation generalizes to higher dimensions [Bader & Kolda 2007]. Its resemblance to database tables will help us expose some interesting connections between databases and sparse matrices. Section 13.4 reviews the most commonly used compressed storage formats for general sparse matrices, namely CSR and CSC. Section 13.5 presents a case on how sparse matrices are represented in the Star-P programming environment. Section 13.6 concludes the chapter.

13.2 Key primitives

Most of the sparse matrix operations have been motivated by numerical linear algebra. Some of them are also useful for graph algorithms:

1. Sparse matrix indexing and assignment (SpRef/SpAsgn): Corresponds to subgraph selection.

2. Sparse matrix-dense vector multiplication (SpMV): Corresponds to breadth-first or depth-first search.

3. Sparse matrix addition and other pointwise operations (SpAdd): Corresponds to graph merging.

4. Sparse matrix-sparse matrix multiplication (SpGEMM): Corresponds to breadth-first or depth-first search to/from multiply vertices simultaneously.

SpRef is the operation of storing a submatrix of a sparse matrix in another sparse matrix ($\mathbf{B} = \mathbf{A}(\mathbf{p}, \mathbf{q})$), and SpAsgn is the operation of assigning a sparse matrix to a submatrix of another sparse matrix ($\mathbf{B}(\mathbf{p}, \mathbf{q}) = \mathbf{A}$). It is worth noting that SpAsgn is the only key primitive that mutates its sparse matrix operand in the general case.* Sparse matrix indexing can be quite powerful and complex if we allow $\mathbf{p}$ and $\mathbf{q}$ to be arbitrary vectors of indices. Therefore, this chapter limits itself to row wise ($\mathbf{A}(i,:)$), column wise ($\mathbf{A}(:,i)$), and element-wise ($\mathbf{A}(i,j)$) indexing, as they find more widespread use in graph algorithms. SpAsgn also requires the matrix dimensions to match, e.g., if $\mathbf{B}(:,i) = \mathbf{A}$ where $\mathbf{B} \in \mathbb{S}^{M \times N}$, then $\mathbf{A} \in \mathbb{S}^{1 \times N}$.

SpMV is the most widely used sparse matrix kernel since it is the workhorse of iterative linear equation solvers and eigenvalue computations. A sparse matrix can be multiplied by a dense vector either on the right ($\mathbf{y} = \mathbf{A}\mathbf{x}$) or on the left ($\mathbf{y}' = \mathbf{x}'\mathbf{A}$). This chapter concentrates on the multiplication on the right. It is generally straightforward to reformulate algorithms that use multiplication on the left so that they use multiplication on the right. Some representative graph computations that use SpMV are page ranking (an eigenvalue computation), breadth-first search, the Bellman–Ford shortest paths algorithm, and Prim's minimum spanning tree algorithm.

SpAdd, $\mathbf{C} = \mathbf{A} \oplus \mathbf{B}$, computes the sum of two sparse matrices of dimensions $M \times N$. SpAdd is an abstraction that is not limited to any summation operator. In general, any pointwise binary scalar operation between two sparse matrices falls into this primitive. Examples include MIN operator that returns the minimum of its operands, logical AND, logical OR, ordinary addition, and subtraction.

SpGEMM computes the sparse product $\mathbf{C} = \mathbf{AB}$, where the input matrices $\mathbf{A} \in \mathbb{S}^{M \times K}$ and $\mathbf{B} \in \mathbb{S}^{K \times N}$ are both sparse. It is a common operation for operating on large graphs, used in graph contraction, peer pressure clustering, recursive

*While $\mathbf{A} = \mathbf{A} \oplus \mathbf{B}$ or $\mathbf{A} = \mathbf{AB}$ may also be considered as mutator operations, these are just special cases when the output is the same as one of the inputs.

formulations of all-pairs shortest path algorithms, and breadth-first search from multiple source vertices.

The computation for matrix multiplication can be organized in several ways, leading to different formulations. One common formulation is the inner product formulation, as shown in Algorithm 13.1. In this case, every element of the product $\mathbf{C}(i,j)$ is computed as the dot product of a row i in $\mathbf{A}$ and a column j in $\mathbf{B}$. Another formulation of matrix multiplication is the outer product formulation (Algorithm 13.2). The product is computed as a sum of n rank-one matrices. Each rank-one matrix is computed as the outer product of column k of $\mathbf{A}$ and row k of $\mathbf{B}$.

Algorithm 13.1. Inner product matrix multiply.
Inner product formulation of matrix multiplication.

$\mathbf{C} : \mathbb{R}^{S(M \times N)} = \textsc{InnerProduct-SpGEMM}(\mathbf{A} : \mathbb{R}^{S(M \times K)}, \mathbf{B} : \mathbb{R}^{S(K \times N)})$

for $i = 1$ **to** M
    do for $j = 1$ **to** N
        do $\mathbf{C}(i,j) = \mathbf{A}(i,:) \cdot \mathbf{B}(:,j)$

Algorithm 13.2. Outer product matrix multiply.
Outer product formulation of matrix multiplication.

$\mathbf{C} : \mathbb{R}^{S(M \times N)} = \textsc{OuterProduct-SpGEMM}(\mathbf{A} : \mathbb{R}^{S(M \times K)}, \mathbf{B} : \mathbb{R}^{S(K \times N)})$

$\mathbf{C} = 0$
for $k = 1$ **to** K
    do $\mathbf{C} = \mathbf{C} + \mathbf{A}(:,k) \cdot \mathbf{B}(k,:)$

SpGEMM can also be set up so that $\mathbf{A}$ and $\mathbf{B}$ are accessed by rows or columns, computing one row/column of the product $\mathbf{C}$ at a time. Algorithm 13.3 shows the column wise formulation where column j of $\mathbf{C}$ is computed as a linear combination of the columns of $\mathbf{A}$ as specified by the nonzeros in column j of $\mathbf{B}$. Figure 13.2 shows the same concept graphically. Similarly, for the row wise formulation, each row i of $\mathbf{C}$ is computed as a linear combination of the rows of $\mathbf{B}$ specified by nonzeros in row i of $\mathbf{A}$ as shown in Algorithm 13.4.

Algorithm 13.3. Column wise matrix multiplication.
Column wise formulation of matrix multiplication.

$\mathbf{C} : \mathbb{R}^{S(M \times N)} = \textsc{ColumnWise-SpGEMM}(\mathbf{A} : \mathbb{R}^{S(M \times K)}, \mathbf{B} : \mathbb{R}^{S(K \times N)})$

for $j = 1$ **to** N
    do for k where $\mathbf{B}(k,j) \neq 0$
        do $\mathbf{C}(:,j) = \mathbf{C}(:,j) + \mathbf{A}(:,k) \cdot \mathbf{B}(k,j)$

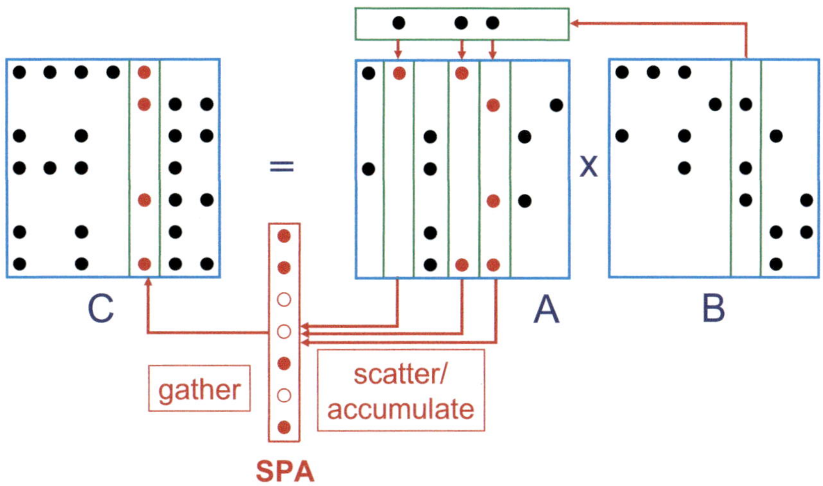

Figure 13.2. Multiply sparse matrices column by column. Multiplication of sparse matrices stored by columns. Columns of **A** are accumulated as specified by the nonzero entries in a column of **B** using a sparse accumulator (SPA). The contents of the SPA are stored in a column of **C** once all required columns are accumulated.

Algorithm 13.4. Row wise matrix multiply.
Row wise formulation of matrix multiplication.

$\mathbf{C} : \mathbb{R}^{S(M \times N)} = \textsc{RowWise-SpGEMM}(\mathbf{A} : \mathbb{R}^{S(M \times K)}, \mathbf{B} : \mathbb{R}^{S(K \times N)})$

```
for i = 1 to M
    do for k where A(i,k) ≠ 0
        do C(i,:) = C(i,:) + A(i,k) · B(k,:)
```

13.3 Triples

The simplest way to represent a sparse matrix is the *triples* (or *coordinates*) format. For each $\mathbf{A}(i,j) \neq 0$, the triple $(i, j, \mathbf{A}(i,j))$ is stored in memory. Each entry in the triple is usually stored in a different array and the whole matrix **A** is represented as three arrays **A**.I (row indices), **A**.J (column indices), and **A**.V (numerical values), as illustrated in Figure 13.3. These separate arrays are called "parallel arrays" by Duff and Reid (see [Duff & Reid 1979]) but we reserve "parallel" for parallel algorithms. Using 8-byte integers for row and column indices, storage cost is $8 + 8 + 8 = 24$ bytes per nonzero.

Modern programming languages offer easier ways of representing an array of tuples than using three separate arrays. An alternative implementation might choose to represent the set of triples as an array of records (or structs). Such an implementation might improve cache performance, especially if the algorithm accesses

$$A = \begin{pmatrix} 19 & 0 & 11 & 0 \\ 0 & 43 & 0 & 0 \\ 0 & 0 & 0 & 0 \\ 0 & 27 & 0 & 35 \end{pmatrix}$$

A.I	A.J	A.V
1	1	19
4	2	27
2	2	43
4	4	35
1	3	11

Figure 13.3. Triples representation.
Matrix **A** (left) and its unordered triples representation (right).

elements of the same index from different arrays. This type of cache optimization is known as array merging [Kowarschik & Weiß 2002]. Some programming languages, such as Python and Haskell, even support tuples as built-in types, and C++ includes tuples support in its standard library [Becker 2006]. In this section, we use the established notation of three separate arrays (**A**.I, **A**.J, **A**.V) for simplicity, but an implementer should keep in mind the other options.

This section evaluates the triples format under different levels of ordering. Unordered triples representation imposes no ordering constraints on the triples. Row ordered triples keep nonzeros ordered with respect to their row indices only. Nonzeros within the same row are stored arbitrarily, irrespective of their column indices. Finally, row-major order keeps nonzeros ordered lexicographically first according to their row indices and then according to their column indices to break ties. It is also possible to order with respect to columns instead of rows, but we analyze the row-based versions. Column-ordered and column-major ordered triples are similar. RAM and I/O complexities of key primitives for unordered and row ordered triples are listed in Tables 13.1 and 13.2.

A theoretically attractive fourth option is to use hashing and store triples in a hash table. In the case of SpGEMM and SpAdd, dynamically managing the output matrix is computationally expensive since dynamic perfect hashing does not yield high performance in practice [Mehlhorn & Näher 1999] and requires $35N$ space [Dietzfelbinger et al. 1994]. A recently proposed dynamic hashing method called Cuckoo hashing is promising. It supports queries in worst-case constant time and updates in amortized expected constant time, while using only $2N$ space [Pagh & Rodler 2004]. Experiments show that it is substantially faster than existing hashing schemes on modern architectures like Pentium 4 and IBM Cell [Ross 2007]. Although hash-based schemes seem attractive, especially for SpAsgn and SpRef primitives [Aspnäs et al. 2006], further research is required to test their efficiency for sparse matrix storage.

13.3.1 Unordered triples

The administrative overhead of the triples representation is low, especially if the triples are not sorted in any order. With unsorted triples, however, there is no

Table 13.1. Unordered and row ordered RAM complexities. Memory access complexities of key primitives on unordered and row ordered triples.

	Unordered	**Row Ordered**
SpRef	$O(nnz(\mathbf{A}))$	$O(\lg nnz(\mathbf{A})+nnz(\mathbf{A}(i,:)))\begin{cases}\mathbf{A}(i,j)\\ \mathbf{A}(i,:)\end{cases}$ $O(nnz(\mathbf{A}))\{\ \mathbf{A}(:,j)$
SpAsgn	$O(nnz(\mathbf{A}))+O(nnz(\mathbf{B}))$	$O(nnz(\mathbf{A}))+O(nnz(\mathbf{B}))$
SpMV	$O(nnz(\mathbf{A}))$	$O(nnz(\mathbf{A}))$
SpAdd	$O(nnz(\mathbf{A})+nnz(\mathbf{B}))$	$O(nnz(\mathbf{A})+nnz(\mathbf{B}))$
SpGEMM	$O(nnz(\mathbf{A})+nnz(\mathbf{B})+\text{flops})$	$O(nnz(\mathbf{A})+\text{flops})$

Table 13.2. Unordered and row ordered I/O complexities. Input/Output access complexities of key primitives on unordered and row ordered triples.

	Unordered	**Row Ordered**
SpRef	$O(scan(\mathbf{A}))$	$O(\lg nnz(\mathbf{A})+scan(\mathbf{A}(i,:))\begin{cases}\mathbf{A}(i,j)\\ \mathbf{A}(i,:)\end{cases}$ $O(scan(\mathbf{A}))\{\ \mathbf{A}(:,j)$
SpAsgn	$O(scan(\mathbf{A})+scan(\mathbf{B}))$	$O(scan(\mathbf{A})+scan(\mathbf{B}))$
SpMV	$O(nnz(\mathbf{A}))$	$O(nnz(\mathbf{A}))$
SpAdd	$O(nnz(\mathbf{A})+nnz(\mathbf{B}))$	$O(nnz(\mathbf{A})+nnz(\mathbf{B}))$
SpGEMM	$O(nnz(\mathbf{A})+nnz(\mathbf{B})+\text{flops})$	$O(\min\{nnz(\mathbf{A})+\text{flops},$ $scan(\mathbf{A})\ \lg(nnz(\mathbf{B}))+\text{flops}\})$

spatial locality when accessing nonzeros of a given row or column.* In the worst case, all indexing operations might require a complete scan of the data structure. Therefore, SpRef has $O(nnz(\mathbf{A}))$ RAM complexity and $O(scan(\mathbf{A}))$ I/O complexity.

SpAsgn is no faster, even though insertions take only constant time per element. In addition to accessing all the elements of the right-hand side matrix $\mathbf{A}$, SpAsgn also invalidates the existing nonzeros that need to be changed in the left-hand side matrix $\mathbf{B}$. Just finding those triples takes time proportional to the number of nonzeros in $\mathbf{B}$ with unordered triples. Thus, RAM complexity of SpAsgn is $O(nnz(\mathbf{A})+nnz(\mathbf{B}))$ and its I/O complexity is $O(scan(\mathbf{A})+scan(\mathbf{B}))$. A simple implementation achieving these bounds performs a single scan of $\mathbf{B}$, outputs only

*A procedure exploits spatial locality if data that are stored in nearby memory locations are likely to be referenced close in time.

the nonassigned triples (e.g., for $\mathbf{B}(:,k) = \mathbf{A}$, those are the triples $(i, j, \mathbf{B}(i,j))$ where $j \neq k$), and finally concatenates the nonzeros in $\mathbf{A}$ to the output.

SpMV has full spatial locality when accessing the elements of $\mathbf{A}$ because the algorithm scans all the nonzeros of $\mathbf{A}$ in the exact order that they are stored. Therefore, $O(scan(\mathbf{A}))$ cache misses are taken for granted as compulsory misses.[*] Although SpMV is optimal in the RAM model without any ordering constraints, its cache performance suffers, as the algorithm cannot exploit any spatial locality when accessing vectors $\mathbf{x}$ and $\mathbf{y}$.

In considering the cache misses involved, for each triple $(i, j, \mathbf{A}(i,j))$, a random access to the jth component of $\mathbf{x}$ is required, and the result of the element-wise multiplication $\mathbf{A}(i,j) \cdot \mathbf{x}(j)$ must be written to the random location $\mathbf{y}(i)$. Assumption 2 implies that the fast memory is not big enough to hold the dense arrays $\mathbf{x}$ and $\mathbf{y}$. Thus, we make up to two extra cache misses per flop. These indirect memory accesses can be clearly seen in the Triples-SpMV code shown in Algorithm 13.5, where the values of $\mathbf{A}.\mathsf{I}(k)$ and $\mathbf{A}.\mathsf{J}(k)$ may change in every iteration. Consequently, I/O complexity of SpMV on unordered triples is

$$nnz(\mathbf{A})/L + 2\ nnz(\mathbf{A}) = O(nnz(\mathbf{A})) \tag{13.1}$$

Algorithm 13.5. Triples matrix vector multiply.
Operation $\mathbf{y} = \mathbf{A}\mathbf{x}$ using triples.

$\mathbf{y} : \mathbb{R}^M = \text{TRIPLES-SPMV}(\mathbf{A} : \mathbb{R}^{S(M \times N)}, \mathbf{x} : \mathbb{R}^N)$

$\mathbf{y} = 0$
for $k = 1$ **to** $nnz(\mathbf{A})$
    do $\mathbf{y}(\mathbf{A}.\mathsf{I}(k)) = \mathbf{y}(\mathbf{A}.\mathsf{I}(k)) + \mathbf{A}.\mathsf{V}(k) \cdot \mathbf{x}(\mathbf{A}.\mathsf{J}(k))$

The SpAdd algorithm needs to identify all (i, j) pairs such that $\mathbf{A}(i,j) \neq 0$ and $\mathbf{B}(i,j) \neq 0$, and to add their values to create a single entry in the resulting matrix. This step can be accomplished by first sorting the nonzeros of the input matrices and then performing a simultaneous scan of sorted nonzeros to sum matching triples. Using linear time counting sort, SpAdd is fast in the RAM model with $O(nnz(\mathbf{A}) + nnz(\mathbf{B}))$ complexity.

Counting sort, in its naive form, has poor cache utilization because the total size of the counting array is likely to be bigger than the size of the fast memory. Sorting the nonzeros of a sparse matrix translates into one cache miss per nonzero in the worst case. Therefore, the complexity of SpAdd in the I/O model becomes $O(nnz(\mathbf{A}) + nnz(\mathbf{B}))$. The number of cache misses can be decreased by using cache optimal sorting algorithms (see [Aggarwal & Vitter 1988]), but such algorithms are comparison based. They do $O(n \lg n)$ work as opposed to linear work. Rahman and Raman [Rahman & Raman 2000] gave a counting sort algorithm that has better cache utilization in practice than the naive algorithm, and still does linear work.

SpGEMM needs fast access to columns, rows, or a given particular element, depending on the algorithm. One can also think of $\mathbf{A}$ as a table of i's and k's

[*] Assuming that no explicit data prefetching mechanism is used.

and $\mathbf{B}$ as a table of k's and j's; then $\mathbf{C}$ is their join on k. This database analogy [Kotlyar et al. 1997] may lead to alternative SpGEMM implementations based on ideas from databases. An outer product formulation of SpGEMM on unordered triples has three basic steps (a similar algorithm for general sparse tensor multiplication is given by Bader and Kolda [Bader & Kolda 2007]):

1. For each $k \in \{1, \ldots, K\}$, identify the set of triples that belongs to the kth column of $\mathbf{A}$ and the kth row of $\mathbf{B}$. Formally, find $\mathbf{A}(:,k)$ and $\mathbf{B}(k,:)$.

2. For each $k \in \{1, \ldots, K\}$, compute the Cartesian product of the row indices of $\mathbf{A}(:,k)$ and the column indices of $\mathbf{B}(k,:)$. Formally, compute the sets $\mathbf{C}_k = \{\mathbf{A}(:,k).\mathsf{I}\} \times \{\mathbf{B}(k,:).\mathsf{J}\}$.

3. Find the union of all Cartesian products, summing up duplicates during set union: $\mathbf{C} = \bigcup_{k \in \{1,\ldots,K\}} \mathbf{C}_k$.

Step 1 of the algorithm can be efficiently implemented by sorting the triples of $\mathbf{A}$ according to their column indices and the triples of $\mathbf{B}$ according to their row indices. Computing the Cartesian products in step 2 takes time

$$\sum_{k=1}^{K} nnz(\mathbf{A}(:,k)) \cdot nnz(\mathbf{B}(k,:)) = \text{flops} \tag{13.2}$$

Finally, summing up duplicates can be done by lexicographically sorting the elements from sets $\mathbf{C}_k$, which has a total running time of

$$O(sort(nnz(\mathbf{A})) + sort(nnz(\mathbf{B})) + \text{flops} + sort(\text{flops})) \tag{13.3}$$

As long as the number of nonzeros is more than the dimensions of the matrices (Assumption 1), it is advantageous to use a linear time sorting algorithm instead of a comparison-based sort. Since a lexicographical sort is not required for finding $\mathbf{A}(:,k)$ or $\mathbf{B}(k,:)$ in step 1, a single pass of linear time counting sort [Cormen et al. 2001] suffices for each input matrix. However, two passes of linear time counting sort are required in step 3 to produce a lexicographically sorted output. RAM complexity of this implementation turns out to be

$$nnz(\mathbf{A}) + nnz(\mathbf{B}) + 3 \cdot \text{flops} = O(nnz(\mathbf{A}) + nnz(\mathbf{B}) + \text{flops}) \tag{13.4}$$

However, due to the cache-inefficient nature of counting sort, this algorithm makes $O(nnz(\mathbf{A}) + nnz(\mathbf{B}) + \text{flops})$ cache misses in the worst case.

Another way to implement SpGEMM on unordered triples is to iterate through the triples of $\mathbf{A}$. For each $(i, j, \mathbf{A}(i,j))$, we find $\mathbf{B}(:,j)$ and multiply $\mathbf{A}(i,j)$ with each nonzero in $\mathbf{B}(:,j)$. The duplicate summation step is left intact. The time this implementation takes is

$$nnz(\mathbf{A}) \cdot nnz(\mathbf{B}) + 3 \cdot \text{flops} = O(nnz(\mathbf{A}) \cdot nnz(\mathbf{B})) \tag{13.5}$$

The term flops is dominated by the term $nnz(\mathbf{A}) \cdot nnz(\mathbf{B})$ according to Theorem 13.1. Therefore, the performance is worse than for the previous implementation that sorts the input matrices first.

Theorem 13.1. *For all matrices* $\mathbf{A}$ *and* $\mathbf{B}$, $\text{flops}(\mathbf{AB}) \leq nnz(\mathbf{A}) \cdot nnz(\mathbf{B})$

Proof. Let the vector of column counts of $\mathbf{A}$ be

$$\mathbf{a} = (a_1, a_2, \ldots, a_k) = (nnz(\mathbf{A}(:,1), nnz(\mathbf{A}(:,2)), \ldots, nnz(\mathbf{A}(:,k))$$

and the vector of row counts of $\mathbf{B}$

$$\mathbf{b} = (b_1, b_2, \ldots, b_k) = (nnz(\mathbf{B}(1,:), nnz(\mathbf{B}(2,:)), \ldots, nnz(\mathbf{B}(k,:)).$$

Note that flops $= \mathbf{a}^{\mathsf{T}}\mathbf{b} = \sum_{i=j} a_i b_j$, and $\sum_{i \neq j} a_i b_j \geq 0$ as $\mathbf{a}$ and $\mathbf{b}$ are nonnegative. Consequently,

$$\begin{aligned} nnz(\mathbf{A}) \cdot nnz(\mathbf{B}) &= \left(\sum_{k=1}^{K} a_k\right) \cdot \left(\sum_{k=1}^{K} b_k\right) = \left(\sum_{i=j} a_i b_j\right) + \left(\sum_{i \neq j} a_i b_j\right) \\ &\geq \sum_{i=j} a_i b_j = \mathbf{a}^{\mathsf{T}}\mathbf{b} = \text{flops} \quad \square \end{aligned}$$

It is worth noting that both implementations of SpGEMM using unordered triples have $O(nnz(\mathbf{A}) + nnz(\mathbf{B}) + \text{flops})$ space complexity, due to the intermediate triples that are all present in the memory after step 2. Ideally, the space complexity of SpGEMM should be $O(nnz(\mathbf{A}) + nnz(\mathbf{B}) + nnz(\mathbf{C}))$, which is independent of flops.

13.3.2 Row ordered triples

The second option is to keep the triples sorted according to their rows or columns only. We analyze the row ordered version; column order is symmetric. This section is divided into three subsections. The first one is on indexing and SpMV. The second one is on a fundamental abstract data type that is used frequently in sparse matrix algorithms, namely the sparse accumulator (SPA). The SPA is used for implementing some of the SpAdd and SpGEMM algorithms throughout the rest of this chapter. Finally, the last subsection is on SpAdd and SpGEMM algorithms.

Indexing and SpMV with row ordered triples

Using row ordered triples, indexing still turns out to be inefficient. In practice, even a fast row access cannot be accomplished since there is no efficient way of spotting the beginning of the ith row without using an index.* Row wise referencing can be done by performing binary search on the whole matrix to identify a nonzero belonging to the referenced row, and then by scanning in both directions to find the rest of the nonzeros belonging to that row. Therefore, SpRef for $\mathbf{A}(i,:)$ has $O(\lg nnz(\mathbf{A}) + nnz(\mathbf{A}(i,:))$ RAM complexity and $O(\lg nnz(\mathbf{A}) + scan(\mathbf{A}(i,:)))$ I/O

*That is a drawback of the triples representation in general. The compressed sparse storage formats described in Section 13.4 provide efficient indexing mechanisms for either rows or columns.

complexity. Element-wise referencing also has the same cost, in both models. Column wise referencing, on the other hand, is as slow as it was with unordered triples, requiring a complete scan of the triples.

SpAsgn might incur excessive data movement, as the number of nonzeros in the left-hand side matrix $\mathbf{B}$ might change during the operation. For a concrete example, consider the operation $\mathbf{B}(i,:) = \mathbf{A}$ where $nnz(\mathbf{B}(i,:)) \neq nnz(\mathbf{A})$ before the operation. Since the data structure needs to keep nonzeros with increasing row indices, all triples with row indices bigger than i need to be shifted by distance $|\,nnz(\mathbf{A}) - nnz(\mathbf{B}(i,:))|$.

SpAsgn has RAM complexity $O(nnz(\mathbf{A})) + nnz(\mathbf{B}))$ and I/O complexity $O(scan(\mathbf{A}) + scan(\mathbf{B}))$, where $\mathbf{B}$ is the left-hand side matrix before the operation. While implementations of row wise and element-wise referencing are straightforward, column wise referencing ($\mathbf{B}(:,i) = \mathbf{A}$) seems harder as it reduces to a restricted case of SpAdd. The restriction is that $\mathbf{B} \in \mathbb{S}^{1\times N}$ has, at most, one nonzero in a given row. Therefore, a similar scanning-based implementation suffices.

Row ordered triples format allows an SpMV implementation that makes, at most, one extra cache miss per flop. The reason is that references to vector $\mathbf{y}$ show good spatial locality: they are ordered with monotonically increasing values of $\mathbf{A}.\mathsf{I}(k)$, avoiding scattered memory referencing on vector $\mathbf{y}$. However, accesses to vector $\mathbf{x}$ are still irregular as the memory stride when accessing $\mathbf{x}$ ($|\mathbf{A}.\mathsf{J}(k+1) - \mathbf{A}.\mathsf{J}(k)|$) might be as big as the matrix dimension N. Memory strides can be reduced by clustering the nonzeros in every row. More formally, this clustering corresponds to reducing the bandwidth of the matrix, which is defined as $\beta(\mathbf{A}) = \max\{|i-j| : \mathbf{A}(i,j) \neq 0\}$. Toledo experimentally studied different methods of reordering the matrix to reduce its bandwidth, along with other optimizations like blocking and prefetching, to improve the memory performance of SpMV [Toledo 1997]. Overall, row ordering does not improve the asymptotic I/O complexity of SpMV over unordered triples, although it cuts the cache misses by nearly half. Its I/O complexity becomes

$$nnz(\mathbf{A})/L + N/L + nnz(\mathbf{A}) = O(nnz(\mathbf{A})) \tag{13.6}$$

The sparse accumulator

Most operations that output a sparse matrix generate it one row at a time. The current active row is stored temporarily on a special structure called the sparse accumulator (SPA) [Gilbert et al. 1992] (or expanded real accumulator [Pissanetsky 1984]). The SPA helps merging unordered lists in linear time.

There are different ways of implementing the SPA as it is an abstract data type not a concrete data structure. In our SPA implementation, $\mathbf{w}$ is the dense vector of values, $\mathbf{b}$ is the boolean dense vector that contains "occupied" flags, and LS is the list that keeps an unordered list of indices, as Gilbert, Moler and Schreiber described [Gilbert et al. 1992].

`Scatter-SPA` function, given in Algorithm 13.6, adds a scalar (*value*) to a specific position (*pos*) of the SPA. Scattering is a constant time operation. Gathering the SPA's nonzeros to the output matrix $\mathbf{C}$ takes $O(nnz(\mathsf{SPA}))$ time. The pseudocode for the `Gather-SPA` is given in Algorithm 13.7. It is crucial to initialize SPA only once at the beginning as this takes $O(N)$ time. Resetting it later for the next active row takes only $O(nnz(\mathsf{SPA}))$ time by using LS to reach all the nonzero elements and resetting only those indices of $\mathbf{w}$ and $\mathbf{b}$.

Algorithm 13.6. Scatter SPA.
Scatters/accumulates the nonzeros in the SPA.

Scatter-SPA(SPA, *value*, *pos*)
if (SPA.$\mathbf{b}(pos) = 0$)
    then SPA.$\mathbf{w}(pos) \leftarrow value$
        SPA.$\mathbf{b}(pos) \leftarrow 1$
        Insert(SPA.LS, *pos*)
    else SPA.$\mathbf{w}(pos) \leftarrow$ SPA.$\mathbf{w}(pos) + value$

Algorithm 13.7. Gather SPA.
Gathers/outputs the nonzeros in the SPA.

nzi = Gather-SPA(SPA, **val**, **col**, *nzcur*)
cptr $\leftarrow$ *head*(SPA.LS)
nzi $\leftarrow 0$ ▷ number of nonzeros in the ith row of $\mathbf{C}$
while *cptr* $\neq$ NIL
    do
        col$(nzcur + n) \leftarrow element(cptr)$ ▷ Set column index
        val$(nzcur + n) \leftarrow$ SPA.$\mathbf{w}(element(cptr))$ ▷ Set value
        $nzi \leftarrow nzi + 1$
        ADVANCE(*cptr*)

The cost of resetting the SPA can be completely avoided by using the *multiple switch* technique (also called the *phase counter* technique) described by Gustavson (see [Gustavson 1976, Gustavson 1997]). Here, $\mathbf{b}$ becomes a dense *switch vector* of integers instead of a dense boolean vector. For computing each row, we use a different switch value. Every time a nonzero is introduced to position *pos* of the SPA, we set the switch to the current active row index (SPA.$\mathbf{b}(pos) = i$). During the computation of subsequent rows $j = \{i+1, \ldots, M\}$, the switch value being less than the current active row index (SPA.$\mathbf{b}(pos) \leq j$) means that the position *pos* of the SPA is "free." Therefore, the need to reset $\mathbf{b}$ for each row is avoided.

SpAdd and SpGEMM with row ordered triples

Using the SPA, we can implement SpAdd with $O(nnz(\mathbf{A}) + nnz(\mathbf{B}))$ RAM complexity. The full procedure is given in Algorithm 13.8. The I/O complexity of this SpAdd implementation is also $O(nnz(\mathbf{A}) + nnz(\mathbf{B}))$ because for each nonzero scanned from inputs, the algorithm checks and updates an arbitrary position of the SPA. From Assumption 2, these arbitrary accesses are likely to incur cache misses every time.

Algorithm 13.8. Row ordered matrix add.
Operation $\mathbf{C} = \mathbf{A} \oplus \mathbf{B}$ using row ordered triples.

$\mathbf{C} : \mathbb{R}^{S(M \times N)} = \text{RowTriples-SpAdd}(\mathbf{A} : \mathbb{R}^{S(M \times N)}, \mathbf{B} : \mathbb{R}^{S(M \times N)})$

1 Set-SPA(SPA) ▷ Set $\mathbf{w} = 0$, $\mathbf{b} = 0$ and create empty list LS
2 $ka \leftarrow kb \leftarrow kc \leftarrow 1$ ▷ Initialize current indices to one
3 **for** $i \leftarrow 1$ **to** M
4   **do while** $(ka \leq nnz(A)$ and $\mathbf{A}.\mathsf{I}(ka) = i)$
5   **do** Scatter-SPA(SPA, $\mathbf{A}.\mathsf{V}(ka), \mathbf{A}.\mathsf{J}(ka)$)
6            $ka \leftarrow ka + 1$
7   **while** $(kb \leq nnz(B)$ and $\mathbf{B}.\mathsf{I}(kb) = i)$
8   **do** Scatter-SPA(SPA, $\mathbf{B}.\mathsf{V}(kb), \mathbf{B}.\mathsf{J}(kb)$)
9            $kb \leftarrow kb + 1$
10       $nznew \leftarrow$ Gather-SPA(SPA, $\mathbf{C}.\mathsf{V}, \mathbf{C}.\mathsf{J}, kc$)
11   **for** $j \leftarrow 0$ **to** $nznew - 1$
12   **do** $\mathbf{C}.\mathsf{I}(kc + j) \leftarrow i$ ▷ Set row index
13       $kc \leftarrow kc + nznew$
14   Reset-SPA(SPA) ▷ Reset $\mathbf{w} = 0$, $\mathbf{b} = 0$ and empty LS

It is possible to implement SpGEMM by using the same outer product formulation described in Section 13.3.1, with a slightly better asymptotic RAM complexity of $O(nnz(\mathbf{A}) + \text{flops})$, as the triples of $\mathbf{B}$ are already sorted according to their row indices. Instead, we describe a row wise implementation, similar to the CSR-based algorithm described in Section 13.4. Because of inefficient row wise indexing support of row ordered triples, however, the operation count is higher than the CSR version. A SPA of size N is used to accumulate the nonzero structure of the current active row of $\mathbf{C}$. A direct scan of the nonzeros of $\mathbf{A}$ allows enumeration of nonzeros in $\mathbf{A}(i,:)$ for increasing values of $i \in \{1, \ldots, M\}$. Then, for each triple $(i, k, \mathbf{A}(i, k))$ in the ith row of $\mathbf{A}$, the matching triples $(k, j, \mathbf{B}(k, j))$ of the kth row of $\mathbf{B}$ need to be found using the SpRef primitive. This way, the nonzeros in $\mathbf{C}(i,:)$ are accumulated. The whole procedure is given in Algorithm 13.9. Its RAM complexity is

$$\sum_{\mathbf{A}(i,k) \neq 0} \left(nnz(\mathbf{B}(i,:)) \lg(nnz(\mathbf{B}))\right) + \text{flops} = O(nnz(\mathbf{A}) \lg(nnz(\mathbf{B})) + \text{flops}) \quad (13.7)$$

where the $\lg(nnz(\mathbf{B}))$ factor per each nonzero in $\mathbf{A}$ comes from the row wise SpRef operation in line 5. Its I/O complexity is

$$O(scan(\mathbf{A}) \lg(nnz(\mathbf{B})) + \text{flops}) \quad (13.8)$$

While the complexity of row wise implementation is asymptotically worse than the outer product implementation in the RAM model, it has the advantage of using only $O(nnz(\mathbf{C}))$ space as opposed to the $O(\text{flops})$ space used by the outer product implementation. On the other hand, the I/O complexities of the outer product version and the row wise version are not directly comparable. Which one is faster depends on the cache line size and the number of nonzeros in $\mathbf{B}$.

Algorithm 13.9. Row ordered matrix multiply.
Operation $\mathbf{C} = \mathbf{AB}$ using row ordered triples.

```
C : R^{S(M×N)} = RowTriples-SpGEMM(A : R^{S(M×K)}, B : R^{S(K×N)})
Set-SPA(SPA)                        ▷ Set w = 0, b = 0 and create empty list LS
ka ← kc ← 1                         ▷ Initialize current indices to one
for i ← 1 to M
      do while (ka ≤ nnz(A) and A.I(ka) = i)
               do BR ← B(A.J(ka), :)         ▷ Using SpRef
                  for kb ← 1 to nnz(BR)
                       do value ← A.NUM(ka) · BR.NUM(kb)
                          Scatter-SPA(SPA, value, BR.J(kb))
                  ka ← ka + 1
         nznew ← Gather-SPA(SPA, C.V, C.J, kc)
         for j ← 0 to nznew − 1
               do C.I(kc + j) ← i       ▷ Set row index
         kc ← kc + nznew
         Reset-SPA(SPA)                 ▷ Reset w = 0, b = 0 and empty LS
```

13.3.3 Row-major ordered triples

We now consider the third option of storing triples in lexicographic order, either in column-major or row-major order. Once again, we focus on the row-oriented scheme in this section. RAM and I/O complexities of key primitives for row-major ordered and CSR are listed in Tables 13.3 and 13.4.

In order to reference a whole row, binary search on the whole matrix, followed by a scan on both directions, is used, as with row ordered triples. As the nonzeros in a row are ordered by column indices, it seems there should be a faster way to access a single element than the method used on row ordered triples. A faster way

Table 13.3. Row-major ordered RAM complexities.
RAM complexities of key primitives on row-major ordered triples and CSR.

	Row-Major Ordered Triples	CSR
SpRef	$O(\lg nnz(\mathbf{A}) + \lg nnz(\mathbf{A}(i,:))\{\ \mathbf{A}(i,j)$ $O(\lg nnz(\mathbf{A}) + nnz(\mathbf{A}(i,:))\{\ \mathbf{A}(i,:)$ $O(nnz(\mathbf{A}))\{\ \mathbf{A}(:,j)$	$O(\lg nnz(\mathbf{A}(i,:))\{\ \mathbf{A}(i,j)$ $O(nnz(\mathbf{A}(i,:))\{\ \mathbf{A}(i,:)$ $O(nnz(\mathbf{A}))\{\ \mathbf{A}(:,j)$
SpAsgn	$O(nnz(\mathbf{A})) + O(nnz(\mathbf{B}))$	$O(nnz(\mathbf{A}) + nnz(\mathbf{B}))$
SpMV	$O(nnz(\mathbf{A}))$	$O(nnz(\mathbf{A}))$
SpAdd	$O(nnz(\mathbf{A}) + nnz(\mathbf{B}))$	$O(nnz(\mathbf{A}) + nnz(\mathbf{B}))$
SpGEMM	$O(nnz(\mathbf{A})+\text{flops})$	$O(nnz(\mathbf{A})+\text{flops})$

Table 13.4. Row-major ordered I/O complexities.
I/O complexities of key primitives on row-major ordered triples and CSR.

	Row-Major Ordered Triples	**CSR**
SpRef	$O(\lg nnz(\mathbf{A}) + search(\mathbf{A}(i,:))\{\ \mathbf{A}(i,j)$ $O(\lg nnz(\mathbf{A}) + scan(\mathbf{A}(i,:))\{\ \mathbf{A}(i,:)$ $O(scan(\mathbf{A}))\{\ \mathbf{A}(:,j)$	$O(search(\mathbf{A}(i,:)))\{\ \mathbf{A}(i,j)$ $O(scan(\mathbf{A}(i,:)))\{\ \mathbf{A}(i,:)$ $O(scan(\mathbf{A}))\{\ \mathbf{A}(:,j)$
SpAsgn	$O(scan(\mathbf{A}) + scan(\mathbf{B}))$	$O(scan(\mathbf{A}) + scan(\mathbf{B}))$
SpMV	$O(nnz(\mathbf{A}))$	$O(nnz(\mathbf{A}))$
SpAdd	$O(scan(\mathbf{A}) + scan(\mathbf{B}))$	$O(scan(\mathbf{A}) + scan(\mathbf{B}))$
SpGEMM	$O(\min\{nnz(\mathbf{A}) + \text{flops},$ $scan(\mathbf{A})\lg(nnz(\mathbf{B}))+\text{flops}\})$	$O(scan(\mathbf{A}) + \text{flops})$

indeed exists, but ordinary binary search would not do it because the beginning and the end of the ith row is not known in advance. The algorithm has three steps:

1. Spot a triple $(i, j, \mathbf{A}(i, j))$ that belongs to the ith row by doing binary search on the whole matrix.

2. From that triple, perform an unbounded binary search [Manber 1989] on both directions. In an unbounded search, the step length is doubled at each iteration. The search terminates at a given direction when it hits a triple that does not belong to the ith row. Those two triples (one from each direction) become the boundary triples.

3. Perform ordinary binary search within the exclusive range defined by the boundary vertices.

The number of total operations is $O(\lg nnz(\mathbf{A}) + \lg nnz(\mathbf{A}(i,:))$. An example is given in Figure 13.4, where $\mathbf{A}(12, 16)$ is indexed.

While unbounded binary search is the preferred method in the RAM model, simple scanning might be faster in the I/O model. Searching an element in an ordered set of N elements can be achieved with $\Theta(\log_L N)$ cost in the I/O model, using B-trees (see [Bayer & McCreight 1972]). However, using an ordinary array, search incurs $\lg N$ cache misses. This may or may not be less than $scan(N)$. Therefore, we define the cost of searching within an ordered row as follows:

$$search(\mathbf{A}(i,:)) = \min\{\lg nnz(\mathbf{A}(i,:)), scan(\mathbf{A}(i,:))\} \tag{13.9}$$

For column wise referencing as well as for SpAsgn operations, row-major ordered triples format does not provide any improvement over row ordered triples.

In SpMV, the only array that does not show excellent spatial locality is $\mathbf{x}$, since $\mathbf{A}.\mathsf{I}, \mathbf{A}.\mathsf{J}, \mathbf{A}.\mathsf{V}$, and $\mathbf{y}$ are accessed with mostly consecutive, increasing index

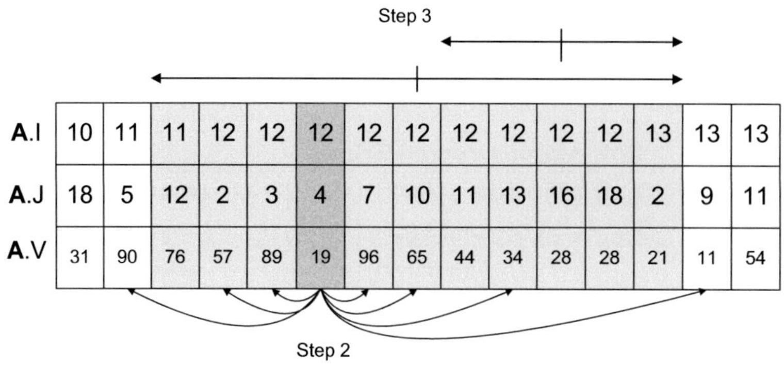

Figure 13.4. Indexing row-major triples.
Element-wise indexing of $\mathbf{A}(12, 16)$ on row-major ordered triples.

values. Accesses to $\mathbf{x}$ are also with increasing indices, which is an improvement over row ordered triples. However, memory strides when accessing $\mathbf{x}$ can still be high, depending on the number of nonzeros in each row and the bandwidth of the matrix. In the worst case, each access to $\mathbf{x}$ might incur a cache miss.

Bender et al. came up with cache-efficient algorithms for SpMV, using the column-major layout, that have an optimal number of cache misses [Bender et al. 2007]. From a high-level view, their method first generates all the intermediate triples of $\mathbf{y}$, possibly with repeating indices. Then, the algorithm sorts those intermediate triples with respect to their row indices, performing additions on the triples with same row index as they occur. I/O optimality of their SpMV algorithm relies on the existence of an I/O optimal sorting algorithm. Their complexity measure assumes a fixed k number of nonzeros per column, leading to I/O complexity of

$$O\left(scan(\mathbf{A}) \log_{Z/L} \frac{N}{\max\{Z, k\}} \right) \tag{13.10}$$

SpAdd is now more efficient even without using any auxiliary data structure. A scanning-based array-merging algorithm is sufficient as long as we do not forget to sum duplicates while merging. Such an implementation has $O(nnz(\mathbf{A})+nnz(\mathbf{B}))$ RAM complexity and $O(scan(\mathbf{A}) + scan(\mathbf{B}))$ I/O complexity.*

Row-major ordered triples allow outer product and row wise SpGEMM implementations at least as efficiently as row ordered triples. Indeed, some finer improvements are possibly by exploiting the more specialized structure. In the case of row wise SpGEMM, a technique called *finger search* [Brodal 2005] can be used to improve the RAM complexity. While enumerating all triples $(i, k, \mathbf{A}) \in \mathbf{A}(i, :)$, they are naturally sorted with increasing k values. Therefore, accesses to $\mathbf{B}(k, :)$ are also with increasing k values. Instead of restarting the binary search from the beginning of $\mathbf{B}$, one can use fingers and only search the yet unexplored subsequence. Note

*These bounds are optimal only if $nnz(\mathbf{A}) = \Theta(nnz(\mathbf{B}))$; see [Brown & Tarjan 1979].

that finger search uses the unbounded binary search as a subroutine when searching the unexplored subsequence. Row wise SpGEMM using finger search has a RAM complexity of

$$O(\text{flops}) + \sum_{i=1}^{m} O\left(nnz(\mathbf{A}(i,:)) \lg \frac{nnz(\mathbf{B})}{nnz(\mathbf{A}(i,:))} \right) \tag{13.11}$$

which is asymptotically faster than the $O(nnz(\mathbf{A}) \lg(nnz(\mathbf{B})) + \text{flops})$ cost of the same algorithm on row ordered triples.

Outer product SpGEMM can be modified to use only $O(nnz(\mathbf{C}))$ space during execution by using multiway merging [Buluç & Gilbert 2008b]. However, this comes at the price of an extra $\lg k$ factor in the asymptotical RAM complexity where k is the number of indices i for which $\mathbf{A}(:,i) \neq \emptyset$ and $\mathbf{B}(i,:) \neq \emptyset$.

Although both of these refined algorithms are asymptotically slower than the naive outer product method, they might be faster in practice because of the cache effects and difference in constants in the asymptotic complexities. Further research is required in algorithm engineering of SpGEMM to find the best performing algorithm in real life.

13.4 Compressed sparse row/column

The most widely used storage schemes for sparse matrices are compressed sparse column (CSC) and compressed sparse row (CSR). For example, MATLAB uses CSC format to store its sparse matrices [Gilbert et al. 1992]. Both are dense collections of sparse arrays. We examine CSR, which is introduced by Gustavson [Gustavson 1972] under the name of sparse row wise representation; CSC is symmetric.

CSR can be seen as a concatenation of sparse row arrays. On the other hand, it is also very close to row ordered triples with an auxiliary index of size $\Theta(N)$. In this section, we assume that nonzeros within each sparse row array are ordered with increasing row indices. This is not a general requirement though. Tim Davis CSparse package [Davis et al. 2006], for example, does not impose any ordering within the sparse arrays.

13.4.1 CSR and adjacency lists

In principle, CSR is almost identical to the adjacency list representation of a directed graph [Tarjan 1972]. In practice, however, it has much less overhead and much better cache efficiency. Instead of storing an array of linked lists as in the adjacency list representation, CSR is composed of three arrays that store whole rows contiguously. The first array (IR) of size $M + 1$ stores the row pointers as explicit integer values, the second array (JC) of size nnz stores the column indices, and the last array (NUM) of size nnz stores the actual numerical values. Observe that column indices stored in the JC array indeed come from concatenating the edge indices of the adjacency lists. Following the sparse matrix/graph duality, it is also meaningful to call the first array the vertex array and the second array the edge

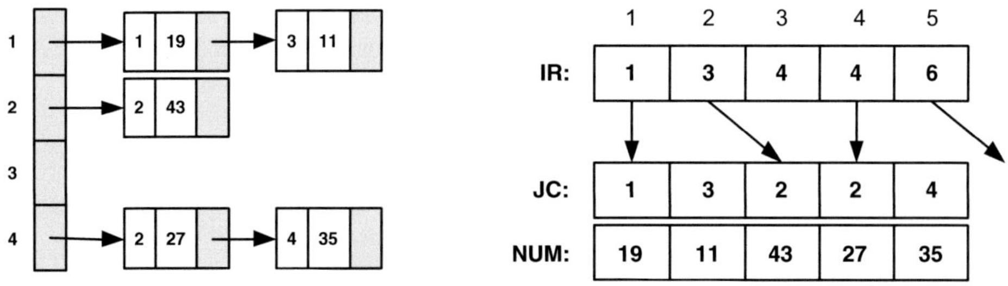

Figure 13.5. CSR format.
Adjacency list (left) and CSR (right) representations of matrix **A**.

array. The vertex array holds the offsets to the edge array, meaning that the nonzeros in the ith row are stored from $\mathsf{NUM}(\mathsf{IR}(i))$ to $\mathsf{NUM}(\mathsf{IR}(i+1)-1)$, and their respective positions within that row are stored from $\mathsf{JC}(\mathsf{IR}(i))$ to $\mathsf{JC}(\mathsf{IR}(i+1)-1)$. Also note that $\mathsf{JC}(i) = \mathsf{JC}(i+1)$ means there are no nonzeros in the ith row.

Figure 13.5 shows the adjacency list and CSR representations of matrix **A**. While the arrows in the adjacency-based representation are actual pointers to memory locations, the arrows in CSR are not. Any data type that supports unsigned integers fulfills the desired purpose of holding the offsets to the edge array.

The efficiency advantage of the CSR data structure compared with the adjacency list can be explained by the memory architecture of modern computers. In order to access all the nonzeros in a given row i, which is equivalent to traversing all the outgoing edges of a given vertex v_i, CSR makes at most $\lceil nnz(\mathbf{A}(i,:))/L \rceil$ cache misses. A similar access to the adjacency list representation incurs $nnz(\mathbf{A}(i,:))$ cache misses in the worst case, worsening as the memory becomes more and more fragmented. In an experiment published in 1998, an array-based representation was found to be 10 times faster to traverse than a linked-list-based representation [Black et al. 1998]. This performance gap is due to the high cost of pointer chasing that happens frequently in linked data structures. The efficiency of CSR comes at a price though: introducing new nonzero elements or deleting a nonzero element is computationally inefficient and best avoided [Gilbert et al. 1992]. Therefore, CSR is best suited for representing static graphs. All the key primitives but SpAsgn work on static graphs.

13.4.2 CSR on key primitives

Contrary to triples storage formats, CSR allows constant-time random access to any row of the matrix. Its ability to enumerate all the elements in the ith row with $O(nnz(\mathbf{A}(i,:))$ RAM complexity and $O(scan(\mathbf{A}(i,:))$ I/O complexity makes it an excellent data structure for row wise SpRef. Element-wise referencing takes, at most, $O(\lg nnz(\mathbf{A}(i,:))$ time in the RAM model as well as the I/O model, using a binary search. Considering column wise referencing, however, CSR does not provide any improvement over the triples format.

On the other hand, even row wise SpAsgn operations are inefficient if the number of elements in the assigned row changes. In that general case, $O(nnz(\mathbf{B}))$ elements might need to be moved. This is also true for column wise and element-wise SpAsgn as long as not just existing nonzeros are reassigned to new values.

The code in Algorithm 13.10 shows how to perform SpMV when matrix $\mathbf{A}$ is represented in CSR format. This code and SpMV with row-major ordered triples have similar performance characteristics except for a few subtleties. When some rows of $\mathbf{A}$ are all zeros, those rows are effectively skipped in row-major ordered triples, but still need to be examined in CSR. On the other hand, when $M \ll nnz$, CSR has a clear advantage since it needs to examine only one index ($\mathbf{A}.\mathsf{JC}(k)$) per inner loop iteration while row-major ordered triples needs to examine two ($\mathbf{A}.\mathsf{I}(k)$ and $\mathbf{A}.\mathsf{J}(k)$). Thus, CSR may have up to a factor of two difference in the number of cache misses. CSR also has some advantages over CSC when the SpMV primitive is considered (especially in the case of $\mathbf{y} = \mathbf{y} + \mathbf{Ax}$), as experimentally shown by Vuduc [Vuduc 2003].

Algorithm 13.10. CSR matrix vector multiply.
Operation $\mathbf{y} = \mathbf{Ax}$ using CSR.

$\mathbf{y} : \mathbb{R}^M = \text{CSR-SpMV}(\mathbf{A} : \mathbb{R}^{S(M)\times N}, \mathbf{x} : \mathbb{R}^N)$

$\mathbf{y} = 0$
for $i = 1$ **to** M
  do for $k = \mathbf{A}.\mathsf{IR}(i)$ **to** $\mathbf{A}.\mathsf{IR}(i+1) - 1$
    do $\mathbf{y}(i) = \mathbf{y}(i) + \mathbf{A}.\mathsf{NUM}(k) \cdot \mathbf{x}(\mathbf{A}.\mathsf{JC}(k))$

Blocked versions of CSR and CSC try to take advantage of clustered nonzeros in the sparse matrix. While blocked CSR (BCSR) achieves superior performance for SpMV on matrices resulting from finite element meshes [Vuduc 2003], mostly by using loop unrolling and register blocking, it is of little use when the matrix itself does not have its nonzeros clustered. Pinar and Heath [Pinar & Heath 1999] proposed a reordering mechanism to cluster those nonzeros to get dense subblocks. However, it is not clear whether such mechanisms are successful for highly irregular matrices from sparse real-world graphs.

Except for the additional bookkeeping required for getting the row pointers right, SpAdd can be implemented in the same way as is done with row-major ordered triples. Luckily, the extra bookkeeping of row pointers does not affect the asymptotic complexities.

One subtlety overlooked in the SpAdd implementations throughout this chapter is management of the memory required by the resulting matrix $\mathbf{C}$. We implicitly assumed that the data structure holding $\mathbf{C}$ has enough space to accommodate all of its elements. Repeated doubling of memory whenever necessary is one way of addressing this issue. Another conservative way is to reserve $nnz(\mathbf{A}) + nnz(\mathbf{B})$ space for $\mathbf{C}$ at the beginning of the procedure and shrink any unused portion after the computation, right before the procedure returns.

The efficiency of accessing and enumerating rows in CSR makes the row wise SpGEMM formulation, described in Algorithm 13.4, the preferred matrix multiplication formulation. An efficient implementation of the row wise SpGEMM using CSR was first given by Gustavson [Gustavson 1978]. It had a RAM complexity of

$$O(M + N + nnz(\mathbf{A}) + \text{flops}) = O(nnz(\mathbf{A}) + \text{flops}) \tag{13.12}$$

where the equality follows from Assumption 1. Recent column wise implementations with similar RAM complexities are provided by Davis in his CSparse software [Davis et al. 2006] and by MATLAB [Gilbert et al. 1992]. The algorithm, presented in Algorithm 13.11 uses the SPA described in Section 13.3.2. Once again, the multiple switch technique can be used to avoid the cost of resetting the SPA for every iteration of the outermost loop. As in the case of SpAdd, generally the space required to store $\mathbf{C}$ cannot be determined quickly. Repeated doubling or more sophisticated methods such as Cohen's algorithm [Cohen 1998] may be used. Cohen's algorithm is a randomized iterative algorithm that does $\Theta(1)$ SpMV operations over a semiring to estimate the row and column counts. It can be efficiently implemented even on unordered triples.

Algorithm 13.11. CSR matrix multiply.
Operation $\mathbf{C} = \mathbf{AB}$ using CSR.

$\mathbf{C} : \mathbb{R}^{S(M)\times N}$ = CSR-SpGEMM($\mathbf{A} : \mathbb{R}^{S(M)\times K}, \mathbf{B} : \mathbb{R}^{S(K)\times N}$)

```
Set-SPA(SPA)                                ▷ Set w = 0, b = 0 and create empty list LS
C.IR(1) ← 0
for i ← 1 to M
      do for k ← A.IR(i) to A.IR(i + 1)
             do for j ← B.IR(A.JC(k)) to B.IR(A.JC(k) + 1)
                    do
                         value ← A.NUM(k) · B.NUM(j)
                         Scatter-SPA(SPA, value, B.JC(j))
         nznew ← Gather-SPA(SPA, C.NUM, C.JC, C.IR(i))
         C.IR(i + 1) ← C.IR(i) + nznew
         Reset-SPA(SPA)                     ▷ Reset w = 0, b = 0 and empty LS
```

The row wise SpGEMM implementation does $O(scan(\mathbf{A}) + \text{flops})$ cache misses in the worst case. Due to the size of the SPA and Assumption 2, the algorithm makes a cache miss for every flop. As long as no cache interference occurs between the nonzeros of $\mathbf{A}$ and the nonzeros of $\mathbf{C}(i, :)$, only $scan(\mathbf{A})$ additional cache misses are made instead of $nnz(\mathbf{A})$.

13.5 Case study: Star-P

This section summarizes how sparse matrices are represented in Star-P. It also includes how the key primitives are implemented in this real-world software solution.

13.5.1 Sparse matrices in Star-P

The current Star-P implementation includes *dsparse* (distributed sparse) matrices, which are distributed across processors by blocks of rows. This layout makes the CSR data structure a logical choice to store the sparse matrix slice on each processor.

The design of sparse matrix algorithms in STAR-P follows the same design principles as in MATLAB [Gilbert et al. 1992].

1. Storage required for a sparse matrix should be $O(nnz)$, proportional to the number of nonzero elements.

2. Running time for a sparse matrix algorithm should be $O(\text{flops})$. It should be proportional to the number of floating-point operations required to obtain the result.

The CSR data structure satisfies the requirement for storage as long as $M < nnz$. The second principle is difficult to achieve exactly in practice. Typically, most implementations achieve running time close to $O(\text{flops})$ for commonly used sparse matrix operations. For example, accessing a single element of a sparse matrix should be a constant-time operation. With a CSR data structure, it typically takes time proportional to the logarithm of the length of the row to access a single element. Similarly, insertion of single elements into a CSR data structure generates extensive data movement. Such operations are efficiently performed with the `sparse/find` routines, which work with triples rather than individual elements.

Sparse matrix-dense vector multiplication (SpMV)

The CSR data structure used in STAR-P is efficient for multiplying a sparse matrix by a dense vector: $\mathbf{y} = \mathbf{A}\mathbf{x}$. It is efficient for communication and it shows good cache behavior for the sequential part of the computation. Our choice of the CSR data structure was heavily influenced by our desire to have good SpMV performance since it forms the core computational kernel for many iterative methods.

The matrix $\mathbf{A}$ and vector $\mathbf{x}$ are distributed across processors by rows. The submatrix of $\mathbf{A}$ on each processor will need some subset of $\mathbf{x}$ depending upon its sparsity structure. When SpMV is invoked for the first time on a dsparse matrix $\mathbf{A}$, STAR-P computes a communication schedule for $\mathbf{A}$ and caches it. When later matvecs are performed using the same $\mathbf{A}$, this communication schedule does not need to be recomputed, thus saving some computing and communication overhead at the cost of extra space required to save the schedule. We experimented with overlapping computation and communication in SpMV. It turns out in many cases that this is less efficient than simply performing the communication first, followed by the computation. As computer architectures evolve, this decision may need to be revisited.

When multiplying from the left, $\mathbf{y} = \mathbf{x}'\mathbf{A}$, the communication is not as efficient. Instead of communicating the required subpieces of the source vector, each processor computes its own destination vector. All partial destination vectors are then summed up into the final destination vector. The communication required is always $O(N)$. The choice of the CSR data structure, while making the communication more efficient when multiplying from the right, makes it more difficult to multiply on the left.

Sparse matrix-sparse matrix multiplication

STAR-P stores its matrices in CSR form. Clearly, computing inner products (Algorithm 13.1) is inefficient, since rows of **A** cannot be efficiently accessed without searching. Similarly, in the case of computing outer products (Algorithm 13.2), rows of **B** have to be extracted. The process of accumulating successive rank-one updates is also inefficient, as the structure of the result changes with each successive update. Therefore, STAR-P uses the row wise formulation of matrix multiplication described in Section 13.2, in which the computation is set up so that only rows of **A** and **B** are accessed, producing a row of **C** at a time.

The performance of sparse matrix multiplication in parallel depends upon the nonzero structures of **A** and **B**. A well-tuned implementation may use a polyalgorithm. Such a polyalgorithm may use different communication schemes for different matrices. For example, it may be efficient to broadcast the local part of a matrix to all processors, but in other cases, it may be efficient to send only the required rows. On large clusters, it may be efficient to interleave communication and computation. On shared memory architectures, however, most of the time is spent in accumulating updates, rather than in communication. In such cases, it may be more efficient to schedule the communication before the computation.

13.6 Conclusions

In this chapter, we gave a brief survey on sparse matrix infrastructure for doing graph algorithms. We focused on implementation and analysis of key primitives on various sparse matrix data structures. We tried to complement the existing literature in two directions. First, we analyzed sparse matrix indexing and assignment operations. Second, we gave I/O complexity bounds for all operations. Taking I/O complexities into account is key to achieving high performance in modern architectures with multiple levels of cache.

References

[Aggarwal & Vitter 1988] A. Aggarwal and J.S. Vitter. The input/output complexity of sorting and related problems. *Communications of the ACM*, 31:1116–1127, 1988.

[Aho et al. 1974] A.V. Aho, J.E. Hopcroft, and J.D. Ullman. *The Design and Analysis of Computer Algorithms.* Boston: Addison–Wesley Longman, 1974.

[Aspnäs et al. 2006] M. Aspnäs, A. Signell, and J. Westerholm. Efficient assembly of sparse matrices using hashing. In B. Kgström, E. Elmroth, J. Dongarra, and J. Wasniewski, eds., *PARA*, volume 4699 of *Lecture Notes in Computer Science*, 900–907. Berlin Heidelberg: Springer, 2006.

[Bader & Kolda 2007] B.W. Bader and T.G. Kolda. Efficient MATLAB computations with sparse and factored tensors. *SIAM Journal on Scientific Computing*, 30:205–231, 2007.

[Bayer & McCreight 1972] R. Bayer and E.M. McCreight. Organization and maintenance of large ordered indexes. *Acta Informatica*, 1:173–189, 1972.

[Becker 2006] P. Becker. *The C++ Standard Library Extensions: A Tutorial and Reference*. Boston: Addison–Wesley Professional, 2006.

[Bender et al. 2007] M.A. Bender, G.S. Brodal, R. Fagerberg, R. Jacob, and E. Vicari. Optimal sparse matrix dense vector multiplication in the I/O-model. In *SPAA '07: Proceedings of the Nineteenth Annual ACM Symposium on Parallel Algorithms and Architectures*, 61–70, New York: Association for Computing Machinery, 2007.

[Black et al. 1998] J.R. Black, C.U. Martel, and H. Qi. Graph and hashing algorithms for modern architectures: Design and performance. In *Algorithm Engineering*, 37–48, 1998.

[Brodal 2005] G.S. Brodal. Finger search trees. In D. Mehta and S. Sahni, eds., *Handbook of Data Structures and Applications*, Chapter 11. Boca Raton, Fla.: CRC Press, 2005.

[Brown & Tarjan 1979] M.R. Brown and R.E. Tarjan. A fast merging algorithm. *Journal of the ACM*, 26:211–226, 1979.

[Buluç & Gilbert 2008b] A. Buluç and J.R. Gilbert. On the representation and multiplication of hypersparse matrices. In *IEEE International Parallel and Distributed Processing Symposium (IPDPS 2008)*, 1–11, Miama, FL, 2008.

[Cohen 1998] E. Cohen. Structure prediction and computation of sparse matrix products. *Journal of Combinatorial Optimization*, 2:307–332, 1998.

[Cormen et al. 2001] T.H. Cormen, C.E. Leiserson, R.L. Rivest, and C. Stein. *Introduction to Algorithms*, Second Edition. Cambridge, Mass.: The MIT Press, 2001.

[Davis et al. 2006] T.A. Davis. *Direct Methods for Sparse Linear Systems*. Philadelphia, SIAM, 2006.

[Dietzfelbinger et al. 1994] M. Dietzfelbinger, A. Karlin, K. Mehlhorn, F. Meyer Auf Der Heide, H. Rohnert, and R.E. Tarjan. Dynamic perfect hashing: upper and lower bounds. *SIAM Journal on Computing*, 23:738–761, 1994.

[Duff & Reid 1979] I.S. Duff and J.K. Reid. Some design features of a sparse matrix code. *ACM Transactions on Mathematical Software*, 5:18–35, 1979.

[Feo et al. 2005] J. Feo, D. Harper, S. Kahan, and P. Konecny. Eldorado. In *CF '05: Proceedings of the 2nd Conference on Computing Frontiers*, 28–34, New York: Association for Computing Machinery, 2005.

[Gilbert et al. 1992] J.R. Gilbert, C. Moler, and R. Schreiber. Sparse matrices in MATLAB: Design and implementation. *SIAM Journal on Matrix Analysis and Applications*, 13:333–356, 1992.

[Gustavson 1972] F.G. Gustavson. Some basic techniques for solving sparse systems of linear equations. In D.J. Rose and R.A. Willoughby, eds., *Sparse Matrices and Their Applications*, 41–52, New York: Plenum Press, 1972.

[Gustavson 1976] F.G. Gustavson. Finding the block lower triangular form of a matrix. In J.R. Bunch and D.J. Rose, eds., *Sparse Matrix Computations*, 275–289. New York: Academic Press, 1976.

[Gustavson 1997] F.G. Gustavson. Efficient algorithm to perform sparse matrix multiplication. *IBM Technical Disclosure Bulletin*, 20:1262–1264, 1977.

[Gustavson 1978] F.G. Gustavson. Two fast algorithms for sparse matrices: Multiplication and permuted transposition. *ACM Transactions on Mathematical Software*, 4:250–269, 1978.

[Hennessy & Patterson 2006] J.L. Hennessy and D.A. Patterson. *Computer Architecture, Fourth Edition: A Quantitative Approach.* San Francisco: Morgan Kaufmann Publishers, 2006.

[Kotlyar et al. 1997] V. Kotlyar, K. Pingali, and P. Stodghill. A relational approach to the compilation of sparse matrix programs. In *European Conference on Parallel Processing*, 318–327, 1997.

[Kowarschik & Weiß 2002] M. Kowarschik and C. Weiß. An overview of cache optimization techniques and cache-aware numerical algorithms. In U. Meyer, P. Sanders, and J. Sibeyn, eds., *Algorithms for Memory Hierarchies*, 213–232, Berlin Heidelberg: Springer-Verlag, 2002.

[Manber 1989] U. Manber. *Introduction to Algorithms: A Creative Approach.* Boston: Addison–Wesley Longman, 1989.

[Mehlhorn & Näher 1999] K. Mehlhorn and S. Näher. *LEDA: A platform for combinatorial and geometric computing.* New York: Cambridge University Press, 1999.

[Meyer et al. 2002] U. Meyer, P. Sanders, and J.F. Sibeyn, eds. *Algorithms for Memory Hierarchies, Advanced Lectures (Dagstuhl Research Seminar, March 10–14, 2002)*, volume 2625 of *Lecture Notes in Computer Science.* Berlin Heidelberg: Springer-Verlag, 2003.

[Pagh & Rodler 2004] R. Pagh and F.F. Rodler. Cuckoo hashing. *Journal of Algorithms*, 51:122–144, 2004.

[Pinar & Heath 1999] A. Pinar and M.T. Heath. Improving performance of sparse matrix-vector multiplication. In *Supercomputing '99: Proceedings of the 1999 ACM/IEEE conference on Supercomputing*, p. 30, New York: Association Association for Computing Machinery, 1999.

[Pissanetsky 1984] S. Pissanetsky. *Sparse Matrix Technology.* London: Academic Press, 1984.

[Rahman & Raman 2000] N. Rahman and R. Raman. Analysing cache effects in distribution sorting. *Journal of Experimental Algorithmics*, 5:14, 2000.

[Ross 2007] K.A. Ross. Efficient hash probes on modern processors. In *Proceedings of the IEEE 23rd International Conference on Data Engineering*, 1297–1301. IEEE, 2007.

[Samet 1984] H. Samet. The quadtree and related hierarchical data structures. *ACM Computing Surveys*, 16:187–260, 1984.

[Tarjan 1972] R.E. Tarjan. Depth-first search and linear graph algorithms. *SIAM Journal on Computing*, 1:146–160, 1972.

[Toledo 1997] S. Toledo. Improving the memory-system performance of sparse-matrix vector multiplication. *IBM Journal of Research and Development*, 41:711–726, 1997.

[Vuduc 2003] R.W. Vuduc. *Automatic performance tuning of sparse matrix kernels.* PhD thesis, University of California, Berkeley, 2003.

[Wise & Franco 1990] D.S. Wise and J.V. Franco. Costs of quadtree representation of nondense matrices. *Journal of Parallel and Distributed Computing*, 9:282–296, 1990.

Chapter 14

New Ideas in Sparse Matrix Matrix Multiplication

*Aydın Buluç** and *John Gilbert*†

Abstract

Generalized sparse matrix matrix multiplication is a key primitive for many high performance graph algorithms as well as some linear solvers such as multigrid. We present the first parallel algorithms that achieve increasing speedups for an unbounded number of processors. Our algorithms are based on the two-dimensional (2D) block distribution of sparse matrices where serial sections use a novel hypersparse kernel for scalability.

14.1 Introduction

Development and implementation of large-scale parallel graph algorithms pose numerous challenges in terms of scalability and productivity [Lumsdaine et al. 2007, Yoo et al. 2005]. Linear algebra formulations of many graph algorithms already exist in the literature; see [Aho et al. 1974, Maggs & Poltkin 1988, Tarjan 1981]. By exploiting the duality between matrices and graphs, linear algebraic formula-

*High Performance Computing Research, Lawrence Berkeley National Laboratory, 1 Cyclotron Road, Berkeley, CA 94720 (abuluc@lbl.gov).

†Computer Science Department, University of California, Santa Barbara, CA 93106-5110 (gilbert@cs.ucsb.edu).

This work is sponsored by the Department of Energy under award number DE-FG02-04ER25632, in part by MIT Lincoln Laboratory under contract number 7000012980.

Preliminary versions of this paper have appeared in the proceedings of the 22nd International Parallel and Distributed Processing Symposium (IPDPS '08) and the 37th International Conference on Parallel Processing (ICPP' 08).

tions aim to apply the existing knowledge on parallel matrix algorithms to parallel graph algorithms. One of the key linear algebraic primitives for graph algorithms is computing the product of two sparse matrices (SpGEMM) over a semiring. It serves as a building block for many algorithms including graph contraction algorithm [Gilbert et al. 2008], breadth-first search from multiple-source vertices, peer pressure clustering [Shah 2007], recursive formulations of all-pairs shortest paths algorithms [D'Alberto & Nicolau 2007], matching algorithms [Rabin & Vazirani 1989], and cycle detection [Yuster & Zwick 2004], as well as for some other applications such as multigrid interpolation/restriction [Briggs et al. 2000] and parsing context-free languages [Penn 2006].

Most large graphs in applications, such as the WWW graph, finite element meshes, planar graphs, and trees are sparse. In this chapter, we consider a graph to be sparse if $nnz = O(N)$, where nnz is the number of edges and N is the number of vertices. Dense matrix multiplication algorithms are inefficient for SpGEMM since they require $O(N^3)$ space and the current fastest dense matrix multiplication algorithm runs in $O(N^{2.38})$ time; see [Coppersmith & Winograd 1987, Seidel 1995]. Furthermore, fast dense matrix multiplication algorithms operate on a ring instead of a semiring, making them unsuitable for many algorithms on general graphs. For example, it is possible to embed the semiring into the ring of integers for the all-pairs shortest paths problem on unweighted and undirected graphs [Seidel 1995], but the same embedding does not work for weighted or directed graphs.

Let $\mathbf{A} \in \mathbb{S}^{M \times N}$ be a sparse rectangular matrix of elements from an arbitrary semiring $\mathbb{S}$. We use $nnz(\mathbf{A})$ to denote the number of nonzero elements in $\mathbf{A}$. When the matrix is clear from context, we drop the parentheses and simply use nnz. For sparse matrix indexing, we use the convenient MATLAB colon notation, where $\mathbf{A}(:, i)$ denotes the ith column, $\mathbf{A}(i, :)$ denotes the ith row, and $\mathbf{A}(i, j)$ denotes the element at the (i, j)th position of matrix $\mathbf{A}$. For one-dimensional arrays, $\mathbf{a}(i)$ denotes the ith component of the array. Sometimes, we abbreviate and use $nnz(j)$ to denote the number of nonzero elements in the jth column of the matrix in context. Array indices are 1-based throughout this chapter. We use flops($\mathbf{A}\, op\, \mathbf{B}$), pronounced "flops," to denote the number of nonzero arithmetic operations required by the operation $\mathbf{A}\, op\, \mathbf{B}$. Again, when the operation and the operands are clear from context, we simply use flops.

The most widely used data structures for sparse matrices are the Compressed Sparse Columns (CSC) and Compressed Sparse Rows (CSR). The previous chapter gives concise descriptions of common SpGEMM algorithms operating both on CSC/CSR and triples. The SpGEMM problem was recently reconsidered in [Yuster & Zwick 2005] over a ring, where the authors used a fast dense matrix multiplication such as arithmetic progression [Coppersmith & Winograd 1987] as a subroutine. Their algorithm uses $O(nnz^{0.7} N^{1.2} + N^{2+o(1)})$ arithmetic operations, which is theoretically close to optimal only if we assume that the number of nonzeros in the resulting matrix $\mathbf{C}$ is $\Theta(N^2)$. This assumption rarely holds in reality. Instead, we provide a work-sensitive analysis by expressing the computation complexity of our SpGEMM algorithms in terms of flops.

Practical sparse algorithms have been proposed by different researchers over the years (see, e.g., [Park et al. 1992, Sulatycke & Ghose 1998]) using various data structures. Although they achieve reasonable performance on some classes of matrices, none of these algorithms outperforms the classical sparse matrix matrix multiplication algorithm for general sparse matrices, which was first described by Gustavson [Gustavson 1978] and was used in MATLAB [Gilbert et al. 1992] and CSparse [Davis et al. 2006]. The classical algorithm runs in $O(\text{flops}+nnz+N)$ time.

In Section 14.2, we present a novel algorithm for sequential SpGEMM that is geared toward computing the product of two *hypersparse* matrices. A matrix is hypersparse if the ratio of nonzeros to its dimension is asymptotically 0. The algorithm is used as the sequential building block of our parallel 2D algorithms described in Section 14.3. Our HYPERSPARSE_GEMM algorithm uses a new $O(nnz)$ data structure, called *DCSC* for *doubly compressed sparse columns*, which is explained in Section 14.2.2. The HYPERSPARSE_GEMM is based on the outer product formulation and has time complexity $O(nzc(\mathbf{A})+nzr(\mathbf{B})+\text{flops}\cdot\lg ni)$, where $nzc(\mathbf{A})$ is the number of columns of $\mathbf{A}$ that contain at least one nonzero, $nzr(\mathbf{B})$ is the number of rows of $\mathbf{B}$ that contain at least one nonzero, and ni is the number of indices i for which $\mathbf{A}(:,i) \neq \emptyset$ and $\mathbf{B}(i,:) \neq \emptyset$. The overall space complexity of our algorithm is only $O(nnz(\mathbf{A}) + nnz(\mathbf{B}) + nnz(\mathbf{C}))$. Notice that the time complexity of our algorithm does not depend on N, and the space complexity does not depend on flops.

Section 14.3 presents parallel algorithms for SpGEMM. We propose novel algorithms based on 2D block decomposition of data in addition to giving the complete description of an existing 1D algorithm. To the best of our knowledge, parallel algorithms using a 2D block decomposition have not earlier been developed for sparse matrix matrix multiplication.

Irony, Toledo, and Tiskin [Irony et al. 2004] proved that 2D dense matrix multiplication algorithms are optimal with respect to the communication volume, making 2D sparse algorithms likely to be more scalable than their one-dimensional (1D) counterparts. In Section 14.4, we show that this intuition is indeed correct by providing a theoretical analysis of the parallel performance of 1D and 2D algorithms.

In Section 14.5, we model the speedup of parallel SpGEMM algorithms using realistic simulations and projections. Our results show that existing 1D algorithms are not scalable to thousands of processors. By contrast, 2D algorithms have the potential for scaling up indefinitely, albeit with decreasing parallel efficiency, which is defined as the ratio of speedup to the number of processors.

14.2 Sequential sparse matrix multiply

We first analyze different formulations of sparse matrix matrix multiplication by using the layered-graph model in Section 14.2.1. This graph theoretical explanation gives insights on the suitability of the outer product formulation for multiplying hypersparse matrices (HYPERSPARSE_GEMM). Section 14.2.2 defines hypersparse matrices and Section 14.2.3 introduces the DCSC data structure that is suitable to store hypersparse matrices. We present our HYPERSPARSE_GEMM algorithm in Section 14.2.2.

14.2.1 Layered graphs for different formulations of SpGEMM

Matrix multiplication can be organized in many different ways. The inner product formulation that usually serves as the definition of matrix multiplication is well known. Given two matrices $\mathbf{A} \in \mathbb{R}^{M\times K}$ and $\mathbf{B} \in \mathbb{R}^{K\times N}$, each element in the product $\mathbf{C} \in \mathbb{R}^{M\times N}$ is computed by the following formula:

$$\mathbf{C}(i,j) = \sum_{k=1}^{K} \mathbf{A}(i,k)\mathbf{B}(k,j) \tag{14.1}$$

This formulation is rarely useful for multiplying sparse matrices since it requires $\Omega(M \cdot N)$ operations regardless of the sparsity of the operands.

We represent the multiplication of two matrices $\mathbf{A}$ and $\mathbf{B}$ as a three-layered graph, following Cohen [Cohen 1998]. The layers have M, K, and N vertices, in that order. The first layer of vertices (U) represents the rows of $\mathbf{A}$, and the third layer of vertices (V) represents the columns of $\mathbf{B}$. The second layer of vertices (W) represents the dimension shared between matrices. Every nonzero $\mathbf{A}(i,l) \neq 0$ in the ith row of $\mathbf{A}$ forms an edge (u_i, w_l) between the first and second layers, and every nonzero in $\mathbf{B}(l,j) \neq 0$ in the jth column of $\mathbf{B}$ forms an edge (w_l, v_j) between the second and third layers.

We perform different operations on the layered graph depending on the way we formulate the multiplication. In all cases, though, the goal is to find pairs of vertices (u_i, v_j) sharing an adjacent vertex $w_k \in W$, and if any pair shares multiple adjacent vertices, to merge their contributions.

Using inner products, we analyze each pair (u_i, v_j) to find the set of vertices in $\widetilde{W_{ij}} \subseteq W = \{w_1, w_2, \ldots, w_k\}$ that are connected to both u_i and v_j in the graph shown in Figure 14.1. The algorithm then accumulates contributions $a_{il} \cdot b_{lj}$ for all $w_l \in \widetilde{W_{ij}}$. The result becomes the value of $\mathbf{C}(i,j)$ in the output. In general, this inner product subgraph is sparse, and a contribution from w_l happens only when both edges a_{il} and b_{lj} exist. However, this sparsity is not exploited using inner products as it needs to examine each (u_i, v_j) pair, even when the set $\widetilde{W_{ij}}$ is empty.

In the outer product formulation, the product is written as the summation of k rank-one matrices:

$$\mathbf{C} = \sum_{k=1}^{K} \mathbf{A}(:,k)\mathbf{B}(k,:) \tag{14.2}$$

A different subgraph results from this formulation as it is the set of vertices W that represent the shared dimension that plays the central role. Note that the edges are traversed in the outward direction from a node $w_i \in W$, as shown in Figure 14.2. For sufficiently sparse matrices, this formulation may run faster because this traversal is performed only for the vertices in W (size K) instead of the inner product traversal that had to be performed for every pair (size MN). The problem with outer product traversal is that it is hard to accumulate the intermediate results into the final matrix.

A row-by-row formulation of matrix multiplication performs a traversal starting from each of the vertices in U towards V, as shown in Figure 14.3 for u_i. Each

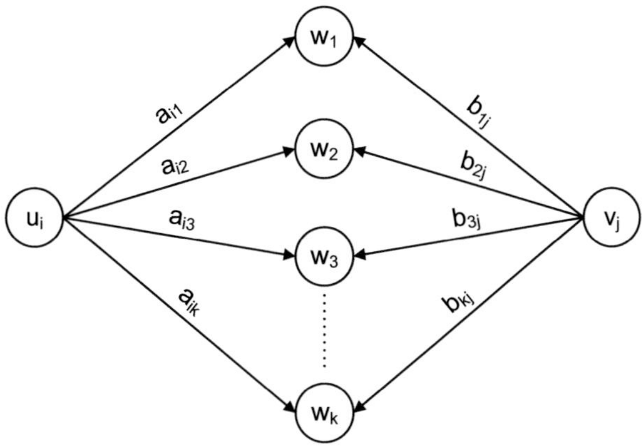

Figure 14.1. Graph representation of the inner product $\mathbf{A}(i,:) \cdot \mathbf{B}(:,j)$.

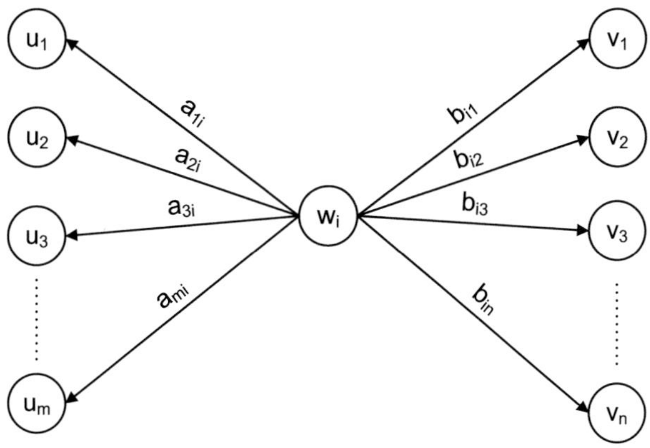

Figure 14.2. Graph representation of the outer product $\mathbf{A}(:,i) \cdot \mathbf{B}(i,:)$.

traversal is independent from each other because they generate different rows of $\mathbf{C}$. Finally, a column-by-column formulation creates an isomorphic traversal, in the reverse direction (from V to U).

14.2.2 Hypersparse matrices

Recall that a matrix is hypersparse if $nnz < N$. Although CSR/CSC is a fairly efficient storage scheme for general sparse matrices having $nnz = \Omega(N)$, it is asymptotically suboptimal for hypersparse matrices. Hypersparse matrices are fairly rare in numerical linear algebra (indeed, a nonsingular square matrix must have $nnz \geq N$), but they occur frequently in computations on graphs, particularly in parallel.

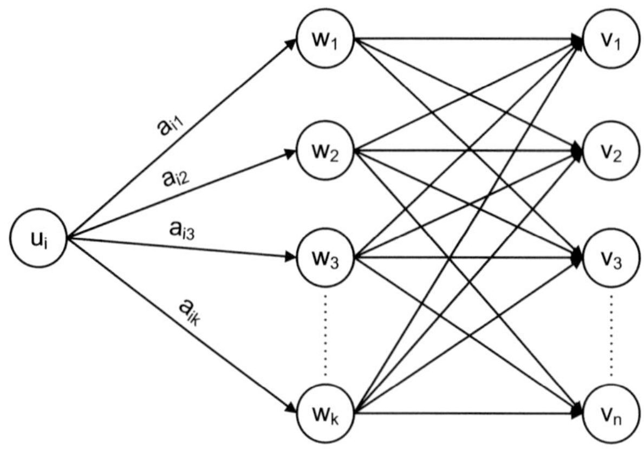

Figure 14.3. Graph representation of the sparse row times matrix product $\mathbf{A}(i,:) \cdot \mathbf{B}$.

Our main motivation for hypersparse matrices comes from parallel processing. Hypersparse matrices arise after the 2D block data decomposition of ordinary sparse matrices for parallel processing. Consider a sparse matrix with c nonzero elements in each column. After the 2D decomposition of the matrix, each processor locally owns a submatrix with dimensions $(N/\sqrt{p}) \times (N/\sqrt{p})$. Storing each of those submatrices in CSC format takes $O(N\sqrt{p} + nnz)$ space, whereas the amount of space needed to store the whole matrix in CSC format on a single processor is only $O(N + nnz)$. As the number of processors increases, the $N\sqrt{p}$ term dominates the nnz term.

Figure 14.4 shows that the average number of nonzeros in a single column of a submatrix, $nnz(j)$, goes to zero as p increases. Storing a graph using CSC is similar to using adjacency lists. The column pointers array represents the vertices, and the row indices array represents their adjacencies. In that sense, CSC is a vertex-based data structure, making it suitable for 1D (vertex) partitioning of the graph. On the other hand, 2D partitioning is based on edges. Therefore, using CSC with 2D distributed data is forcing a vertex-based representation on edge distributed data. The result is unnecessary replication of column pointers (vertices) on each processor along the processor column.

The inefficiency of CSC leads to a more fundamental problem: any algorithm that uses CSC and scans all the columns is not scalable for hypersparse matrices. Even without any communication at all, such an algorithm cannot scale for $n\sqrt{p} \geq max\{\text{flops}, nnz\}$. SpMV and SpGEMM are algorithms that scan column indices. For these operations, any data structure that depends on the matrix dimension (such as CSR or CSC) is asymptotically too wasteful for submatrices.

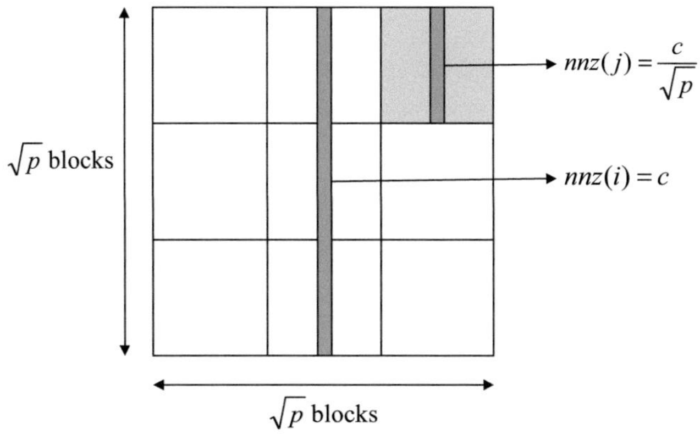

Figure 14.4. 2D sparse matrix decomposition.

14.2.3 DCSC data structure

We use a new data structure for our sequential hypersparse matrix matrix multiplication. This structure, called *DCSC* for doubly compressed sparse columns, has the following properties.

1. It uses $O(nnz)$ storage.

2. It lets the hypersparse algorithm scale with increasing sparsity.

3. It supports fast access to columns of the matrix.

For an example, consider the 9×9 matrix with 4 nonzeros whose triples representation is given in Figure 14.6. Figure 14.5 shows its CSC storage, which includes repetitions and redundancies in the column pointers array (JC). Our new data structure compresses the JC array to avoid repetitions, giving the CP(column pointers) array of DCSC as shown in Figure 14.7. DCSC is essentially a sparse array of sparse columns, whereas CSC is a dense array of sparse columns.

After removing repetitions, CP[i] does no longer refer to the ith column. A new JC array, which is parallel to CP, gives us the column numbers. Although our HYPERSPARSE_GEMM algorithm does not need column indexing, DCSC supports fast column indexing for completeness. Whenever column indexing is needed, we construct an AUX array that contains pointers to nonzero columns (columns that have at least one nonzero element). Each entry in AUX refers to a $\lceil n/\,nzc \rceil$-sized chunk of columns, pointing to the first nonzero column in that chunk (there might be none). The storage requirement of DCSC is $O(nnz)$ since $|\mathsf{NUM}| = |\mathsf{IR}| = nnz$, $|\mathsf{JC}| = nzc$, $|\mathsf{CP}| = nzc + 1$, and $|\mathsf{AUX}| \approx nzc$.

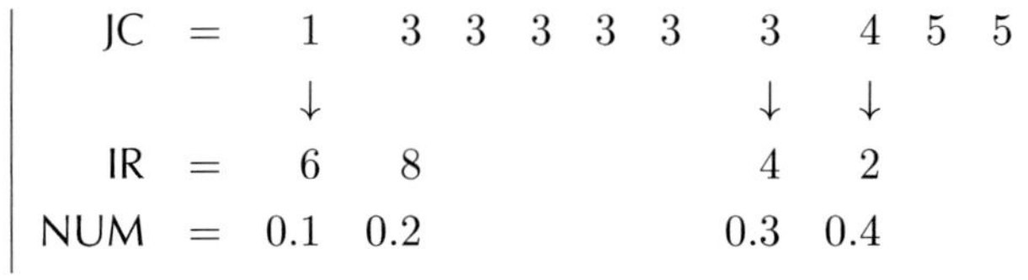

Figure 14.5. Matrix A in CSC format.

A.1	**A**.J	**A**.V
6	1	0.1
8	1	0.2
4	7	0.3
2	8	0.4

Figure 14.6. Matrix A in triples format.

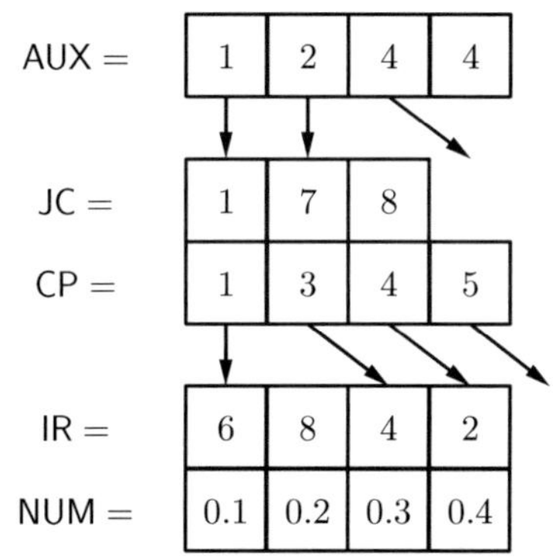

Figure 14.7. Matrix A in DCSC format.

In our implementation, the AUX array is a temporary work array that is constructed on demand, only when an operation requires repetitive use of it. This keeps the storage and copying costs low. The time to construct AUX is only $O(nzc)$, which is subsumed by the cost of multiplication.

14.2.4 A sequential algorithm to multiply hypersparse matrices

The sequential hypersparse algorithm (HYPERSPARSE_GEMM) is based on outer product multiplication. Therefore, it requires fast access to rows of matrix **B**. This could be accomplished by having each input matrix represented in DCSC and also in DCSR (doubly compressed sparse rows), which is the same as the transpose in DCSC. This method, which we described in an early version of this work [Buluç & Gilbert 2008b], doubles the storage but does not change the asymptotic space and time complexities. Here, we describe a more practical version where **B** is transposed as a preprocessing step, at a cost of trans(**B**). The actual cost of transposition is either $O(N + nnz(\mathbf{B}))$ or $O(nnz(\mathbf{B}) \lg nnz(\mathbf{B}))$, depending on the implementation.

The idea behind the HYPERSPARSE_GEMM algorithm is to use the outer product formulation of matrix multiplication efficiently. The first observation about DCSC is that the JC array is already sorted. Therefore, **A**.JC is the sorted indices

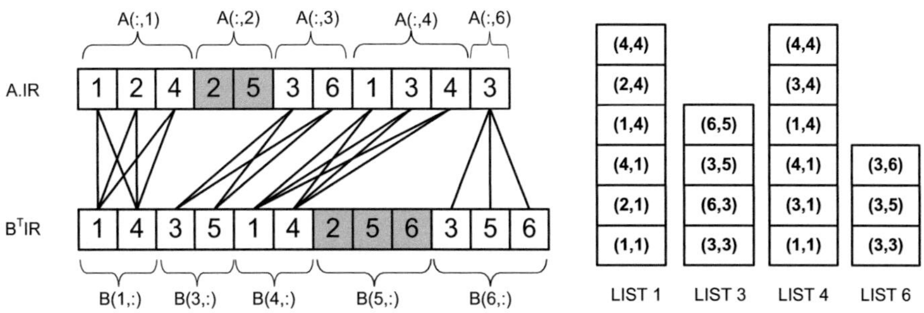

Figure 14.8. Cartesian product and the multiway merging analogy.

$$
\mathbf{A} = \begin{array}{c} \\ 1\\2\\3\\4\\5\\6 \end{array}
\begin{array}{c} \begin{array}{cccccc} 1 & 2 & 3 & 4 & 5 & 6 \end{array} \\
\begin{pmatrix}
\times & & & \times & & \\
\times & \times & & & & \\
 & & \times & \times & & \times \\
\times & & & \times & & \\
 & \times & & & & \\
 & & \times & & &
\end{pmatrix} \end{array},
\quad
\mathbf{B} = \begin{array}{c} \\ 1\\2\\3\\4\\5\\6 \end{array}
\begin{array}{c} \begin{array}{cccccc} 1 & 2 & 3 & 4 & 5 & 6 \end{array} \\
\begin{pmatrix}
\times & & & \times & & \\
 & & & & & \\
 & & \times & & \times & \\
\times & & & \times & & \\
 & \times & & & \times & \times \\
 & & \times & & \times & \times
\end{pmatrix} \end{array}
$$

Figure 14.9. Nonzero structures of operands A and B.

of the columns that contain at least one nonzero, and similarly $\mathbf{B^T}$.JC is the sorted indices of the rows that contain at least one nonzero. In this formulation, the ith column of $\mathbf{A}$ and the ith row of $\mathbf{B}$ are multiplied to form a rank-one matrix. The naive algorithm does the same procedure for all values of i and gets n different rank-one matrices, adding them to the resulting matrix $\mathbf{C}$ as they become available. Our algorithm has a preprocessing step that finds intersection Isect $= \mathbf{A}$.JC $\cap$ $\mathbf{B^T}$.JC, which is the set of indices that participate nontrivially in the outer product.

The preprocessing takes $O(nzc(\mathbf{A}) + nzr(\mathbf{B}))$ time as $|\mathbf{A}.\mathrm{JC}| = nzc(\mathbf{A})$ and $|\mathbf{B^T}.\mathrm{JC}| = nzr(\mathbf{B})$. The next phase of our algorithm performs |Isect| Cartesian products, each of which generates a fictitious list of size $nnz(\mathbf{A}(:,i)) \cdot nnz(\mathbf{B}(i,:))$. The lists can be generated sorted because all the elements within a given column are sorted according to their row indices (i.e., $\mathrm{IR}(\mathrm{JC}(i)) \ldots \mathrm{IR}(\mathrm{JC}(i) + 1)$ is a sorted range). The algorithm merges those sorted lists, summing up the intermediate entries having the same (row_id, col_id) index pair, to form the resulting matrix $\mathbf{C}$. Therefore, the second phase of HYPERSPARSE_GEMM is similar to multiway merging [Knuth 1997]. The only difference is that we never explicitly construct the lists; we compute their elements one by one on demand.

Figure 14.8 shows the setup for the matrices from Figure 14.9. Since $\mathbf{A}.\mathrm{JC} = \{1, 2, 3, 4, 6\}$ and $\mathbf{B^T}.\mathrm{JC} = \{1, 3, 4, 5, 6\}$, Isect $= \{1, 3, 4, 6\}$ for this product. The algorithm does not touch the shaded elements since they do not contribute to the output.

The merge uses a priority queue (represented as a heap) of size ni, which is the size of Isect, the number of indices i for which $\mathbf{A}(:,i) \neq \emptyset$ and $\mathbf{B}(i,:) \neq \emptyset$. The value in a heap entry is its NUM value and the key is a pair of indices (i,j) in column-major order. The idea is to repeatedly extract the entry with minimum key from the heap and insert another element from the list that the extracted element originally came from. If there are multiple elements in the lists with the same key, then their values are added on the fly. If we were to explicitly create ni lists instead of doing the computation on the fly, we would get the lists shown in the right side of Figure 14.8, which are sorted from bottom to top. For further details of multiway merging, consult Knuth [Knuth 1997].

The time complexity of this phase is $O(\text{flops} \cdot \lg ni)$, and the space complexity is $O(nnz(\mathbf{C}) + ni)$. The output is a stack of NUM values in column-major order. The $nnz(\mathbf{C})$ term in the space complexity comes from the output, and the flops term in the time complexity comes from the observation that

$$\sum_{i \in \text{Isect}} nnz(\mathbf{A}(:,i)) \cdot nnz(\mathbf{B}(i,:)) = \text{flops}$$

The final phase of the algorithm constructs the DCSC structure from this column-major ordered stack. This requires $O(nnz(\mathbf{C}))$ time and space.

The overall time complexity of our algorithm is $O(nzc(\mathbf{A}) + nzr(\mathbf{B}) + \text{flops} \cdot \lg ni)$, plus the preprocessing time to transpose matrix $\mathbf{B}$. Note that $nnz(\mathbf{C})$ does not appear in this bound since $nnz(\mathbf{C}) \leq \text{flops}$. We opt to keep the cost of transposition separate because our parallel 2D block SpGEMM will amortize this transposition of each block over $\sqrt{p}$ uses of that block. Therefore, the cost of transposition will be negligible in practice. The space complexity is $O(nnz(\mathbf{A}) + nnz(\mathbf{B}) + nnz(\mathbf{C}))$. The time complexity does not depend on N, and the space complexity does not depend on flops.

Algorithm 14.1 gives the pseudocode for the whole algorithm. It uses two subprocedures: CARTMULT-INSERT generates the next element from the ith fictitious list and inserts it to the heap PQ, and INCREMENT-LIST increments the pointers of the ith fictitious list or deletes the list from the intersection set if it is empty.

To justify the extra logarithmic factor in the flops term, we briefly analyze the complexity of each submatrix multiplication in the parallel 2D block SpGEMM. Our parallel 2D block SpGEMM performs $p\sqrt{p}$ submatrix multiplications since each submatrix of the output is computed using $\mathbf{C}_{ij} = \sum_{k=1}^{\sqrt{p}} \mathbf{A}_{ik}\mathbf{B}_{kj}$. Therefore, with increasing number of processors and under perfect load balance, flops scales with $1/p\sqrt{p}$, nnz scales with $1/p$, and N scales with $1/\sqrt{p}$. Figure 14.10 shows the trends of these three complexity measures as p increases. The graph shows that the N term becomes the bottleneck after around 50 processors and flops becomes the lower-order term. In contrast to the classical algorithm, our HYPERSPARSE_GEMM algorithm becomes independent of N, by putting the burden on the flops instead.

Algorithm 14.1. Hypersparse matrix multiply.
Pseudocode for hypersparse matrix matrix multiplication algorithm.

$\mathbf{C} : \mathbb{R}^{S(M \times N)} = \text{HyperSparseGEMM}(\mathbf{A} : \mathbb{R}^{S(M \times K)}, \mathbf{B}^{\mathbf{T}} : \mathbb{R}^{S(N \times K)})$

```
isect ← Intersection(A.JC, B^T.JC)
for j ← 1 to |isect|
      do CartMult-Insert(A, B^T, PQ, isect, j)
         Increment-List(isect, j)
while IsNotFinished(isect)
      do (key, value) ← Extract-Min(PQ)
         (product, i) ← UnPair(value)
         if key ≠ Top(Q)
            then Enqueue(Q, key, product)
            else UpdateTop(Q, product)
         if IsNotEmpty(isect(i))
            then CartMult-Insert(A, B^T, PQ, lists, isect, i)
                 Increment-List(isect, i)
Construct-Dcsc(Q)
```

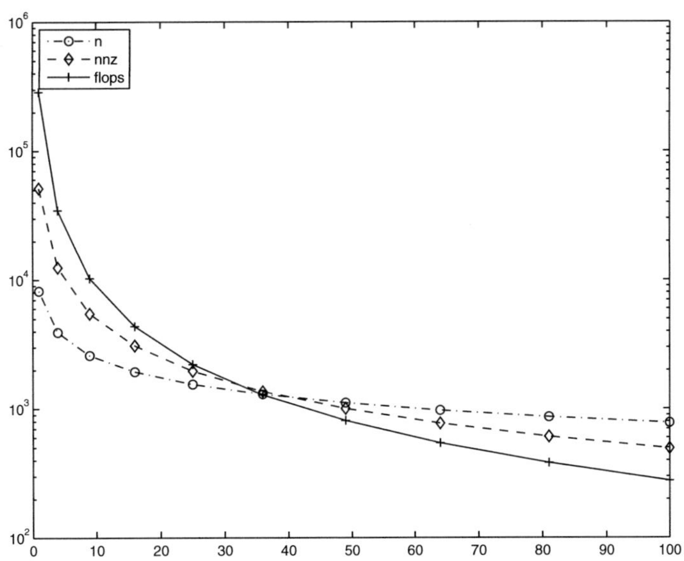

Figure 14.10. Trends of different complexity measures for submatrix multiplications as p increases. The inputs are randomly permuted R-MAT matrices (scale 15 with an average of 8 nonzeros per column) that are successively divided into $(N/\sqrt{p}) \times (N/\sqrt{p})$. The counts are averaged over all submatrix multiplications.

14.3 Parallel algorithms for sparse GEMM

This section describes parallel algorithms for multiplying two sparse matrices in parallel on p processors, which we call PSpGEMM. The design of our algorithms is motivated by distributed memory systems, but we expect them to perform well in shared memory too, as they avoid hot spots and load imbalances by ensuring proper work distribution among processors. Like most message passing algorithms, they can be implemented in the partitioned global address space (PGAS) model as well.

14.3.1 1D decomposition

We assume the data is distributed to processors in block rows, where each processor receives M/p consecutive rows. We write $\mathbf{A}_i = \mathbf{A}(ip : (i+1)p - 1, :)$ to denote the block row owned by the ith processor. To simplify the algorithm description, we use $\mathbf{A}_{ij}$ to denote $\mathbf{A}_i(:, jp : (j+1)p - 1)$, the jth block column of $\mathbf{A}_i$, although block rows are not physically partitioned:

$$\mathbf{A} = \begin{pmatrix} \mathbf{A}_1 \\ \hline \dots \\ \hline \mathbf{A}_p \end{pmatrix} = \left(\begin{array}{c|c|c} \mathbf{A}_{11} & \dots & \mathbf{A}_{1p} \\ \hline \dots & & \dots \\ \hline \mathbf{A}_{p1} & \dots & \mathbf{A}_{pp} \end{array}\right), \mathbf{B} = \begin{pmatrix} \mathbf{B}_1 \\ \hline \dots \\ \hline \mathbf{B}_p \end{pmatrix} \tag{14.3}$$

For each processor $P(i)$, the computation is

$$\mathbf{C}_i = \mathbf{C}_i + \mathbf{A}_i\,\mathbf{B} = \mathbf{C}_i = \mathbf{C}_i + \sum_{j=1}^{p} \mathbf{A}_{ij}\,\mathbf{B}_j$$

14.3.2 2D decomposition

Our 2D parallel algorithms, Sparse Cannon and Sparse SUMMA, use the hypersparse algorithm, which has complexity $O(nzc(\mathbf{A}) + nzr(\mathbf{B}) + \text{flops} \cdot \lg ni)$, as shown in Section 14.2.2, for multiplying submatrices. Processors are logically organized on a square $\sqrt{p} \times \sqrt{p}$ mesh, indexed by their row and column indices so that the (i,j)th processor is denoted by $P(i,j)$. Matrices are assigned to processors according to a 2D block decomposition. Each node gets a submatrix of dimensions $(N/\sqrt{p}) \times (N/\sqrt{p})$ in its local memory. For example, $\mathbf{A}$ is partitioned as shown below and $\mathbf{A}_{ij}$ is assigned to processor $P(i,j)$:

$$\mathbf{A} = \left(\begin{array}{c|c|c} \mathbf{A}_{11} & \dots & \mathbf{A}_{1\sqrt{p}} \\ \hline \dots & & \dots \\ \hline \mathbf{A}_{\sqrt{p}1} & \dots & \mathbf{A}_{\sqrt{p}\sqrt{p}} \end{array}\right) \tag{14.4}$$

For each processor $P(i)$, the computation is

$$\mathbf{C}_{ij} = \sum_{k=1}^{\sqrt{p}} \mathbf{A}_{ik}\,\mathbf{B}_{kj}$$

14.3.3 Sparse 1D algorithm

The row wise SpGEMM forms one row of $\mathbf{C}$ at a time, and each processor may potentially need to access all of $\mathbf{B}$ to form a single row of $\mathbf{C}$. However, only a portion of $\mathbf{B}$ is locally available at any time in parallel algorithms. The algorithm thus performs multiple iterations to fully form one row of $\mathbf{C}$. We use a SPA to accumulate the nonzeros of the current active row of $\mathbf{C}$. Algorithm 14.2 shows the pseudocode of the algorithm. Loads and unloads of SPA, which is not amortized by the number of nonzero arithmetic operations in general, dominate the computational time.

Algorithm 14.2. Matrix matrix multiply.
Operation $\mathbf{C} \leftarrow \mathbf{AB}$ using block row Sparse1D.

$\mathbf{C} : \mathbb{R}^{P(S(N)\times N)} = \text{BLOCK1D-PSPGEMM}(\mathbf{A} : \mathbb{R}^{P(S(N)\times N)}, \mathbf{B} : \mathbb{R}^{P(S(N)\times N)})$

for all processors $P(i)$ in parallel
    do INITIALIZE(SPA)
        for $j \leftarrow 1$ **to** p
            do BROADCAST($\mathbf{B}_j$)
                for $k \leftarrow 1$ **to** N/p
                    do LOAD(SPA, $\mathbf{C}_i(k,:)$)
                        SPA $\leftarrow$ SPA $+ \mathbf{A}_{ij}(k,:)\,\mathbf{B}_j$
                        UNLOAD(SPA, $\mathbf{C}_i(k,:)$)

14.3.4 Sparse Cannon

Our first 2D algorithm is based on Cannon's algorithm for dense matrices (see [Cannon 1969]). The pseudocode of the algorithm is given in Algorithm 14.5. Sparse Cannon, although elegant, is not our choice of algorithm for the final implementation, as it is hard to generalize to nonsquare grids, nonsquare matrices, and matrices whose dimensions are not perfectly divisible by grid dimensions.

Algorithm 14.3. Circular shift left.
Circularly shift left by s along the processor row.

LEFT-CIRCULAR-SHIFT(**Local** : $\mathbb{R}^{S(N\times N)}, s$)

SEND(**Local**, $P(i, (j-s) \bmod \sqrt{p})$) $\triangleright$ This is processor $P(i,j)$
RECEIVE(**Temp**, $P(i, (j+s) \bmod \sqrt{p})$)
Local $\leftarrow$ **Temp**

Algorithm 14.4. Circular shift up.
Circularly shift up by s along the processor column.

UP-CIRCULAR-SHIFT(**Local** : $\mathbb{R}^{S(N\times N)}, s$)

SEND(**Local**, $P((i-s) \bmod \sqrt{p}, j)$) $\triangleright$ This is processor $P(i,j)$
RECEIVE(**Temp**, $P((i+s) \bmod \sqrt{p}, j)$)
Local $\leftarrow$ **Temp**

Algorithm 14.5. Cannon matrix multiply.
Operation $\mathbf{C} \leftarrow \mathbf{AB}$ using Sparse Cannon.

$\mathbf{C} : \mathbb{R}^{P(S(N\times N))} = \text{CANNON-PSPGEMM}(\mathbf{A} : \mathbb{R}^{P(S(N\times N))}, \mathbf{B} : \mathbb{R}^{P(S(N\times N))})$

for all processors $P(i, j)$ in parallel
  do LEFT-CIRCULAR-SHIFT($\mathbf{A}_{ij}, i - 1$)
    UP-CIRCULAR-SHIFT($\mathbf{B}_{ij}, j - 1$)
for all processors P_{ij} in parallel
  do for $k \leftarrow 1$ **to** $\sqrt{p}$
    do $\mathbf{C}_{ij} \leftarrow \mathbf{C}_{ij} + \mathbf{A}_{ij}\,\mathbf{B}_{ij}$
      LEFT-CIRCULAR-SHIFT($\mathbf{A}_{ij}, 1$)
      UP-CIRCULAR-SHIFT($\mathbf{B}_{ij}, 1$)

14.3.5 Sparse SUMMA

SUMMA [Van De Geijn & Watts 1997] is a memory efficient, easy to generalize algorithm for parallel dense matrix multiplication. It is the algorithm used in parallel BLAS (see [Chtchelkanova et al. 1997]). As opposed to Cannon's algorithm, it allows a tradeoff to be made between latency cost and memory by varying the degree of blocking. The algorithm, illustrated in Figure 14.11, proceeds in k/b stages. At each stage, $\sqrt{p}$ active row processors broadcast b columns of $\mathbf{A}$ simultaneously along their rows and $\sqrt{p}$ active column processors broadcast b rows of $\mathbf{B}$ simultaneously along their columns.

Sparse SUMMA is our algorithm of choice for our final implementation because it is easy to generalize to nonsquare matrices, matrices whose dimensions are not perfectly divisible by grid dimensions.

14.4 Analysis of parallel algorithms

In this section, we analyze the parallel performance of our algorithms and show that they scale better than existing 1D algorithms in theory. We begin by introducing

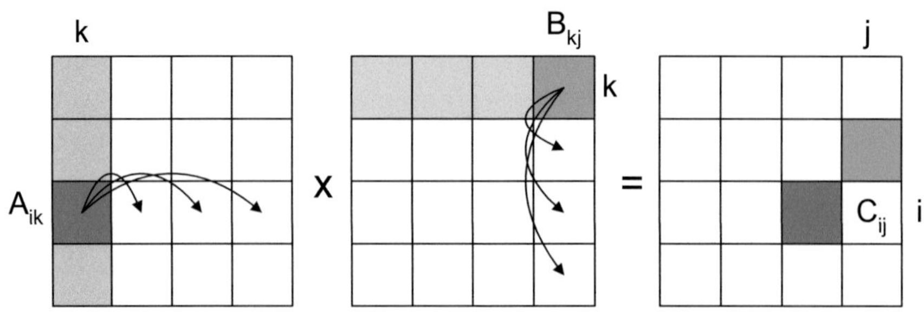

Figure 14.11. Sparse SUMMA execution ($b = N/\sqrt{p}$).

our parameters and model of computation. Then, we present a theoretical analysis showing that 1D decomposition, at least with the current algorithm, is not sufficient for PSpGEMM to scale. Finally, we analyze our 2D algorithms in depth.

In our analysis, the cost of one floating-point operation, along with the cost of cache misses and memory indirections associated with the operation, is denoted by γ, measured in nanoseconds. The latency of sending a message over the communication interconnect is α, and the inverse bandwidth is β, measured in nanoseconds and nanoseconds per word transferred, respectively. The running time of a parallel algorithm on p processors is given by

$$T_p = T_{comm} + T_{comp}$$

where T_{comm} denotes the time spent in communication and T_{comp} is the time spent during local computation phases. T_{comm} includes both the latency (delay) costs and the actual time it takes to transfer the data words over the network. Hence, the cost of transmitting h data words in a communication phase is

$$T_{comm} = \alpha + h\beta$$

The sequential work of SpGEMM, unlike dense GEMM, depends on many parameters. This makes parallel scalability analysis a tough process. Therefore, we restrict our analysis to sparse matrices following the Erdős–Rényi graph model. Consequently, the analysis is probabilistic, exploiting the independent and identical distribution of nonzeros. When we talk about quantities such as nonzeros per subcolumn, we mean the expected number of nonzeros. Our analysis assumes that there are $c > 0$ nonzeros per row/column. The sparsity parameter c, albeit oversimplifying, is useful for analysis purposes since it makes different parameters comparable to each other. For example, if $\mathbf{A}$ and $\mathbf{B}$ both have sparsity c, then $nnz(\mathbf{A}) = cN$ and $\text{flops}(\mathbf{AB}) = c^2N$. It also allows us to decouple the effects of load imbalances from the algorithm analysis because the nonzeros are assumed to be evenly distributed across processors.

The lower bound on sequential SpGEMM is $\Omega(\text{flops}) = \Omega(c^2N)$. This bound is achieved by some row wise and column wise implementations (see [Gilbert et al. 1992, Gustavson 1978]), provided that $c \geq 1$. The row wise implementation of Gustavson that uses CSR is the natural kernel to be used in the 1D algorithm where data is distributed by rows. As shown in the previous chapter, it has an asymptotic complexity of

$$O(N + nnz(\mathbf{A}) + \text{flops}) = O(N + cN + c^2N) = \Theta(c^2N)$$

Therefore, we take the sequential work (W) to be $\gamma c^2 N$ in our analysis.

14.4.1 Scalability of the 1D algorithm

We begin with a theoretical analysis whose conclusion is that 1D decomposition is not sufficient for PSpGEMM to scale. In BLOCK1D_PSPGEMM, each processor sends and receives $p-1$ point-to-point messages of size $nnz(\mathbf{B})/p$. Therefore,

$$T_{comm} = (p-1)\left(\alpha + \beta\,\frac{nnz(\mathbf{B})}{p}\right) = \Theta(p\,\alpha + \beta\,c\,N) \tag{14.5}$$

We previously showed that the BLOCK1D_PSPGEMM algorithm is unscalable with respect to both communication and computation costs (see, for instance, [Buluç & Gilbert 2008a]). In fact, it gets slower as the number of processors grows. The current STAR-P implementation (see [Shah 2007]) bypasses this problem by all-to-all broadcasting nonzeros of the $\mathbf{B}$ matrix, so that the whole $\mathbf{B}$ matrix is essentially assembled at each processor. This avoids the cost of loading and unloading SPA at every stage, but it uses $nnz(\mathbf{B})$ memory at each processor.

14.4.2 Scalability of the 2D algorithms

In this section, we provide an in-depth theoretical analysis of our parallel 2D SpGEMM algorithms and conclude that they scale significantly better than their 1D counterparts. Although our analysis is limited to the Erdős–Rényi model, its conclusions are strong enough to be convincing.

In CANNON_PSPGEMM, each processor sends and receives $\sqrt{p}-1$ point-to-point messages of size $nnz(\mathbf{A})/p$, and $\sqrt{p}-1$ messages of size $nnz(\mathbf{B})/p$. Therefore, the communication cost per processor is

$$T_{comm} = \sqrt{p}\left(2\,\alpha + \beta\left(\frac{nnz(\mathbf{A}) + nnz(\mathbf{B})}{p}\right)\right) = \Theta\left(\alpha\sqrt{p} + \frac{\beta\, c\, N}{\sqrt{p}}\right) \tag{14.6}$$

The average number of nonzeros in a column of a local submatrix $\mathbf{A}_{ij}$ is $c/\sqrt{p}$. Therefore, for a submatrix multiplication $\mathbf{A}_{ik}\mathbf{B}_{kj}$,

$$ni(\mathbf{A}_{ik}, \mathbf{B}_{kj}) = \min\left\{1, \frac{c^2}{p}\right\}\frac{N}{\sqrt{p}} = \min\left\{\frac{N}{\sqrt{p}}, \frac{c^2\, N}{p\sqrt{p}}\right\}$$

$$\text{flops}(\mathbf{A}_{ik}\mathbf{B}_{kj}) = \frac{\text{flops}(\mathbf{AB})}{p\sqrt{p}} = \frac{c^2\, N}{p\sqrt{p}}$$

$$T_{mult} = \sqrt{p}\left(2\,\min\left\{1, \frac{c}{\sqrt{p}}\right\}\frac{N}{\sqrt{p}} + \frac{c^2 N}{p\sqrt{p}}\lg\left(\min\left\{\frac{N}{\sqrt{p}}, \frac{c^2 N}{p\sqrt{p}}\right\}\right)\right)$$

The probability of a single column of $\mathbf{A}_{ik}$ (or a single row of $\mathbf{B}_{kj}$) having at least one nonzero is $\min\{1, c/\sqrt{p}\}$, where 1 covers the case $p \le c^2$ and $c/\sqrt{p}$ covers the case $p > c^2$.

The overall cost of additions, using p processors and Brown and Tarjan's $O(m\,\lg n/m)$ algorithm [Brown & Tarjan 1979] for merging two sorted lists of size m and n (for $m < n$), is

$$T_{add} = \sum_{i=1}^{\sqrt{p}}\left(\frac{\text{flops}}{p\sqrt{p}}\,\lg i\right) = \frac{\text{flops}}{p\sqrt{p}}\,\lg\prod_{i=1}^{\sqrt{p}} i = \frac{\text{flops}}{p\sqrt{p}}\,\lg\left(\sqrt{p}!\right)$$

Note that we might be slightly overestimating since we assume flops/ $nnz(\mathbf{C}) \approx 1$ for simplicity. From Stirling's approximation and asymptotic analysis, we know that $\lg(n!) = \Theta(n \lg n)$ [Cormen et al. 2001]. Thus, we get

$$T_{add} = \Theta\left(\frac{\text{flops}}{p\sqrt{p}}\sqrt{p}\lg\sqrt{p}\right) = \Theta\left(\frac{c^2 N \lg\sqrt{p}}{p}\right)$$

There are two cases to analyze: $p > c^2$ and $p \leq c^2$. Since scalability analysis is concerned with the asymptotic behavior as p increases, we just provide results for the $p > c^2$ case. The total computation cost $T_{comp} = T_{mult} + T_{add}$ is

$$T_{comp} = \gamma\left(\frac{cn}{\sqrt{p}} + \frac{c^2 n}{p}\lg\left(\frac{c^2 n}{p\sqrt{p}}\right) + \frac{c^2 n \lg\sqrt{p}}{p}\right) = \gamma\left(\frac{cn}{\sqrt{p}} + \frac{c^2 n}{p}\lg\left(\frac{c^2 n}{p}\right)\right) \quad (14.7)$$

In this case, parallel efficiency is

$$E = \frac{W}{p\left(T_{comp} + T_{comm}\right)} = \frac{\gamma c^2 n}{(\gamma+\beta)\, c n\sqrt{p} + \gamma c^2 n \lg\left(\frac{c^2 n}{p}\right) + \alpha\, p\sqrt{p}} \quad (14.8)$$

Scalability is not perfect and efficiency deteriorates as p increases due to the first term. Speedup is, however, not bounded, as opposed to the 1D case. In particular, $\lg(c^2 n/p)$ becomes negligible as p increases, and scalability due to latency is achieved when $\gamma c^2 n \propto \alpha\, p\sqrt{p}$, where it is sufficient for n to grow on the order of $p^{1.5}$. The biggest bottleneck for scalability is the first term in the denominator, which scales with $\sqrt{p}$. Consequently, two different scaling regimes are likely to be present: a close to linear scaling regime until the first term starts to dominate the denominator and a $\sqrt{p}$-scaling regime afterwards.

Compared to the 1D algorithms, Sparse Cannon both lowers the degree of unscalability due to bandwidth costs and mitigates the bottleneck of computation. This makes overlapping communication with computation more promising.

Sparse SUMMA, like dense SUMMA, incurs an extra cost over Cannon for using row wise and column wise broadcasts instead of nearest-neighbor communication, which might be modeled as an additional $O(\lg p)$ factor in communication cost. Other than that, the analysis is similar to Sparse Cannon and we omit the details. Using the DCSC data structure, the expected cost of fetching b consecutive columns of a matrix $\mathbf{A}$ is b plus the size (number of nonzeros) of the output [Buluç & Gilbert 2008b]. Therefore, the algorithm asymptotically has the same computation cost for all values of b.

14.5 Performance modeling of parallel algorithms

In this section, we project the estimated speedup of 1D and 2D algorithms in order to evaluate their prospects in practice. We use a quasi-analytical performance model in which we first obtain realistic values for the parameters (γ, β, α) of the algorithm performance, then use them in our projections.

In order to obtain a realistic value for γ, we performed multiple runs on an AMD Opteron 8214 (Santa Rosa) processor using matrices of various dimensions

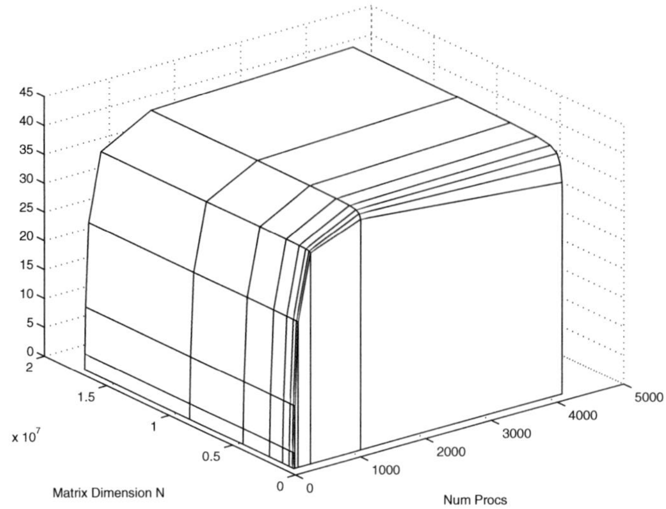

Figure 14.12. Modeled speedup of synchronous sparse 1D algorithm.

and sparsity, estimating the constants using nonlinear regression. One surprising result is the order of magnitude difference in the constants between sequential kernels. The classical algorithm, which is used as the 1D SpGEMM kernel, has $\gamma = 293.6$ nsec, whereas HYPERSPARSE_GEMM, which is used as the 2D kernel, has $\gamma = 19.2$ nsec. We attribute the difference to cache friendliness of the hypersparse algorithm. The interconnect supports $1/\beta = 1$ GB/sec point-to-point bandwidth and a maximum of $\alpha = 2.3$ microseconds latency, both of which are achievable on TACC's Ranger Cluster. The communication parameters ignore network contention.

Figures 14.12 and 14.13 show the modeled speedup of BLOCK1D_PSPGEMM and CANNON_PSPGEMM for matrix dimensions from $N = 2^{17}$ to 2^{24} and number of processors from $p = 1$ to 4096. The inputs are Erdős–Rényi graphs.

We see that BLOCK1D_PSPGEMM's speedup does not go beyond 50x, even on larger matrices. For relatively small matrices, having dimensions $N = 2^{17} - 2^{20}$, it starts slowing down after a thousand processors, where it achieves less than 40x speedup. On the other hand, CANNON_PSPGEMM shows increasing and almost linear speedup for up to 4096 processors even though the slope of the curve is less than one. It is crucial to note that the projections for the 1D algorithm are based on the memory inefficient implementation that performs an all-to-all broadcast of **B**. This is because the original memory efficient algorithm given in Section 14.3.1 actually slows down as p increases.

It is worth explaining one peculiarity. The modeled speedup turns out to be higher for smaller matrices than for bigger matrices. Remember that communication requirements are on the same order as computational requirements for parallel SpGEMM. Intuitively, the speedup should be independent of the matrix dimension in the absence of load imbalance and network contention, but since we are estimating the speedup with respect to the optimal sequential algorithm, the

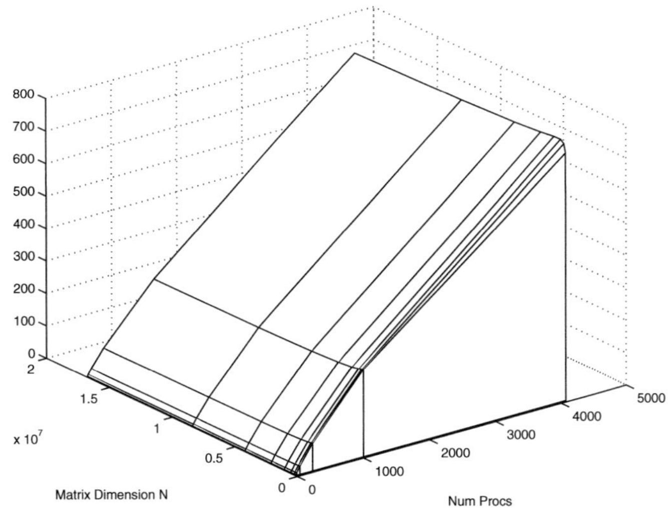

Figure 14.13. Modeled speedup of synchronous Sparse Cannon.

overheads associated with the hypersparse algorithm are bigger for larger matrices. The bigger the matrix dimension, the slower the hypersparse algorithm is with respect to the optimal algorithm, due to the extra logarithmic factor. Therefore, speedup is better for smaller matrices in theory. This is not the case in practice because the peak bandwidth is usually not achieved for small-sized data transfers and load imbalances are severer for smaller matrices.

We also evaluate the effects of overlapping communication with computation. Following Krishnan and Nieplocha [Krishnan & Nieplocha 2004], we define the nonoverlapped percentage of communication as

$$w = 1 - \frac{T_{comp}}{T_{comm}} = \frac{T_{comm} - T_{comp}}{T_{comm}}$$

The speedup of the asynchronous implementation is

$$S = \frac{W}{T_{comp} + w(T_{comm})}$$

Figure 14.14 shows the modeled speedup of asynchronous SpCannon assuming truly one-sided communication. For smaller matrices with dimensions $N = 2^{17}$–2^{20}, speedup is about 25% more than the speedup of the synchronous implementation.

The modeled speedup plots should be interpreted as upper bounds on the speedup that can be achieved on a real system using these algorithms. Achieving these speedups on real systems requires all components to be implemented and working optimally. The conclusion we derive from those plots is that no matter how hard we try, it is impossible to get good speedup with the current 1D algorithms.

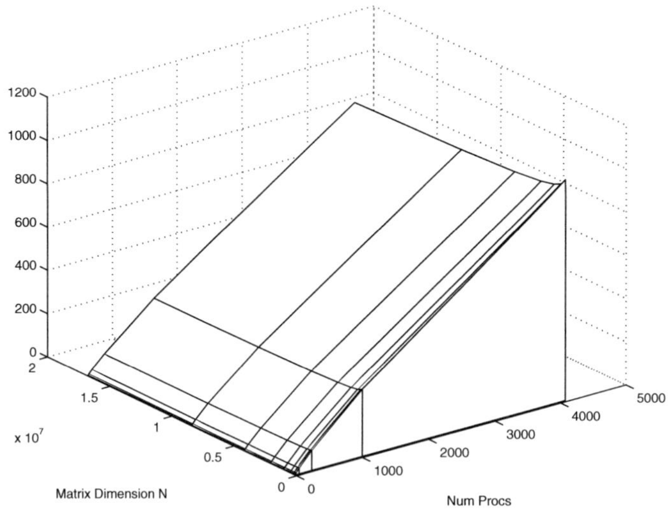

Figure 14.14. Modeled speedup of asynchronous Sparse Cannon.

References

[Aho et al. 1974] A.V. Aho, J.E. Hopcroft, and J.D. Ullman. *The Design and Analysis of Computer Algorithms.* Boston: Addison–Wesley Longman, 1974.

[Bader et al. 2004] D. Bader, J. Feo, J. Gilbert, J. Kepner, D. Koester, E. Loh, K. Madduri, B. Mann, and T. Meuse. HPCS scalable synthetic compact applications #2. version 1.1.

[Briggs et al. 2000] W.L. Briggs, V.E. Henson, and S.F. McCormick. *A multigrid tutorial, Second Edition.* Philadelphia: SIAM, 2000.

[Brown & Tarjan 1979] M.R. Brown and R.E. Tarjan. A fast merging algorithm. *Journal of the ACM*, 26:211–226, 1979.

[Buluç & Gilbert 2008a] A. Buluç and J.R. Gilbert. Challenges and advances in parallel sparse matrix-matrix multiplication. In *The 37th International Conference on Parallel Processing (ICPP '08)*, 503–510, Portland, Oregon, 2008.

[Buluç & Gilbert 2008b] A. Buluç and J.R. Gilbert. On the representation and multiplication of hypersparse matrices. In *IEEE International Parallel and Distributed Processing Symposium (IPDPS 2008)*, 1530–2075, Miami, FL, 2008.

[Cannon 1969] L.E. Cannon. *A cellular computer to implement the Kalman filter algorithm.* Ph.D. thesis, Montana State University, 1969.

[Chtchelkanova et al. 1997] A. Chtchelkanova, J. Gunnels, G. Morrow, J. Overfelt, and R.A. van de Geijn. Parallel implementation of BLAS: General techniques for Level 3 BLAS. *Concurrency: Practice and Experience*, 9:837–857, 1997.

[Cohen 1998] E. Cohen. Structure prediction and computation of sparse matrix products. *Journal of Combinatorial Optimization*, 2:307–332, 1998.

[Coppersmith & Winograd 1987] D. Coppersmith and S. Winograd. Matrix multiplication via arithmetic progressions. In *STOC '87: Proceedings of the 19th Annual ACM Conference on Theory of Computing*, 1–6, New York: ACM Press.

[Cormen et al. 2001] T.H. Cormen, C.E. Leiserson, R.L. Rivest, and C. Stein. *Introduction to Algorithms*, Second Edition. Cambridge, MA: The MIT Press, 2001.

[D'Alberto & Nicolau 2007] P. D'Alberto and A. Nicolau. R-Kleene: A high-performance divide-and-conquer algorithm for the all-pair shortest path for densely connected networks. *Algorithmica*, 47:203–213, 2007.

[Davis et al. 2006] T.A. Davis. *Direct Methods for Sparse Linear Systems*. Philadelphia: SIAM, 2006.

[Gilbert et al. 1992] J.R. Gilbert, C. Moler, and R. Schreiber. Sparse matrices in MATLAB: Design and implementation. *SIAM Journal on Matrix Analysis and Applications*, 13:333–356, 1992.

[Gilbert et al. 2008] J.R. Gilbert, S. Reinhardt, and V.B. Shah. A unified framework for numerical and combinatorial computing. *Computing in Science and Engineering*, 10:20–25, 2008.

[Gustavson 1978] F.G. Gustavson. Two fast algorithms for sparse matrices: multiplication and permuted transposition. *ACM Transactions on Mathematical Software*, 4:250–269, 1978.

[Irony et al. 2004] D. Irony, S. Toledo, and A. Tiskin. Communication lower bounds for distributed-memory matrix multiplication. *Journal of Parallel and Distributed Computing*, 64:1017–1026, 2004.

[Irwin et al. 1997] J. Irwin, J.-M. Loingtier, J.R. Gilbert, G. Kiczales, J. Lamping, A. Mendhekar, and T. Shpeisman. Aspect-oriented programming of sparse matrix code. In *ISCOPE '97: Proceedings of the Scientific Computing in Object-Oriented Parallel Environments*, 249–256, London: Springer-Verlag, 1997.

[Kleinberg et al. 1979] J.M. Kleinberg, R. Kumar, P. Raghavan, S. Rajagopalan, and A. Tomkins. The web as a graph: Measurements, models, and methods. In *Proceedings of the 5th Annual Conference on Computing and Combinatorics (COCOON '99)*, 1–17, 1999.

[Knuth 1997] D.E. Knuth. *The art of computer programming, volume 1 (3rd ed.): Fundamental algorithms*. Redwood City, CA: Addison–Wesley Longman, 1997.

[Krishnan & Nieplocha 2004] M. Krishnan and J. Nieplocha. SRUMMA: A matrix multiplication algorithm suitable for clusters and scalable shared memory systems. In *Proceedings of 18th International Parallel and Distributed Processing Symposium*, 70, IEEE Computer Society, 2004.

[Leskovec et al. 2005] J. Leskovec, D. Chakrabarti, J.M. Kleinberg, and C. Faloutsos. Realistic, mathematically tractable graph generation and evolution, using Kronecker multiplication. In *Proceedings of the 9th European Conference on Principles and Practice of Knowledge Discovery in Databases (PKDD 2005)*, 133–145, 2005.

[Lumsdaine et al. 2007] A. Lumsdaine, D. Gregor, B. Hendrickson, and J. Berry. Challenges in parallel graph processing. *Parallel Processing Letters*, 17:5–20, 2007.

[Maggs & Poltkin 1988] B.M. Maggs and S.A. Poltkin. Minimum-cost spanning tree as a path-finding problem. *Information Processing Letters*, 26:291–293, 1988.

[Park et al. 1992] S.C. Park, J.P. Draayer, and S.-Q. Zheng. Fast sparse matrix multiplication. *Computer Physics Communications*, 70:557–568, 1992.

[Penn 2006] G. Penn. Efficient transitive closure of sparse matrices over closed semirings. *Theoretical Computer Science*, 354:72–81, 2006.

[Rabin & Vazirani 1989] M.O. Rabin and V.V. Vazirani. Maximum matchings in general graphs through randomization. *Journal of Algorithms*, 10:557–567, 1989.

[Sanders 2000] P. Sanders. Fast priority queues for cached memory. *Journal of Experimental Algorithmics*, 5:7, 2000.

[Seidel 1995] R. Seidel. On the all-pairs-shortest-path problem in unweighted undirected graphs. *Journal of Computer and System Sciences*, 51:400–403, 1995.

[Shah 2007] V.B. Shah. *An Interactive System for Combinatorial Scientific Computing with an Emphasis on Programmer Productivity*. Ph.D. thesis, University of California, Santa Barbara, June 2007.

[Sleator & Tarjan 1985] D.D. Sleator and R.E. Tarjan. Self-adjusting binary search trees. *Journal of the ACM*, 32:652–686, 1985.

[Sulatycke & Ghose 1998] P. Sulatycke and K. Ghose. Caching-efficient multithreaded fast multiplication of sparse matrices. In *Proceedings of the First Merged International Parallel Processing Symposium and Symposium on Parallel and Distributed Processing (IPPS '98)*, 117–123, IEEE Computer Society, 1998.

[Tarjan 1981] R.E. Tarjan. A unified approach to path problems. *Journal of the ACM*, 28:577–593, 1981.

[Van De Geijn & Watts 1997] R.A. Van De Geijn and J. Watts. SUMMA: Scalable universal matrix multiplication algorithm. *Concurrency: Practice and Experience*, 9:255–274, 1997.

[Yoo et al. 2005] A. Yoo, E. Chow, K. Henderson, W. McLendon, B. Hendrickson, and U. Catalyurek. A scalable distributed parallel breadth-first search algorithm on bluegene/L. In *Proceedings of the 2005 ACM/IEEE Conference on Supercomputing*, 25, IEEE Computer Society, 2005.

[Yuster & Zwick 2004] R. Yuster and U. Zwick. Detecting short directed cycles using rectangular matrix multiplication and dynamic programming. In *Proceedings of the 15th Annual ACM-SIAM Symposium on Discrete Algorithms*, 254–260, Philadelphia: SIAM, 2004.

[Yuster & Zwick 2005] R. Yuster and U. Zwick. Fast sparse matrix multiplication. *ACM Transactions on Algorithms*, 1:2–13, 2005.

Chapter 15
Parallel Mapping of Sparse Computations

*Eric Robinson**, *Nadya Bliss**, *and Sanjeev Mohindra**

Abstract

Expressing graph algorithms in the language of linear algebra aids strongly in automatically mapping algorithms onto parallel architectures. Matrices extend naturally to parallel mapping schemes. They allow for schemes not strictly based around the source or destination vertex of the edge, but rather the pairwise combination of the two. The problem of mapping over sparse data, especially data distributed according to a power law, is difficult. Automated techniques are best for achieving optimal parallel throughput.

15.1 Introduction

Previous chapters have defined many common data layouts, or maps, used for parallel arrays. These maps are most appropriate for processing dense data. Sparse data, especially sparse power law data, requires more complex data mapping techniques.

For sparse data, the maps yielding high performance are intricate and nonintuitive, making them next to impossible for a human to discover. Figure 15.1 shows the scaling as the number of processors increases for an edge betweenness

*MIT Lincoln Laboratory, 244 Wood Street, Lexington, MA 02420 (erobinson@ll.mit.edu, nt@ll.mit.edu, smohindra@ll.mit.edu).

This work is sponsored by the Department of the Air Force under Air Force Contract FA8721-05-C-0002. Opinions, interpretations, conclusions, and recommendations are those of the author and are not necessarily endorsed by the United States Government.

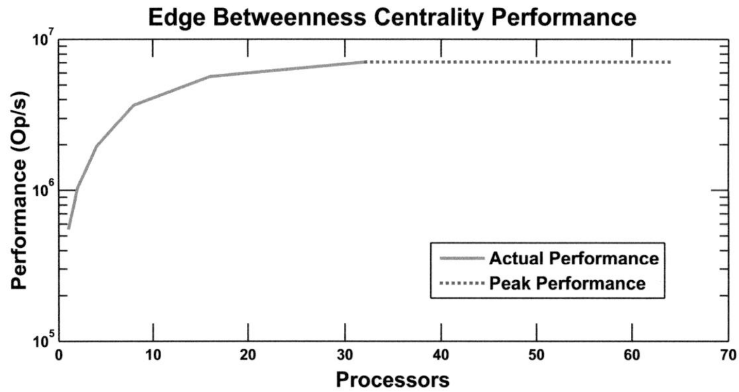

Figure 15.1. Performance scaling.
Scaling of edge betweenness centrality algorithm on a cluster. The algorithm's peak performance is obtained after using only 32 machines.

centrality algorithm that uses a sparse adjacency matrix. The maps in this case were manually chosen on the basis of algorithm analysis. While the maps do yield initial performance benefits, this quickly flattens out after around 8 to 16 processors are added. The actual performance falls well short of the desired performance for the algorithm.

While it is impossible to find an optimal data layout for many sparse problems in a tractable amount of time, a good layout can commonly be found using automated search techniques. The computational cost of finding an ideal map for a particular set of inputs to a program is typically high. In order for data mapping to be beneficial, the reward for finding a good map must be correspondingly high. In general, the benefit of data mapping can be seen for two different types of problems:

- problems where at least one input is used repeatedly, and
- problems where input content may vary, but data layout remains constant.

In both of these cases, the cost of mapping is amortized over multiple calls to the mapped program. This, in many cases, allows the performance gains to outweigh the upfront cost for optimization.

Automated mapping can be done in a variety of ways. Oftentimes the process is not completely automated, but may have a human in the loop to point the mapping program in the correct direction. The rest of this chapter presents a detailed description of one software approach to this problem.

15.2 Lincoln Laboratory mapping and optimization environment

Linear algebra operations have long been in use in front-end processing for sensor processing applications. There is an increasing demand for linear algebra in back-end processing, where graph applications dominate. Many graph algorithms have linear algebraic equivalents using a sparse adjacency matrix. These algorithms, such as breadth-first search, clustering, and centrality, form the backbone of many typical back-end applications. By leveraging this representation, it becomes possible to process both front- and back-end data in the same manner, using algebraic operations on matrices.

With growing demands on the performance of sensor applications, parallel implementations of these algorithms become necessary. For front-end processing, which typically operates on dense matrices, this yields large rewards for very little effort. Parallel strategies for dense matrix operations are well studied and yield good performance results. However, there is no clear method for efficiently parallelizing the sparse matrix operations typical to back-end processing.

The difference between sparse and dense algorithm efficiency arises due to differences in their data-access patterns. While dense matrix algorithms naturally access large continuous chunks of data, allowing them to take advantage of streaming access, this is not true for sparse matrices. Naive sparse matrix algorithms typically require random access, which, on modern architectures, incurs a high cost. This cost is more pronounced for parallel processing, where remote random access requires use of both the memory and the network. Sparse matrix algorithms typically underperform their dense matrix counterparts in terms of *op/s* (operations per second) due to the large memory and network latencies they incur.

15.2.1 LLMOE overview

To overcome memory and network latencies, the software-hardware interaction must be co-optimized. The Lincoln Laboratory Mapping and Optimization Environment (LLMOE) simplifies the co-optimization process by abstracting both the algorithm and hardware implementations and provides tools for analyzing both. Our system, implemented in MATLAB using pMatlab [Bliss & Kepner 2007] for parallelization, is shown in Figure 15.2. This figure illustrates the four main components of LLMOE.

- The *program analysis* component is responsible for converting the user program, taken as input, into a *parse graph*, a description of the high-level operations and their dependencies on one another.

- The *data mapping* component is responsible for distributing the data of each variable specified in the user code across the processors in the architecture.

- The *operations analysis* component is responsible for taking the parse graph and data maps and forming the *dependency graph*, a description of the low-level operations and their dependencies on one another.

- The *architecture simulation* component is responsible for taking the dependency graph and a model of a hardware architecture and simulating it on that architecture. Once the simulation is finished, the results can either be returned to the user or fed back into the data mapping component in order to further optimize the data distribution.

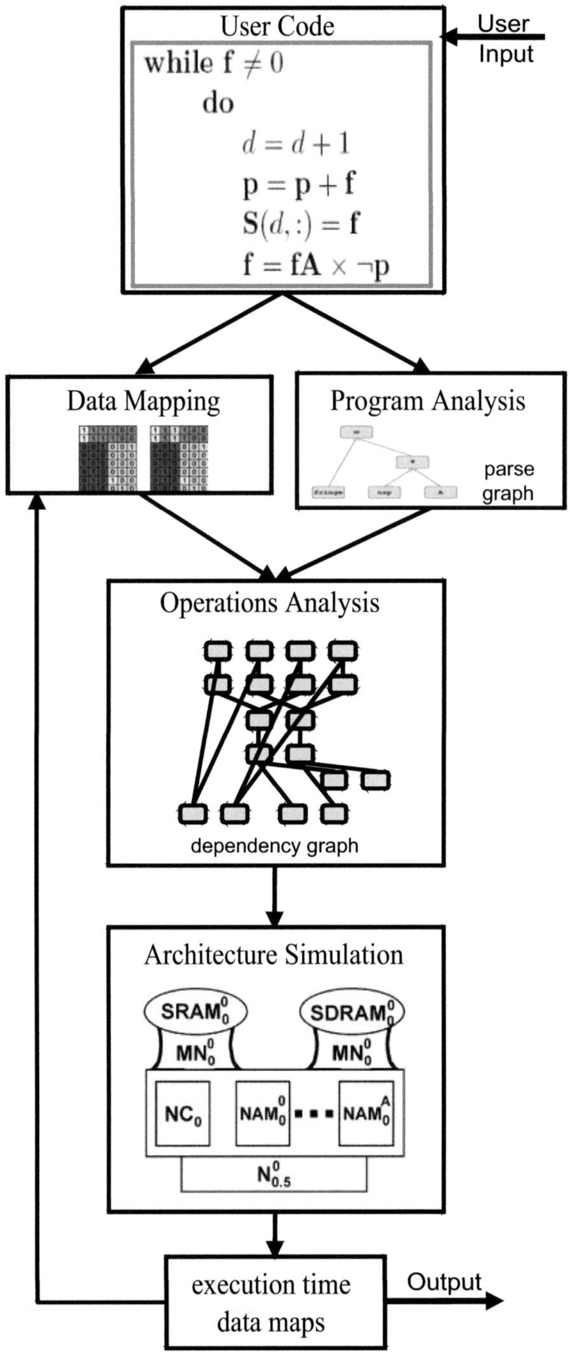

Figure 15.2. LLMOE.
Overview of the LLMOE framework and its components.

LLMOE's modular design allows for various hardware and software implementations to be evaluated relative to one another. The data mapping and architecture simulation are designed to be pluggable components to provide flexibility. This design lets users determine the best software-hardware pairing for their application. In addition, individual operator implementations (e.g., matrix multiplication) can be analyzed because LLMOE supports the definitions of operations at the dependency graph level. LLMOE provides a standard interface in the program and operations analysis components via the parse tree and dependency graph, respectively.

Currently, LLMOE supports three data mappers. These mappers typically specify maps using a PITFALLS data structure [Ramaswamy & Banerjee 1995], though other formats are possible. The *manual mapper* accepts fully specified data maps from the user. The *genetic mapper* uses a genetic algorithm and the feedback from the architecture simulation to optimize the map selection. Finally, the *atlas mapper* looks up the appropriate data maps for certain types of operations and sparsity patterns in a predefined atlas of maps.

In addition, LLMOE currently defines two architecture simulators. The input to each of these simulators includes the dependency graph and an architectural model that considers memory and network bandwidth/latency, CPU rate, and network topology. The *topological simulator* partitions the dependency graph into stages based on the dependencies and then executes each stage sequentially, with the instructions in the stage being executed in parallel. The *event-driven simulator* uses PhoenixSim [Chan et al. 2010], an event-driven simulator primarily for photonic networks built on top of OMNeT [Ponger], to simulate the dependency graph operations. An operation is scheduled as soon as all of its dependencies have been fulfilled.

The LLMOE system eases the efforts involved in co-optimization of matrix operations, of key importance in the intelligence, surveillance, and reconnaissance (ISR) domain. It uses a front-end MATLAB interface to allow users to input their application in a language familiar to them. In addition to the pluggable components for data mapping and architecture simulation, the user may also specify new low-level operation implementations. The system is itself parallel, leveraging pMatlab to speed up mapping and simulation time.

15.2.2 Mapping in LLMOE

In order to find the best mapping for a sparse matrix multiplication operation, the efficiency of computation and communication at a fine-grain level are co-optimized. Note that while data maps provide the distribution information for the matrices, they do not provide routing information. Once the maps for the arrays are defined, the set of communication operations that must occur can be enumerated. However, in order to evaluate the performance of a given mapping, a route for each communication operation must be chosen. These routes cannot be enumerated until a map is defined. Figure 15.3 illustrates a simplified mapping example over a distributed addition. In the figure, the dependency graph communication operations are highlighted. For each highlighted operation, a number of routing options exists. The number of possible routes is dependent on the topology of the underlying

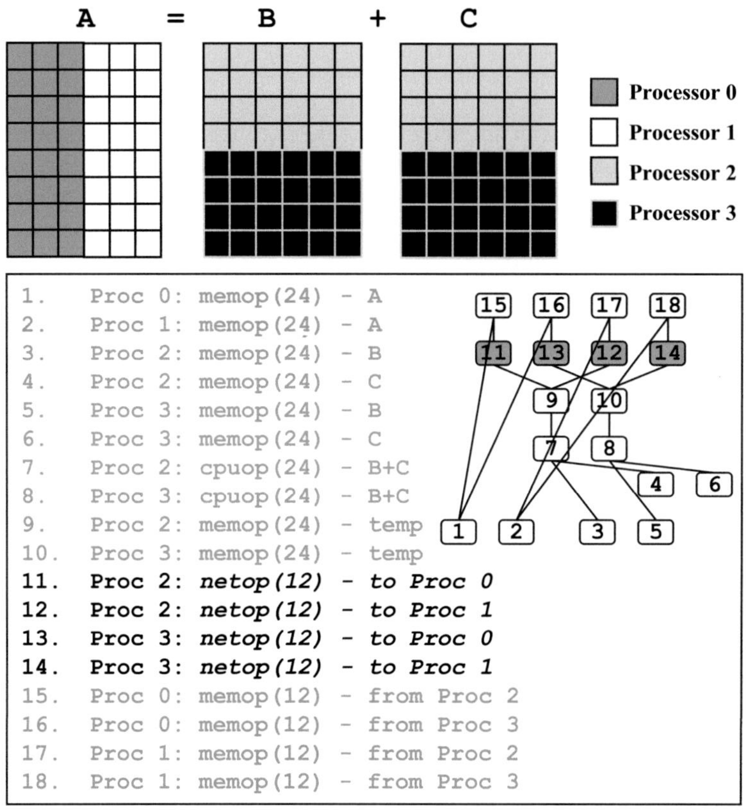

Figure 15.3. Parallel addition with redistribution.
Top: illustrates how a parallel addition is mapped. Bottom: presents the dependency graph for the addition. For each communication operation (highlighted) a number of possible routes exist.

hardware architecture. Since the best map depends on chosen routes and routes cannot be enumerated until a map is chosen, the problem is combinatorial and no closed form solution exists. This type of problem can be solved by a stochastic search method. Evaluation of the quality of a solution requires creation and simulation of fine-grained dependency graphs (Figure 15.3) on a machine or hardware model. A stochastic search technique well suited for parallelization is chosen. The next section describes the genetic algorithm [Mitchell 1998] problem formulation in greater detail.

Co-optimization of mapping and routing

Figure 15.4 presents an overview of the nested genetic algorithm (GA) for mapping and routing.

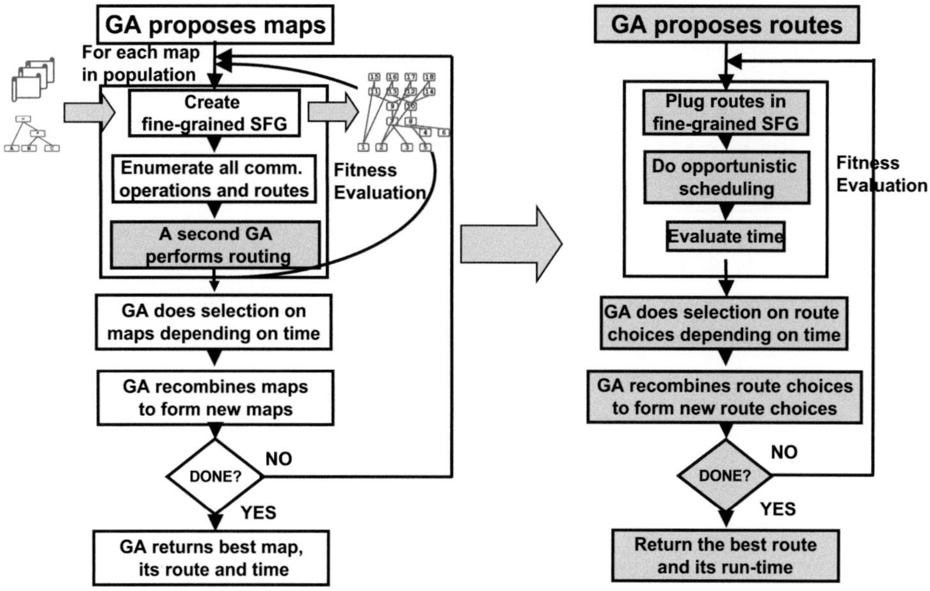

Figure 15.4. Nested genetic algorithm (GA).
The outer GA searches over maps, while the inner GA searches over routes for all communication operations given a map set.

The outer GA searches over maps, while the inner GA searches over route options for a given set of maps. One input to the GA is a parse graph. The outer GA generates a population of maps for the arrays in the parse graph. For the matrix-multiply operation $\mathbf{A} = \mathbf{A} * \mathbf{B}$, there are two arrays for which maps must be generated, A and B. The result, in this case, is declared to have the same mapping as the left operand. For each <maps, parse graph> pair, a dependency graph is generated. While the parse graph contains operations such as matrix multiplication and assignment, the dependency graph contains only computation, communication, and memory operations, as illustrated in Figure 15.3 (bottom).

Once the dependency graph is constructed, the communication operations and corresponding route choices can be enumerated. At this point, the inner GA assigns route choices to communication operations and evaluates the performance of the given <maps, routes> pair against a hardware model. Fitness evaluation is performed using opportunistic scheduling, and operations in the dependency graph are overlapped whenever possible.

Outer GA

The outer GA iterates over sets of maps for arrays in the computation. The minimum block size for a matrix is chosen based on the size of the matrix being mapped. An individual in the outer GA is represented as a set of matrices blocked according to minimum block size. Processors are assigned to each block, and mutation

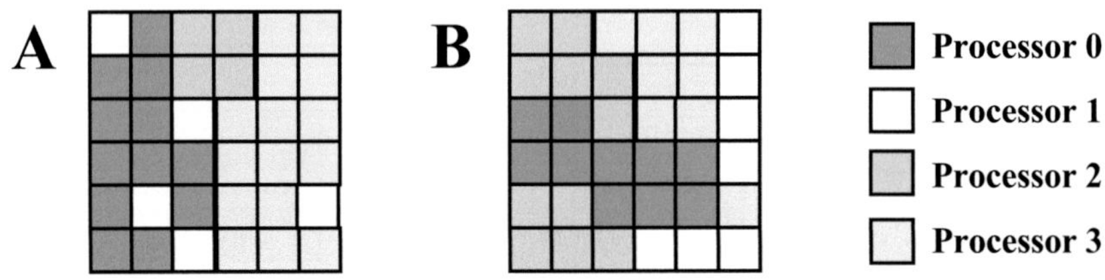

Figure 15.5. Outer GA individual.
Here, there are 36 blocks in each of the two matrices in the individual. Different shades of gray indicate different processors assigned to blocks.

1	9	7	5	3	1

Figure 15.6. Inner GA individual.
The set of communication operations is represented as a linear array, with each entry in the array containing the index of a route chosen for the given communication operation.

and crossover operators manipulate the processor assignments for each block. Figure 15.5 illustrates a typical individual for a matrix multiplication operation.

Inner GA

The inner GA iterates over routes. The representation for the inner GA is simply the list of communication operations. The length of the list is equal to the number of communication operations for a given set of maps. Each entry in the list represents the index of a route chosen for a particular communication operation. Figure 15.6 illustrates an individual for the inner GA.

Search space

When one is performing a stochastic search, it is helpful to characterize the size of the search space. The search space, S, for the nested GA formulation of the mapping and routing problem is given by

$$S = P^B r^C$$

where

P = number of processors
B = number of blocks
C = number of communication operations
r = average number of route options per communication operations.

Table 15.1. Individual fitness evaluation times.

Time to evaluate for some sample sparsity patterns of 1024×1024 matrices. The time shown is the average over 30,000 evaluations.

Sparsity pattern	Evaluation time (min)
toroidal	0.45
power law	1.77

Consider an architecture with 32 processors, where for any given pair of processors there are four possible routes between them. For a mapping scheme involving just 64 blocks, or two blocks owned by each processor, and 128 communication operations, or four communication operations performed by each processor, the search space size is already extremely large, greater than 2×10^{173}.

Parallelization of the nested GA

Fitness evaluation of a <maps, routes> pair requires building a dependency graph consisting of all communication, memory, and computation operations, performing opportunistic scheduling of operations, and simulating the operations on a machine model. This evaluation is computationally expensive, as illustrated by Table 15.1.

Since the mapping and optimization environment is written in MATLAB, we used pMatlab to parallelize the nested GA and run it on LLGrid: Lincoln Laboratory cluster computing capability [Reuther et al. 2007]. Figure 15.7 illustrates the parallelization process.

As indicated by Figure 15.7, parallelization required minimal changes to the code (Table 15.2). A GA is well suited to parallelization since each fitness evaluation can be performed independently of all other fitness evaluations.

15.2.3 Mapping performance results

This section discusses the performance results of the maps found by using LLMOE. It shows that LLMOE assists in finding efficient ways of distributing sparse computations onto parallel architectures and gaining insight into the type of mappings that perform well.

Machine model

The results presented here are simulated results on a hardware or machine model. LLMOE allows for the machine model to be altered freely. With this, focus may be placed on various architectural properties that affect the performance of sparse computations. Table 15.3 describes the parameters of the model used for the results presented.

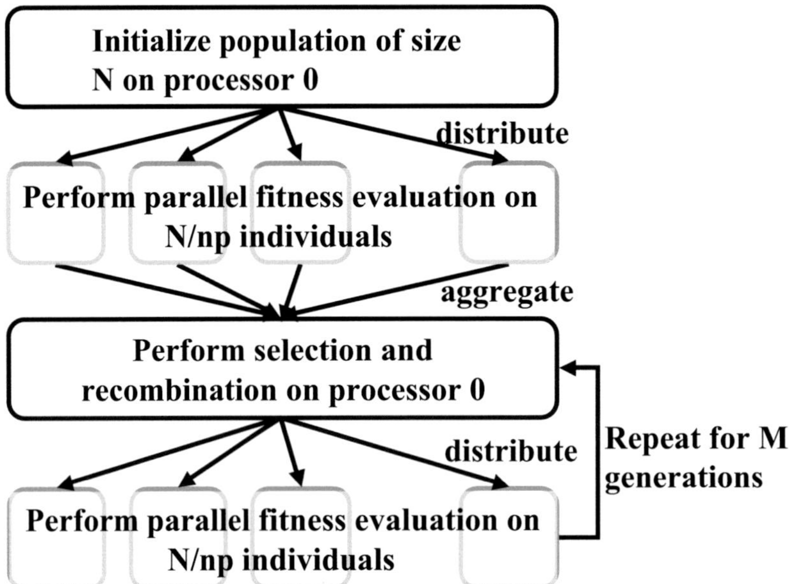

Figure 15.7. Parallelization process.
Fitness evaluation is performed in parallel on np processors. Selection and recombination are performed on the leader processor.

Table 15.2. Lines of code.

Parallelization with pMatlab requires minimal changes to the code.

Serial program	1400
Parallel program	1420
% Increase	1.4

Table 15.3. Machine model parameters.

Parameter	Value
Topology	ring
Processors	8
CPU rate	28 GFLOPS
CPU efficiency	30%
Memory rate	256 GBytes/sec
Memory latency	10^{-8} sec
Network rate	256 GBytes/sec
Network latency	10^{-8} sec

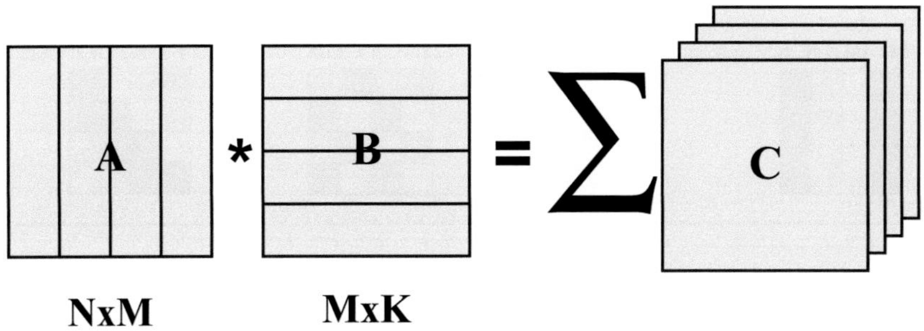


```
for i = 1:M
  C = C+A(:,i)*B(i,:);
end
```

Figure 15.8. Outer product matrix multiplication.

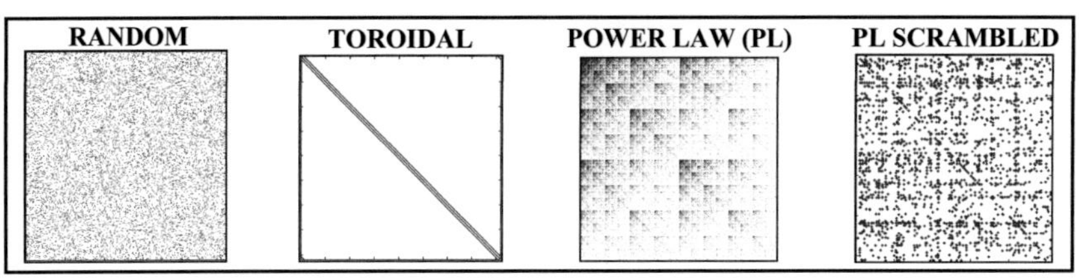

Figure 15.9. Sparsity patterns.

Matrix multiplication algorithm

LLMOE was applied to an outer product matrix multiplication algorithm (see [Golub & Van Loan 1996]). Figure 15.8 illustrates the algorithm and the corresponding pseudocode. This algorithm was chosen because of the independent computation of slices of matrix **C**. This property makes the algorithm well suited for parallelization.

Sparsity patterns

LLMOE solutions should apply to general sparse matrices so LLMOE was tested on a number of different sparsity patterns. Figure 15.9 illustrates the sparsity patterns mapped in increasing order of load-balancing complexity, from random sparse to scrambled power law.

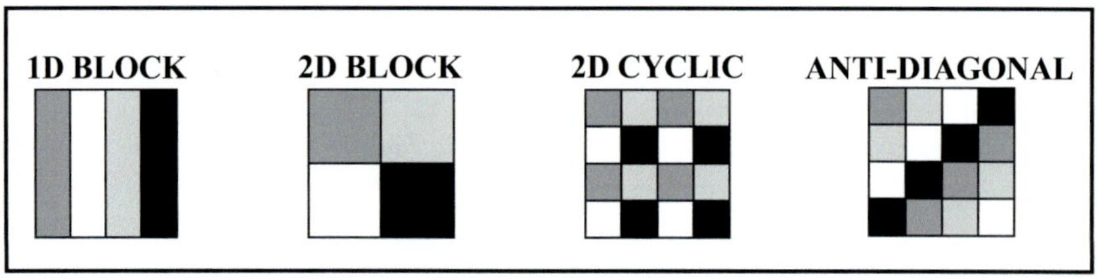

Figure 15.10. Benchmark maps.
We compared our results with the results using standard mappings.

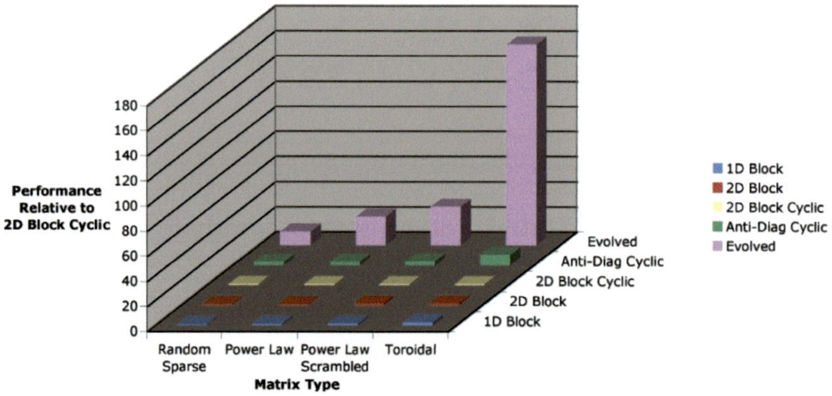

Figure 15.11. Mapping performance results.

Benchmarks

Figure 15.10 illustrates a number of standard mappings that the results obtained with LLMOE were compared against.

Results

Figure 15.11 presents performance results achieved by LLMOE. That performance outperforms standard maps by more than an order of magnitude. The results are normalized with regard to the performance achieved using a 2D block-cyclic map, as that is the most commonly used map for sparse computations. In order to show that the results were repeatable and statistically significant over a number of runs of the GA, multiple runs were performed. Figure 15.12 shows statistics for 30 runs of the GA on a power law matrix. Observe that there is good consistency in terms of solution found between multiple runs of the mapping framework.

Note that while a large number of possible solutions were considered, only a small fraction of the search space has been explored. For the statistics runs in Figure 15.12, the outer GA was run for 30 generations with 1000 individuals. The

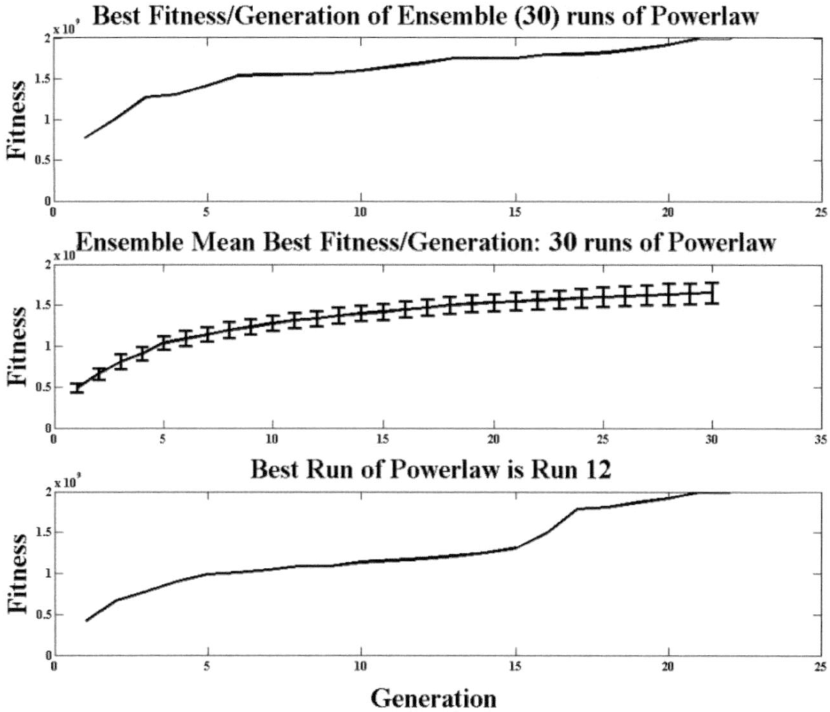

Figure 15.12. Run statistics.
The top plot shows best overall fitness found for each generation. The middle plot shows average fitness for each generation. Finally, the bottom plot shows the behavior of the best of the 30 runs over 30 generations.

inner GA used a greedy heuristic to pick the shortest route between two nodes whenever possible. Thus, the total number of solutions considered was

$$30 \times 1000 = 30,000$$

The size of the search space per equation is

$$S = P^B r^C = 8^{128} \times 2^{100} = 5 \times 10^{145}$$

where 8 is the number of processors in the machine model; 128 is the number of blocks used for 256×256 matrices; O(100) is the number of communication operations; 2 is the number of possible routes for each communication operation given a ring topology. Thus, the GA performs well in this optimization space, as it is able to find good solutions while exploring a rather insignificant fraction of the search space.

Conclusion

LLMOE provides a tool to analyze and co-optimize problems requiring the partitioning sparse arrays and graphs. It allows an efficient partition of the data used in these problems to be obtained in a reasonable amount of time. This is possible even in circumstances where the search space of potential mappings is so large as to make the problem unapproachable by typical methods.

References

[Bliss & Kepner 2007] N.T. Bliss and J. Kepner. pMatlab parallel MATLAB library. *International Journal of High Performance Computing Applications (IJHPCA)*. Special Issue on High-Productivity Programming Languages and Models, 21: 336–359, 2007.

[Chan et al. 2010] J. Chan, G. Hendry, A. Biberman, K. Bergman, and L. Carloni. PhoenixSim: A simulator for physical-layer analysis of chip-scale photonic interconnection networks. In *Proceedings of DATE: Design, Automation and Test in Europe Conference and Exhibition*, 691–696, 2010.

[Golub & Van Loan 1996] G.H. Golub and C.F. Van Loan. *Matrix Computations*, 3rd edition. Baltimore: Johns Hopkins University Press, Baltimore, MD, 1996.

[Mitchell 1998] M. Mitchell. *An Introduction to Genetic Algorithms.* Cambridge, Mass.: MIT Press, 1998.

[Ponger] G. Ponger. OMNeT: Objective modular network testbed. In *Proceedings of the International Workshop on Modeling, Analysis, and Simulation on Computer and Telecommunication Systems,* 323–326, 1993.

[Ramaswamy & Banerjee 1995] S. Ramaswamy and P. Banerjee. Automatic generation of efficient array redistribution routines for distributed memory multicomputers. In *Proceedings of the Fifth Symposium on the Frontiers of Massively Parallel Computation,* 342–349, 1995.

[Reuther et al. 2007] A. Reuther, J. Kepner, A. McCabe, J. Mullen, N. Bliss, and H. Kim. Technical challenges of supporting interactive HPC. In *DoD High Performance Computing Modernization Program Users Group Conference,* 403–409, 2007.

Chapter 16

Fundamental Questions in the Analysis of Large Graphs

*Jeremy Kepner**, *David A. Bader*†, *Robert Bond**, *Nadya Bliss**, *Christos Faloutsos*‡, *Bruce Hendrickson*§, *John Gilbert*¶, *and Eric Robinson**

Abstract

Graphs are a general approach for representing information that spans the widest possible range of computing applications. They are particularly important to computational biology, web search, and knowledge discovery. As the sizes of graphs increase, the need to apply advanced mathematical and computational techniques to solve these problems is growing dramatically. Examining the mathematical and computational foundations of the analysis of large graphs generally leads to more questions than answers. This book concludes with a discussion of some of these questions.

*MIT Lincoln Laboratory, 244 Wood Street, Lexington, MA 02420 (kepner@ll.mit.edu, rbond@ll.mit.edu, nt@ll.mit.edu, erobinson@ll.mit.edu).

†College of Computing, Georgia Institute of Technology, Atlanta, GA 30332 (bader@cc.gatech.edu).

‡School of Computer Science, Carnegie Mellon University, Pittsburgh, PA 15213-3891 (christos@cs.cmu.edu).

§Discrete Algorithms & Math Department Sandia National Laboratories Albuquerque, NM 87185 (bahendr@sandia.gov).

¶Computer Science Department, University of California, Santa Barbara, CA 93106-5110 (gilbert@cs.ucsb.edu).

This work is sponsored by the Department of the Air Force under Air Force Contract FA8721-05-C-0002. Opinions, interpretations, conclusions, and recommendations are those of the authors and are not necessarily endorsed by the United States Government.

Of the many questions relating to the the mathematical and computational foundations of the analysis of large graphs, five important classes appear to emerge. These are:

- Ontology, schema, data model.
- Time evolution.
- Detection theory.
- Algorithm scaling.
- Computer architecture.

These questions are discussed in greater detail in the subsequent sections.

16.1 Ontology, schema, data model

Graphs are a highly general structure that can describe complex relationships (edges) between entities (vertices). It would appear self-evident that graphs are a good match for many important problems in computational biology, web search, and knowledge discovery. However, these problems contain far more information than just vertices and edges. Vertices and edges contain metadata describing their inherent properties. Incorporating vertex/edge metadata is critical to analyzing these graphs. Furthermore, the diversity of vertex and edge types often makes it unclear which is which.

Knowledge representation using graphs has emerged, faded, and re-emerged over time. The recent re-emergence is due to the increased interest in very large networks and the data they contain. Revisiting the first order logical basis for knowledge representation using graphs, and finding efficient representations and algorithms for querying graph-based knowledge representation databases is a fundamental question.

The mapping of the data for a specific problem onto a data structure for that problem is given by many names: ontology, schema, data model, etc. Historically, ontologies have been generated by experts by hand and applied to the data. Increasingly large, complex, and dynamic data sets make this approach infeasible. Thus, a fundamental question is how to create ontologies from data sets automatically.

Higher order graphs (complexes, hierarchical, and hypergraphs) that can capture more sophisticated relationships between entities may allow for more useful ontologies. However, higher order graphs raise a number of additional questions. How do we extend graph algorithms to higher order graphs? What are the performance benefits of higher order graphs on specific applications (e.g., pattern detection or matching)? What is the computational complexity of algorithms running on higher order graphs? What approximations can be used to reduce the complexities introduced by higher order graphs?

16.2 Time evolution

Time evolution is an important feature of graphs. In what ways are the spatiotemporal graphs arising from informatics and analytics problems similar to or different from the more static graphs used in traditional scientific computing? Are higher order graphs necessary to capture the behavior of graphs as they grow?

For static graphs, there is a rich set of basic algorithms for connectivity (e.g., s-t connectivity, shortest paths, spanning tree, connected components, biconnected components, planarity) and for centrality (e.g., closeness, degree, betweenness), as well as for flows (max flow, min cut). For dynamic graphs, are there new classes of basic algorithms that have no static analogs? What is the complexity of these? For instance, determining whether a vertex switches clusters over time (i.e., allegiance switching) has no static analog. Another example is detecting the genesis and dissipation of communities.

A related issue is probabilistic graphs. How do we apply graph algorithms to probabilistic graphs, where the edges exist with some temporal probability?

16.3 Detection theory

The goal of many graph analysis techniques is to find items of interest in a graph. Historically, many of these techniques are based on heuristics about topological features of the graph that should be indicative of what is being sought. Massive graphs will need to be analyzed using statistical techniques. The application of statistical approaches and detection theory are used in other domains and should be explored for application to large graphs. For example, how do spectral techniques applied to sparse matrices apply to large graphs? Can detection be enhanced with spectral approaches? Spectral approaches applied to dynamic graphs provide a tensor perspective similar to those used in many signal processing applications that have two spatial and one temporal dimension.

A key element of statistical detection is a mathematical description of the items of interest. The signature of an item may well be another graph that must be embedded in a larger graph while preserving some distance metric and doing so in a computationally tractable way. The optimal mapping (matching or detection) of a subgraph embedded in a graph is an NP-hard problem, but perhaps there are approximation approaches that are within reach for expected problem sizes. A subproblem that is of relevance to visualization and analysis is that of projecting a large graph into a small chosen graph. The inverse problem is constructing a graph from its projections.

The second key element of statistical detection is a mathematical description of the background. Many real-world phenomena exhibit random-like graphical relationships (social networks, etc.). Statistical detection theory relies on an extensive theory of random matrices. Can these results be extended to graphs? How do we construct random-like graphs? What are the basic properties of random-like graphs, and how can one derive one graph property from another? How can we efficiently generate random graphs with properties that mirror graphs of interest?

16.4 Algorithm scaling

As graphs become increasingly large (straining the limits of storage), algorithm scaling becomes increasingly important. The computational complexities of feasible algorithms are narrowing. $O(1)$ algorithms remain trivial, $O(N|M)$ algorithms are tractable along with $O(N|M \log(N|M))$, providing the constants are reasonable. Algorithms that were feasible on smaller graphs are increasingly becoming less feasible on those that $O(N^2)$, $O(NM)$, and $O(M^2)$. Thus, better scaling algorithms or approximations are required to analyze large graphs.

Similar issues exist in bandwidth, where storage retrieval and parallel communication are often larger bottlenecks than single-processor computation. Existing algorithms need to be adapted for both parallel and hierarchical storage environments.

16.5 Computer architecture

Graph algorithms present a unique set of challenges for computing architectures. These include both software algorithm mapping and hardware challenges.

Parallel graph algorithms are very difficult to code and optimize. Can parallel algorithmic components be organized to allow nonexperts to analyze large graphs in diverse and complex ways? What primitives or kernels are critical to supporting graph algorithms? More specifically, what should be the "combinatorial BLAS?" That is, what is a parsimonious but complete set of primitives that (a) are powerful enough to enable a wide range of combinatorial computations, (b) are simple enough to hide low-level details of data structures and parallelism, and (c) allow efficient implementation on a useful range of computer architectures. Semiring sparse matrix multiplication and related operations may form such a set of primitives. Other primitives have been suggested, e.g., the visitor-based approach of Berry/Hendrickson/Lumsdaine. Can these be reduced to a common set, or is there a good way for them to interoperate?

Parallel computing algorithms generally rely on effective algorithm mappings that can minimize the amount of communication required. Do these graphs contain good separators that can be exploited for partitioning? If so, efficient parallel algorithm mappings can be found that will work well on conventional parallel computers. Can efficient mappings be found that can be applied to a wide range of graphs? Can the recursive Kronecker structure of a power law graph be exploited for parallel mapping? How sensitive is parallel performance to changes in the graph (both with additions/removals of edges and as the graph grows in size)?

From a hardware perspective, graph processing typically involves randomly hopping from vertex to vertex in memory, and experiencing bad memory locality. Modern architectures stress fast memory bandwidth, but typically high latency. It is not clear if the shift to multicore may make this problem better or worse, as typical multicore machines only have a single memory subsystem, resulting in multiple processors competing for the same memory. In distributed computing and clusters, bad memory locality translates into bad processor locality for computations where the data (graph) is distributed. Rather than poor performance due to memory latency, clusters typically see poor performance on graph algorithms due to high communication costs.

The underlying issue is data locality. Modern architectures are not optimized for computations with poor data locality, and the problem only looks to get worse. Casting graph algorithms by using an adjacency matrix formulation may help. The linear algebra community has a broad background body of work in distributing matrix operations for better data locality. Fundamentally, what architectural features could be added to complex, heterogeneous many-core chips to make them more suitable for these applications?

Index